T0324894

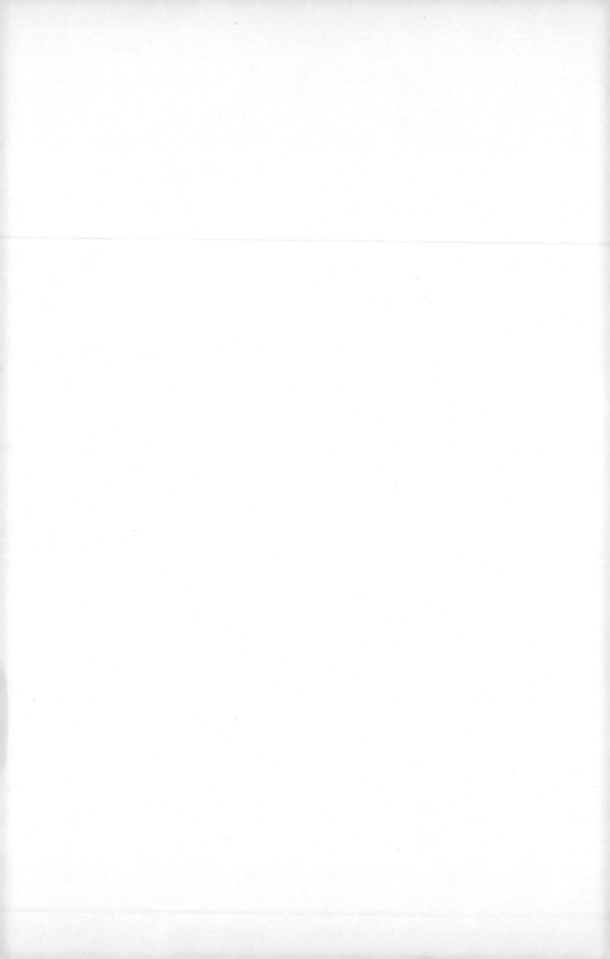

Methods in Molecular Biology

Series Editor
John M. Walker
School of Life and Medical Sciences
University of Hertfordshire
Hatfield, Hertfordshire, AL10 9AB, UK

For further volumes:
http://www.springer.com/series/7651

Computational Design of Ligand Binding Proteins

Edited by

Barry L. Stoddard

Division of Basic Sciences, Fred Hutchinson Cancer Research Center, Seattle, Washington, USA

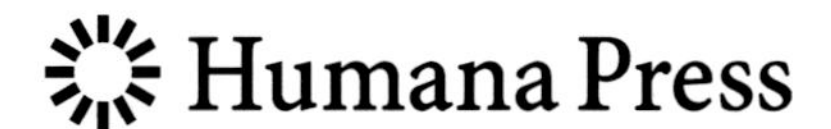

Editor
Barry L. Stoddard
Division of Basic Sciences
Fred Hutchinson Cancer Research Center
Seattle, Washington, USA

ISSN 1064-3745 ISSN 1940-6029 (electronic)
Methods in Molecular Biology
ISBN 978-1-4939-3567-3 ISBN 978-1-4939-3569-7 (eBook)
DOI 10.1007/978-1-4939-3569-7

Library of Congress Control Number: 2016937968

Printed on acid-free paper

This Humana Press imprint is published by Springer Nature
The registered company is Springer Science+Business Media LLC New York

Preface

Introduction: Design and Creation of Ligand-Binding Proteins

The appropriate balance of ligand binding affinity and specificity is a fundamental feature of most if not all biological processes, including immune recognition, cellular metabolism, regulation of gene expression, and cell signaling. The ability to accurately predict and recapitulate the physical basis for ligand binding behavior is therefore a crucial part of understanding and manipulating such biological phenomena. It also represents a critical technical requirement in the reciprocal fields of drug design and protein engineering.

This book provides a collection of protocols and approaches, compiled and described by many of today's leaders in the field of protein engineering, that they apply to the problem of creating ligand-binding proteins that display desirable combinations of target affinity and specificity. The descriptions provided by each chapter's authors also provide a snapshot of their current "belief system" regarding the challenging problem of protein engineering and design, as it is applied to the creation of novel ligand binding functions.

The problem of how to effectively engineer novel binding properties onto protein scaffolds, and how to do so while exploiting the information that is provided by high-resolution protein structures, has been under investigation for almost 40 years if not longer. Such efforts date back at least to the design of small folded peptides and proteins capable of binding individual nucleosides and single-stranded DNA, followed by subsequent attempts to generate additional ligand binding functions using various protein scaffolds (*see* Refs. [1, 2] for early examples of such work). By the early 1990s, some of the first computational algorithms intended to design novel ligand binding sites into proteins of known structure had been described [3], and the field of structure-based protein engineering as it is known today was underway.

Although the field of protein engineering, including the specific problem of designing novel ligand binding capabilities onto engineered protein folds, now comprises an extensive and growing publication record, significant challenges regarding the accurate calculation or prediction of protein–ligand binding affinities (even when provided a high-resolution structure of the actual complex) still represent significant hurdles to the field's advancement. For example:

- Several recent studies have demonstrated that current methods for structure-based calculation of binding affinities display variable accuracies. At least three broad (and somewhat overlapping) classes of scoring functions for predicting binding affinities from high-resolution structures have been developed: *force-field* (formulated by calculating the individual energetic contributions of physical interactions between the protein and ligand) [4, 5], *knowledge-based* (produced by statistical mining of large databases of protein–ligand structures to deduce rules and models that govern binding affinity) [6–9], and *empirical* (in which binding energy is calculated to be a product of a collection of weighted energy terms fit to a training data set of known binding affinities, with the weighting coefficients calculated via linear regression analyses) [10–14]. Even with all these tools, the accuracy of many methods that are intended to calculate structure-based binding affinities (as well as the ability to identify and rank the most tightly bound ligands to a given protein) has been shown to often be somewhat poor

[15–17], leading to the conclusion by one group that "more precise chemical descriptions of the protein–ligand complex do not generally lead to a more accurate prediction of binding affinity" [17]. Therefore, the reliable prediction of affinity remains a significant challenge in biophysical chemistry [15].

- Even for the most thoroughly studied of ligand-binding proteins, the basis for tight, specific binding is not well understood. For example, avidin and streptavidin exhibit some of the highest known affinities to their cognate molecular ligand (Ka ~ 10^{15} M^{-1}). Over 20 years of studies on these proteins have produced a wide range of hypotheses regarding their high affinities, including exceptional shape complementarity across a stabilized network of hydrophobic side chains and precisely arranged hydrogen bond partners [18], the precisely tuned dynamic behavior of the protein [19], a large free energy benefit upon ligand binding due to the strengthening of noncovalent interactions within the protein scaffold [20], or the induction of polarized moieties within the bound complex that create a cooperative effect between neighboring hydrogen bonds [21]. Not surprisingly, attempts to engineer altered binding properties onto avidin or streptavidin have yielded constructs with unexpected and unpredictable properties [22].
- Attempts to computationally engineer novel ligand-binding proteins have either been unsuccessful [23, 24] or have produced computationally designed constructs that display low affinities. Optimization of those designed proteins has then required laborious rounds of random mutagenesis and affinity maturation [25, 26].

The sources of error in calculating and modeling protein–ligand binding interactions and affinities are myriad, and their relative importance is still not entirely clear. These include: (1) Inaccuracies in the treatment of solvent and desolvation effects during binding [27–29]. (2) Limited consideration of protein dynamics [30–32]. (3) Difficulties incorporating the contribution of entropic changes into calculations of binding energies, leading to examples where modifications of ligand binding sites that lead to favorable enthalpic gains are confounded by substantial losses in entropy, with no improvement in overall binding affinity (recently reviewed extensively in Ref. [33]). Even for the most straightforward aspect of a protein–ligand interface (i.e., the observation of direct interatomic interactions and corresponding estimation of their enthalpic contributions to binding), uncertainties exist regarding interatomic distance cutoffs [17] and best strategies for estimating charge and protonation states [34].

Therefore, the creation of novel ligand-binding proteins that display tight binding affinity to their desired target and that also can discriminate between closely related targets remains an important goal, but is plagued by rather poor understanding of how to accurately calculate binding affinities or predict binding specificity, even when armed high structural information of protein–ligand complexes. As a result, the creation of highly specific ligand-binding proteins with high affinity remains extremely challenging and generally requires a substantial investment of time and effort to identify designed protein scaffolds that are actually active, and then to manually optimize their behavior. Nevertheless, studies from groups around the world have recently demonstrated that engineered proteins can, with considerable effort, be created that perform as desired, even in highly demanding in vivo applications. In this book, a series of 21 author groups present individual chapters that describe, in considerable detail, the types of overall thought processes and approaches, as well as very detailed computational and/or experimental protocols, that are used in their research groups as they attempt to address and resolve the difficulties associated with the design and creation of engineered ligand-binding proteins.

The reader will find a wide variety of technical issues and variables described in this volume. The first three chapters are largely concerned with a fundamental challenge that precedes actual protein engineering: identifying, characterizing, and modeling protein–ligand binding sites and predicting their corresponding modes and affinities of molecular interaction. Various strategies are shown to rely on both sequence-based and structure-based methods of analysis, and often utilize evolutionary information to determine the relative importance of positions within individual protein scaffolds that are important for form and function. With the development of controlled, blind binding site prediction challenges within the protein informatics and design community, the number of methods available to perform such analyses has exploded, as summarized in Chapter 2. Virtually all structure-based methods for binding site evaluation rely on accurate modeling of protein–ligand conformational sampling and scoring of individual docked solutions, which is further discussed in Chapters 3 and 4.

Beyond the basic ability to identify and model protein–ligand binding sites and their interactions, the field of protein engineering also now has at its disposal a number of increasingly powerful and robust computational platforms for structure-based engineering, including the widely used and rapidly evolving ROSETTA program suite as well as other programs such as POCKETOPTIMIZER and PROTEUS. Many of the fundamental features of these computational program suites, as well as individual examples of their utility and application for the design of a protein binding site for a defined small molecular ligand, are found in Chapters 5 through 7.

The output of even the most powerful structure-based computational design algorithms is usually augmented by considerable experimental time and effort, generally consisting of the preparation of combinatorial protein libraries or the systematic generation of large numbers of individual protein mutants on top of designed protein constructs, which are then subjected to selections or screens for optimal activity. While the ultimate goal of protein design is to eliminate the need for such manual intervention and effort, at this time many strategies for protein design involve combining information from computational design to the subsequent creation and screening of protein mutational libraries. Several examples of such approaches, which have resulted in particularly notable recent successes in protein engineering and the creation of designed ligand-binding proteins', are outlined and described in Chapters 8–10 and can then be found at various points within the remaining chapters.

Finally, the exact technical hurdles and necessary approaches required for the creation of ligand-binding proteins obviously are dependent upon the chemical and structural nature of the ligand to be recognized and bound with high affinity and specificity. The remaining 12 chapters describe a variety of specific scenarios and methodological approaches, ranging from the design of metal-binding proteins and light-induced ligand-binding proteins, to the creation of binding proteins that also display catalytic activity, to binding of larger peptide, protein, DNA, and RNA ligands.

The continued development of approaches to design and create ligand-binding proteins, beyond enabling the creation of unique protein-based reagents and molecules for biotechnology and medicine, will continue to test and refine the ability of modern biophysical chemistry to fundamentally understand and exploit the forces and principles that drive molecular recognition. The behaviors and properties of designed ligand-binding proteins resulting from the types of methods described in this book (including the "failures"—those constructs that fail to bind their intended targets and those that bind to unintended ligands) will eventually be explained by systematically examining their structures and properties. As has been famously attributed to Richard Feynman, "That which I cannot create, I do not understand." The following volume provides detailed (although by no means complete and total) examples of the current approaches and methods by which the protein engineering and design community attempt to do both.

Seattle, WA, USA ***Barry L. Stoddard***

References

1. Gutte B, Daumigen M, Wittschieber E (1979) Design, synthesis and characterisation of a 34-residue polypeptide that interacts with nucleic acids. Nature 281:650–655
2. Moser R, Thomas RM, Gutte B (1983) Artificial crystalline DDT-binding polypeptide. FEBS 157:247–251
3. Hellinga HW, Richards FM (1991) Construction of new ligand binding sites in proteins of known structure. I. Computer-aided modeling of sites with pre-defined geometry. J Mol Biol 222:763–785
4. Huang N, Kalyanaraman C, Bernacki K et al. (2006) Molecular mechanics methods for predicting protein-ligand binding. Phys Chem Chem Phys 8: 5166–5177
5. Ewing T, Makino S, Skillman A et al. (2001) DOCK 4.0: Search strategies for automated molecular docking of flexible molecule databases. J Comut Mol Des 15:411–428
6. Gehlhaar DK, Verkhivker GM, Rejto PA et al. (1995) Molecular recognition of the inhibitor AG-1343 by HIV-1 protease: conformationally flexible docking by evolutionary programming. Chem Biol 2: 317–324
7. Muegge I, Martin Y (1999) A general and fast scoring function for protein-ligand interactions: a simplified potential approach. J Med Chem 42: 791–804
8. Mooij W, Verdonk M (2005) General and targeted statistical potentials for protein-ligand interactions. Proteins 61: 272–287
9. Hohlke H, Hendlich M, Klebe G (2000) Knowledge-based scoring function to predict protein-ligand interactions. J Mol Biol 295: 337–356
10. Bohm H (1994) The development of a simple empirical scoring function to estimate the binding constant for a protein-ligand complex of known three-dimensional structure. J Comput Mol Des 8: 243–256
11. Eldridge M, Murray C, Auton T et al. (1997) Empirical scoring functions: the development of a fast empirical scoring function to estimate the binding affinity of ligands in receptor complexes. J Comput Aid Mol Des 11: 425–445
12. Friesner R, Al E (2004) Glide: a new approach for rapid, accurate docking and scoring. J Med Chem 47: 1739–1749
13. Krammer A, Kirchhoff P, Jiang X et al. (2005) LigScore: a novel scoring function for predicting binding affinities. J Mol Graphics Model 23: 395–407
14. Wang R, Lai L, Wang S (2002) Further development and validation of empirical scoring functions for structure-based binding affinity prediction. J Comput Mol Des 16: 11–26
15. Ross G, Morris G, Biggin P (2013) One size does not fit all: the limits of structure-based models in drug discovery. J Chem Theory Comput 9: 4266–4274
16. Ashtawy H, Mahapatra N (2012) A comparative assessment of ranking accuracies of conventional and machine-learning-based scoring functions for protein-ligand binding affinity prediction. IEEE/ACM Trans Comput Biol Bioinform 9: 1301–1312
17. Ballester P, Schreyer A, Blundell T (2014) Does a more precise chemical description of protein-ligand complexes lead to more accurate prediction of binding affinity? J Chem Inform Model 54: 944–955
18. Livnah O, Bayer EA, Wilchek M et al. (1993) Three-dimensional structures of avidin and the avidin-biotin complex. Proc Natl Acad Sci U S A 90: 5076–5080
19. Trong I, Wang Z, Hyre D et al. (2011) Streptavidin and its biotin complex at atomic resolution. Acta Crystallogr D Biol Crystallogr 67:813–821
20. Williams D, Stephens E, O'brien D et al. (2004) Understanding noncovalent interactions: ligand binding energy and catalytic efficiency from ligand-induced reductions in motion within receptors and enzymes. Angew Chem Int Ed Engl 43:6596–6616
21. Dechancie J, Houk K (2008) The origins of femtomolar protein–ligand binding: hydrogen bond cooperativity and desolvation energetics in the biotin–(strept)avidin binding site. JACS 129: 5419–5429
22. Aslan FM, Yu Y, Mohr SC et al. (2005) Engineered single-chain dimeric streptavidins with an unexpected strong preference for biotin-4-fluorescein. Proc Natl Acad Sci U S A 102: 8507–8512
23. Schreir B, Stumpp C, Wiesner S et al. (2009) Computational design of ligand binding is not a solved problem. Proc Natl Acad Sci U S A 106: 18491–18496
24. Looger L, Dwyer M, Smith J et al. (2003) Computational design of receptor and sensor

proteins with novel functions. Nature 423: 185–190
25. Procko E, Berguig G, Shen B et al. (2014) A computationally designed inhibitor of an Epstein-Barr viral Bcl-2 protein induces apoptosis in infected cells. Cell 157: 1644–1656
26. Tinberg CE, Khare SD, Dou J et al. (2013) Computational design of ligand-binding proteins with high affinity and selectivity. Nature 501: 212–216
27. Leach A, Shoichet B, Peishoff C (2006) Prediction of protein-ligand interactions. Docking and Scoring: successes and gaps. J Med Chem 49: 5851–5855
28. Schneider G (2010) Virtual screening: an endless staircase? Nat Rev Drug Discov 9: 273–276
29. Huang S, Grinter S, Zou X (2010) Scoring functions and their evaluation methods for protein-ligand docking: recent advances and future directions. Phys Chem Chem Phys 12: 12899–12908
30. Michel J, Esses J (2010) Prediction of protein-ligand binding affinity by free energy simulations: assumptions, pitfalls and expectations. J Comput Aid Mol Des 24: 639–658
31. Mobley D (2012) Let's get honest about sampling. J Comput Aid Mol Des 26: 93–95
32. Guvench O, Mackerell A (2009) Computational evaluation of protein-small molecule binding. Curr Opin Struct Biol 19: 56–61
33. Chodera J, Mobley D (2013) Entropy-enthalpy compensation: role and ramification in biomolecular ligand recognition and design. Ann Rev Biophys 42: 121–142
34. Rocklin GJ, Boyce SE, Fischer M et al. (2013) Blind prediction of charged ligand binding affinities in a model binding site. J Mol Biol 425: 4569–4583

Contents

Contributors

BRITTANY ALLISON • *Department of Chemistry, Vanderbilt University, Nashville, TN, USA; Center for Structural Biology, Vanderbilt University, Nashville, TN, USA*
GEORGIOS ARCHONTIS • *Theoretical and Computational Biophysics Group, Department of Physics, University of Cyprus, Nicosia, Cyprus*
MINKYUNG BAEK • *Department of Chemistry, Seoul National University, Seoul, Republic of Korea*
BRIAN M. BAKER • *Department of Chemistry and Biochemistry and the Harper Cancer Research Institute, University of Notre Dame, South Bend IN USA*
KYLE A. BARLOW • *Graduate Program in Bioinformatics, California Institute for Quantitative Biomedical Research, and Department of Bioengineering and Therapeutic Sciences, University of California, San Francisco, San Francisco, CA, USA*
BRIAN J. BENDER • *Department of Chemistry, Vanderbilt University, Nashville, TN, USA; Department of Pharmacology, Vanderbilt University, Nashville, TN, USA*
STEVE J. BERTOLANI • *Department of Chemistry, University of California Davis, Davis, CA, USA*
JANUSZ M. BUJNICKI • *Laboratory of Bioinformatics and Protein Engineering, International Institute of Molecular and Cell Biology in Warsaw, Warsaw, Poland; Bioinformatics Laboratory, Institute of Molecular Biology and Biotechnology, Faculty of Biology, Adam Mickiewicz University, Poznan, Poland*
DYLAN ALEXANDER CARLIN • *Biophysics Graduate Group, University of California Davis, Davis, CA, USA*
MARINO CONVERTINO • *Department of Biochemistry and Biophysics, University of North Carolina, Chapel Hill, NC, USA*
BRUNO E. CORREIA • *Institute of Bioengineering, Ecole polytechnique fédérale de Lausanne, Lausanne, Switzerland*
WAYNE DAWSON • *Laboratory of Bioinformatics and Protein Engineering, International Institute of Molecular and Cell Biology in Warsaw, Warsaw, Poland*
NIKOLAY V. DOKHOLYAN • *Department of Biochemistry and Biophysics, University of North Carolina, Chapel Hill, NC, USA*
KAREN DRUART • *Department of Biology, Laboratoire de Biochimie (CNRS UMR7654), Ecole Polytechnique, Palaiseau, France*
SANJIB DUTTA • *Department of Biology, Massachusetts Institute of Technology, Cambridge, MA, USA*
GEVORG GRIGORYAN • *Department of Biological Sciences, Dartmouth College, Hanover, NH, USA; Department of Computer Science, Dartmouth College, Hanover, NH, USA*
WILLIAM A. HANSEN • *Computational Biology and Molecular Biophysics Program, Rutgers State University of New Jersey, Piscataway, NJ, USA; Center for Integrative Proteomics Research, Rutgers State University of New Jersey, Piscataway, NJ, USA*
LIM HEO • *Department of Chemistry, Seoul National University, Seoul, Republic of Korea*

BIRTE HÖCKER • *Max Planck Institute for Developmental Biology, Tübingen, Germany; Lehrstuhl für Biochemie, Universität Bayreuth, Bayreuth, Germany*
DANIEL HOERSCH • *California Institute for Quantitative Biomedical Research and Department of Bioengineering and Therapeutic Sciences, University of California, San Francisco, San Francisco, CA, USA; Fachbereich Physik, Freie Universität Berlin, Berlin, Germany*
TIM JACOBS • *University of North Carolina, Chapel Hill, NC, USA*
JOANNA M. KASPRZAK • *Laboratory of Bioinformatics and Protein Engineering, International Institute of Molecular and Cell Biology in Warsaw, Warsaw, Poland; Bioinformatics Laboratory, Institute of Molecular Biology and Biotechnology, Faculty of Biology, Adam Mickiewicz University, Poznan, Poland*
AMY E. KEATING • *Department of Biology, Massachusetts Institute of Technology, Cambridge, MA, USA*
SAGAR D. KHARE • *Department of Chemistry and Chemical Biology, Rutgers State University of New Jersey, Piscataway, NJ, USA; Center for Integrative Proteomics Research, Rutgers State University of New Jersey, Piscataway, NJ, USA*
TANJA KORTEMME • *California Institute for Quantitative Biomedical Research and Department of Bioengineering and Therapeutic Sciences, University of California, San Francisco, San Francisco, CA, USA*
BRIAN KUHLMAN • *Department of Biochemistry and Biophysics, University of North Carolina, Chapel Hill, NC, USA*
HASUP LEE • *Department of Chemistry, Seoul National University, Seoul, Republic of Korea*
TOM LINSKEY • *University of Washington, Seattle, WA, USA*
MARK W. LUNT • *Department of Chemical and Biological Engineering, Colorado State University, Fort Collins, CO, USA*
BHARAT MADAN • *Laboratory of Bioinformatics and Protein Engineering, International Institute of Molecular and Cell Biology in Warsaw, Warsaw, Poland*
MARCIN MAGNUS • *Laboratory of Bioinformatics and Protein Engineering, International Institute of Molecular and Cell Biology in Warsaw, Warsaw, Poland*
LIAM JAMES MCGUFFIN • *School of Biological Sciences, University of Reading, Reading, UK*
JENS MEILER • *Department of Chemistry, Vanderbilt University, Nashville, TN, USA; Center for Structural Biology, Vanderbilt University, Nashville, TN, USA; Department of Pharmacology, Vanderbilt University, Nashville, TN, USA*
ELENI MICHAEL • *Theoretical and Computational Biophysics Group, Department of Physics, University of Cyprus, Nicosia, Cyprus*
DAVID MIGNON • *Department of Biology, Laboratoire de Biochimie (CNRS UMR7654), Ecole Polytechnique, Palaiseau, France*
JEREMY H. MILLS • *Department of Biochemistry, University of Washington, Seattle, WA, USA*
ROCCO MORETTI • *Department of Chemistry, Vanderbilt University, Nashville, TN, USA; Center for Structural Biology, Vanderbilt University, Nashville, TN, USA*
MEHDI NELLEN • *Max Planck Institute for Developmental Biology, Tübingen, Germany*
VINCENT L. PECORARO • *Department of Chemistry, University of Michigan, Ann Arbor, MI, USA*

BRIAN G. PIERCE • *Institute for Bioscience and Biotechnology Research, University of Maryland, Rockville, MD, USA*
JEFFERSON S. PLEGARIA • *Department of Chemistry, University of Michigan, Ann Arbor, MI, USA*
SAVVAS POLYDORIDES • *Theoretical and Computational Biophysics Group, Department of Physics, University of Cyprus, Nicosia, Cyprus*
ERIK PROCKO • *Department of Biochemistry, University of Illinois, Urbana, IL, USA*
LOTHAR "LUTHER" REICH • *Department of Biology, Massachusetts Institute of Technology, Cambridge, MA, USA*
TIMOTHY P. RILEY • *Department of Chemistry and Biochemistry, University of Notre Dame, Notre Dame, IN, USA; Harper Cancer Research Institute, University of Notre Dame, Notre Dame, IN, USA*
RYAN S. RITTERSON • *California Institute for Quantitative Biomedical Research and Department of Bioengineering and Therapeutic Sciences, University of California, San Francisco, San Francisco, CA, USA*
DANIEL BARRY ROCHE • *Institut de Biologie Computationnelle, LIRMM, CNRS, Université de Montpellier, Montpellier, France; Centre de Recherche en Biologie cellulaire de Montpellier, CNRS-UMR 5237, Montpellier, France*
CHAOK SEOK • *Department of Chemistry, Seoul National University, Seoul, Republic of Korea*
JUSTIN B. SIEGEL • *Department of Chemistry, University of California Davis, One Shields Avenue, Davis, CA, USA; Genome Center, University of California Davis, One Shields Avenue, Davis, CA, USA; Department of Biochemistry and Molecular Medicine, University of California Davis, One Shields Avenue, Davis, CA, USA*
DANIEL-ADRIANO SILVA • *Department of Biochemistry, University of Washington, Seattle, WA, USA*
THOMAS SIMONSON • *Department of Biology, Laboratoire de Biochimie (CNRS UMR7654), Ecole Polytechnique, Palaiseau, France*
NISHANT K. SINGH • *Department of Chemistry and Biochemistry, University of Notre Dame, Notre Dame, IN, USA; Harper Cancer Research Institute, University of Notre Dame, Notre Dame, IN, USA*
CHRISTOPHER D. SNOW • *Department of Chemical and Biological Engineering, Colorado State University, Fort Collins, CO, USA*
YIFAN SONG • *Department of Biochemistry, University of Washington, Seattle, WA, USA*
ANDRE C. STIEL • *Max Planck Institute for Developmental Biology, Tübingen, Germany*
KRZYSZTOF SZCZEPANIAK • *Laboratory of Bioinformatics and Protein Engineering, International Institute of Molecular and Cell Biology in Warsaw, Warsaw, Poland*
SUMMER THYME • *Department of Molecular and Cellular Biology, Harvard University, Cambridge, MA, USA*
CHRISTINE E. TINBERG • *Department of Biochemistry, University of Washington, Seattle, WA, USA; Amgen, South San Francisco, CA, USA*

IRINA TUSZYNSKA • *Laboratory of Bioinformatics and Protein Engineering, International Institute of Molecular and Cell Biology in Warsaw, Warsaw, Poland; Institute of Informatics, University of Warsaw, Warsaw, Poland*
MENG WANG • *Department of Chemical and Biomolecular Engineering, University of Illinois at Urbana-Champaign, Urbana, IL, USA*
ZHIPING WENG • *Program in Bioinformatics and Integrative Biology, University of Massachusetts Medical School, Worcester, MA, USA*
HUIMIN ZHAO • *Departments of Chemical and Biomolecular Engineering, Biochemisry, and Chemistry and the Institute for Genomic Biology, University of Illinois at Urbana-Champaign, Urbana, IL USA*
FAN ZHENG • *Department of Biological Sciences, Dartmouth College, Hanover, NH, USA*

Chapter 1

In silico Identification and Characterization of Protein-Ligand Binding Sites

Daniel Barry Roche and Liam James McGuffin

Abstract

Protein–ligand binding site prediction methods aim to predict, from amino acid sequence, protein–ligand interactions, putative ligands, and ligand binding site residues using either sequence information, structural information, or a combination of both. In silico characterization of protein–ligand interactions has become extremely important to help determine a protein's functionality, as in vivo-based functional elucidation is unable to keep pace with the current growth of sequence databases. Additionally, in vitro biochemical functional elucidation is time-consuming, costly, and may not be feasible for large-scale analysis, such as drug discovery. Thus, in silico prediction of protein–ligand interactions must be utilized to aid in functional elucidation. Here, we briefly discuss protein function prediction, prediction of protein–ligand interactions, the Critical Assessment of Techniques for Protein Structure Prediction (CASP) and the Continuous Automated EvaluatiOn (CAMEO) competitions, along with their role in shaping the field. We also discuss, in detail, our cutting-edge web-server method, FunFOLD for the structurally informed prediction of protein–ligand interactions. Furthermore, we provide a step-by-step guide on using the FunFOLD web server and FunFOLD3 downloadable application, along with some real world examples, where the FunFOLD methods have been used to aid functional elucidation.

Key words Protein function prediction, Protein–ligand interactions, Binding site residue prediction, Biochemical functional elucidation, Critical Assessment of Techniques for Protein Structure Prediction (CASP), Continuous Automated EvaluatiOn (CAMEO), Protein structure prediction, Structure-based function prediction, Quality assessment of protein–ligand binding site predictions

1 Introduction

Proteins play an essential role in all cellular activity, which includes: enzymatic catalysis, maintaining cellular defenses, metabolism and catabolism, signaling within and between cells, and the maintenance of the cells' structural integrity. Hence, the identification and characterization of a protein binding site and associated ligands is a crucial step in the determination of a protein's functionality [1–3].

Barry L. Stoddard (ed.), *Computational Design of Ligand Binding Proteins*, Methods in Molecular Biology, vol. 1414, DOI 10.1007/978-1-4939-3569-7_1, © Springer Science+Business Media New York 2016

1.1 Predicting Protein–Ligand Interactions

Protein–ligand interaction prediction methods can be categorized into two broad groups: sequence-based methods and structure-based methods [1, 3, 4]. Sequence-based methods utilize evolutionary conservation to determine residues, which may be structurally or functionally important. These methods include firestar [5, 6], WSsas [7], INTREPID [8], Multi-RELIEF [9], ConSurf [10], ConFunc [11], DISCERN [12], TargetS [13], and LigandRFs [14]. Structure-based methods can additionally be separated into geometric-based methods (FINDSITE [15], Surflex-PSIM [16], LISE [17], Patch-Surfer2.0 [18], CYscore [19], LigDig [20], and EvolutionaryTrace [21, 22]), energetic methods (SITEHOUND [23]), and miscellaneous methods that utilize information from homology modeling (FunFOLD [3], FunFOLD2 [2], COACH [24], COFACTOR [25], GalaxySite [26], and GASS [27]), surface accessibility (LigSite^{CSC} [28]), and physiochemical properties, utilized by methods including SCREEN [29].

1.2 The Role of CASP and CAMEO on the Development of Protein–Ligand Interaction Methods

In recent years, there has been an explosion in the development and availability of protein–ligand binding site prediction methods. This is a direct result of the inclusion of a ligand binding site prediction category in the Critical Assessment of Techniques for Protein Structure Prediction (CASP) competition [30–32], along with the subsequent inclusion of ligand binding site prediction in the Continuous Automated EvaluatiOn (CAMEO) competition [33].

Ligand binding site residue prediction was first introduced in CASP8 [30], where the aim was to predict putative binding site residues, in the target protein, which may interact with a bound biologically relevant ligand. The top methods in CASP8 (LEE [4] and 3DLigandSite [34]) utilized homologous structures with bound biologically relevant ligands in their prediction strategies. In both CASP9 [31] and CASP10 [32], protein–ligand interaction methods converged on similar strategies; the structural superposition of models, onto templates bound to biologically relevant ligands [1].

After the CASP10 competition, the protein–ligand interaction analysis moved to the CAMEO [33] continuous evaluation competition. This was a direct result of a lack of targets for evaluation, over the 3-month prediction period of the CASP competition, although predictions were still accepted for the CASP11 competition. This also resulted in a change of prediction format, where methods not only have to predict potential ligand binding site residues, but also predict the probability that each residue binds to a specific ligand type: I, Ion; O, Organic ligand; N, nucleotide; and P, peptide. In addition, the most likely type that a protein may bind is also predicted [33]. The continuous weekly assessment of CAMEO allows for a much better picture, of how a method performs, on a large diverse data set, containing a wide diversity of ligand types [33].

1.3 Metrics to Assess Protein–Ligand Interactions

Both CASP and CAMEO utilize a number of different metrics to analyze protein–ligand interaction predictions. The first score utilized in CASP8 [30] was the Matthews Correlation Coefficient (MCC) score [35]. The MCC score is a statistical score for the comparison of predicted ligand binding site residues to observed ligand binding site residues, by analyzing the number of residues assigned as true positives, false positives, true negatives, and false negatives, resulting in a score between −1 and 1 (1 is a perfect prediction, 0 is a random prediction). The disadvantage of the MCC score is that it is a statistical measure, which does not take into account the 3D nature of a protein. Additionally, it is often a subjective matter to assign observed ligand binding site residues, even in an experimental structure, which is another disadvantage of using a purely statistical metric.

Thus, we proposed a new scoring metric: the Binding-site Distance Test (BDT) score [36], which addresses some of the problems associated with the MCC score. The BDT score takes into account the distance in 3D space a predicted binding site residue is from an observed binding site residue. The BDT score ranges from 0 to 1 (1 is a perfect prediction, 0 is a random prediction). Binding sites which are predicted close to the observed binding site score higher than binding sites predicted far from the observed site. The BDT score was used in addition to the MCC score in both the CASP9 [31] and CASP10 [32] assessments and is now a standard assessment metric used in CAMEO [33].

1.4 The FunFOLD2 Server for the Prediction of Protein–Ligand Interactions

The FunFOLD server has been developed with the user in mind, providing an intuitive interface (Fig. 1), which allows users to easily predict protein–ligand interactions for their protein of interest [2]. Additionally, for the more expert user, a PDB file of the top IntFOLD2-TS [37] model containing the biologically relevant ligand cluster can be downloaded for further interrogation, along with predicted ligand–protein interaction quality scores. Additionally, the results are available in CASP FN and CAMEO-LB format. The FunFOLD2 server takes as input a protein sequence, and optionally a short name for the target protein. Also, the user has the option to include an email address, to allow for easy results delivery or the submission page can be bookmarked and returned to later, when results are available. The FunFOLD2 server runs the IntFOLD2-TS structure prediction algorithm to produce a set of models and related templates that can be used to predict protein–ligand interactions. The FunFOLD2 [2] method combines the original FunFOLD method [3] for ligand binding site residue prediction, the FunFOLDQA method [1] for ligand binding site quality assessment, and a number of scores to comply with the CAEMO-LB prediction format [33].

The original FunFOLD method [3] was designed based on the following concept: protein structural templates from the PDB con-

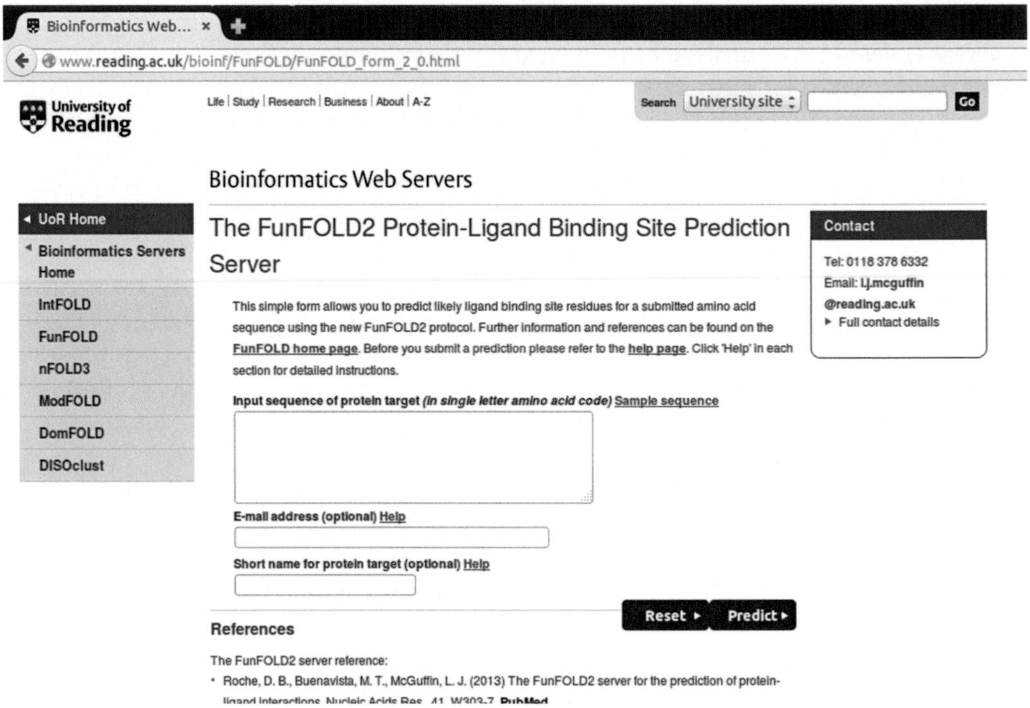

Fig. 1 Submission page for the FunFOLD server

taining biologically relevant ligands, and having the same fold (according to TM-align [38]), as the model built for the target under analysis, may contain similar binding sites. Firstly, the FunFOLD algorithm takes as input a model and a set of template PDB IDs (generated by IntFOLD2-TS [37]). Secondly, TM-align [38] is used to superpose each template determined to contain a biologically relevant ligand onto the target model (originally the method used an in-house curated ligand list, now the latest version, FunFOLD3, described below, makes use of the BioLip database [39]). Template-model superpositions having a TM-score ≥ 0.4 are used in the next step. TM-scores ranging from 0.4 to 0.6 has been shown to mark the transition step of significantly related folds [40]. Thirdly, all retained templates are superposed onto the model and ligands are assigned to clusters using an agglomerative hierarchical clustering algorithm, identifying each continuous mass of contacting ligands, thus locating potential binding pockets. Ligands are determined to be in contact within a cluster if the contact distance is less than or equal to the Van der Waal radius of the contacting atoms plus 0.5 Å. The location of the largest ligand cluster is thus determined to be the putative binding site.

Fourthly, putative ligand binding site residues are determined using a novel residue voting method. The distance between all atoms in the ligand cluster and all atoms in the modeled 3D protein is calculated. Again, residues are determined to be in contact with the ligand cluster, if the contact distance between any atom in the residue and any atom in the ligand cluster is less than or equal to the Van der Waal radius of the contacting atoms plus 0.5 Å. Finally, the next step is "residue voting," where all residues determined to be in contact with the ligand cluster are further analyzed and included in the final prediction if a residue has at least one contact to 2 ligands within the cluster and at least 25 % of the ligands in the cluster [3].

The next tool utilized by the FunFOLD2 server [2] is the FunFOLDQA algorithm [1], which assesses the quality of the FunFOLD prediction [3], outputting a set of quality scores. The FunFOLDQA algorithm produces five feature-based scores: BDTalign, Identity, Rescaled BLOSUM62 score, Equivalent Residue Ligand Distance Score, and 3D Model Quality (using ModFOLDclust2 [41]), which are subsequently combined using a neural network to produce predicted MCC and BDT scores. The predicted MCC and BDT scores can be used to rank the FunFOLD predictions of the top 10 IntFOLD2-TS models, to find the best prediction. This has been shown to provide statistically significant improvements of protein–ligand prediction quality over using FunFOLD alone [1]. The BDTalign score basically determines the fit of the model binding site into the binding sites of the templates used in the prediction. The Identity score assesses the relationship between the binding site residues, which are equivalent in 3D space, between the model and the templates, scoring them according to their amino acid identity. The Rescaled BLOSUM62 score utilizes the same concept as the Identity score, but scores equivalent residues in 3D space according to the BLOSUM62 scoring matrix. Furthermore, the Equivalent Residue Ligand Distance score scores equivalent residues in 3D space between the model and each template according to their distance from the bound ligand.

The final component of the FunFOLD2 server [2] is to score the resultant ligand binding site residues, from the top prediction, based on the CAMEO-LB criteria. The first score is a global functional propensity metric, which calculates the probability that the protein will bind to each ligand type (I, Ion; O, Organic; N, Nucleotide; P, Peptide). The second score is the per-residue functional propensity metric, which determines the propensity that each predicted ligand binding site residue is in contact with each ligand type (I, O, N, & P) [2].

1.5 The FunFOLD3 Algorithm for the Prediction of Protein–Ligand Interactions

The FunFOLD3 algorithm is the latest implementation of FunFOLD. FunFOLD3 was designed to produce predictions to comply with the CAMEO-LB prediction format [33], including the development of new metrics to predict per-atom *P-values*. Another major change in FunFOLD3 is the use of the BioLip database [39], for the determination of biologically relevant ligands at multiple binding sites. In addition to the provision of functional annotations, namely EC [42] numbers and GO terms [43]. The FunFOLD3 algorithm along with FunFOLDQA [1] has been integrated into the latest version of the IntFOLD server pipeline [44] and is available as an executable JAR file. The executable version of FunFOLD3 does not incorporate the FunFOLDQA binding site quality scoring module, however, the FunFOLDQA program may be downloaded as a separate JAR executable if desired.

The FunFOLD2 method and its previous implementations have been benchmarked at CASP9 and CASP10 and were amongst the top performing methods [31, 32]. In addition to CASP, the FunFOLD2 and FunFOLD3 methods are now continuously benchmarked by CAMEO [33] (http://www.cameo3d.org). Furthermore, the FunFOLD algorithms have been utilized in numerous studies, including the investigation of barley powdery mildew proteins [45, 46], calcium binding proteins [47], and olfactory proteins [48], which have resulted in biologically significant findings.

In summary, the use of computational methods for the prediction of protein–ligand interactions is essential in the era of high-throughput next-generation sequencing, as experimental methods are unable to keep pace. The prediction of protein–ligand interactions can lead to the interpretation of a protein's general function. These predictions can be further utilized in subsequent in silico, in vivo and in vitro studies, for the discovery of new functions, as well as in drug discovery, which can impact on issues such as health and disease.

2 Materials and Systems Requirements

2.1 Web Server Requirements

1. For the FunFOLD2 web server [2], internet access and a web browser are required. The server is freely accessible at: http://www.reading.ac.uk/bioinf/FunFOLD/ (*See* Fig. 1 and **Note 1**). The FunFOLD2 server has been extensively tested on Google Chrome and Firefox, which are recommended for proper use. The server also works on other browsers such as Internet Explorer, Safari and Opera, but these browsers have not been tested as extensively.
2. To run your protein–ligand interaction predictions on the FunFOLD2 server you require an amino acid sequence for your protein of interest, in single-letter code format. Additionally, a short name can be given for the target sequence submitted and an email address can be included to inform the user when the prediction is complete. If the length of the target amino acid sequence is longer than 500 amino acids, it is best to divide the

target sequence into domains, using PFAM [49] or SMART [50], then submit each domain sequence separately. For a more detailed explanation along with potential problems that can be encountered at the submission stage *see* **Note 1**.

2.2 Requirements for the FunFOLD3 Downloadable Executable

A downloadable version of the FunFOLD3 method is available as an executable JAR file, which can be run locally. The executable has several dependencies and system requirements which are briefly described below. The executable along with a detailed README file and example input and output data can be downloaded from the following location: http://www.reading.ac.uk/bioinf/downloads/ (*See* **Note 2** for potential errors that may be encountered).

The system requirements are as follows:

1. A linux-based operating system such as Ubuntu.
2. A recent version of Java (www.java.com/getjava/).
3. A recent version of PyMOL (www.pymol.org).
4. The TM-align program [38] (http://zhanglab.ccmb.med.umich.edu/TM-align/). Please ensure the TM-align program is working on your system before attempting to run FunFOLD3. Ensure that you have the correct 32-bit/64-bit version for your hardware and that the TMalign file is made executable: chmod +x TMalign.
5. wget and ImageMagick installed system wide.
6. The CIF chemical components database file [51] should be downloaded from here: ftp://ftp.wwpdb.org/pub/pdb/data/monomers/components.cif.
7. The BioLip databases [39] containing ligand and receptor PDB files are also required (up to 30 GB or disc space may be required). The databases need to be downloaded in two sections: firstly all annotations prior to 2013-03-06 can be downloaded from here for the receptor database: http://zhanglab.ccmb.med.umich.edu/BioLiP/download/receptor_2013-03-6.tar.bz2 (3.6 G) and from here for the ligand database: http://zhanglab.ccmb.med.umich.edu/BioLiP/download/ligand_2013-03-6.tar.bz2 (438 M). The Text File of the BioLip annotations can be downloaded from here: http://zhanglab.ccmb.med.umich.edu/BioLiP/download/BioLiP.tar.bz2. To update the databases to include annotations after 2013-03-6 it is recommended to download and use this perl script which will update the databases: http://zhanglab.ccmb.med.umich.edu/BioLiP/download/download_all_sets.pl. The BioLip text file: http://zhanglab.ccmb.med.umich.edu/BioLiP/download/BioLiP.tar.bz2 and all the weekly update text files should be concatenated to form a large text file containing all of the annotations. Furthermore, it is recommended to regularly update your BioLip and CIF databases.

Additionally, a shell script is available as downloadBioLipdata.sh, which can be downloaded from here: http://www.reading.ac.uk/bioinf/downloads/, in a compressed directory: FunFOLD3Package.tar.gz. To run the shell script simply edit the file paths for the location of the BioLip databases and the executable directory.

8. Please ensure your system environment is set to English, as utilizing other languages may cause problems with the FunFOLD calculations: export LC_ALL=en_US.utf-8.
9. Note the FunFOLD3 executable does not contain the FunFOLDQA code. The FunFOLDQA code is available to download as a separate executable if desired.

3 Methods

In this section we present a step-by-step guide on utilizing the FunFOLD2 server and the FunFOLD3 downloadable executable, to produce protein–ligand interaction predictions for the user's sequence of interest. We also describe interesting case studies of the FunFOLD3 method and its previous implementations.

3.1 The FunFOLD2 Server

1. Navigate to the FunFOLD2 submission page: http://www.reading.ac.uk/bioinf/FunFOLD/FunFOLD_form_2_0.html.
2. The next step is to paste the full single-letter format amino acid sequence of your protein of interest into the text box provided on the submission page (*see* Fig. 1).
3. Optionally, the user can provide a short name for their target sequence.
4. The user has the option to supply their email address, which enables an email to be sent to the user once the results of the target sequences become available.
5. Once all of the required information boxes, on the submission page, have been filled, the user then needs to click on the submit button to enable submission of their prediction.
6. Presently, submissions are limited to one per IP address, to enable the maintenance of speed and server capacity. Upon completion of the user's prediction, their IP address is automatically unlocked and they can then submit their next target sequence. *See* **Note 1** for common problems encountered at the submission step.
7. Upon job completion an email is sent to the user, which contains a link to the prediction results for the target sequence. *See* Fig. 2 for an example results page (FunFOLD3 via the IntFOLD server) and Fig. 3 for example results from CASP11.

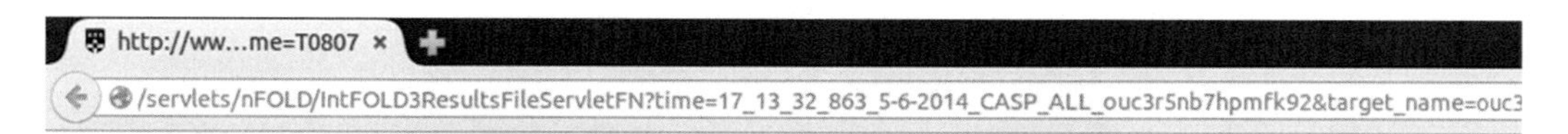

PyMOL generated image of ligand binding residues prediction for T0807

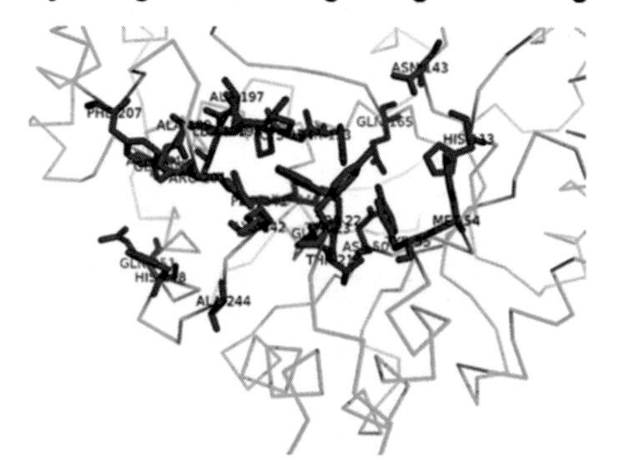

Click here to download PDB file of this model with the superposition of all identified ligands.

Predicted ligand binding residues are shown as blue sticks in the image above.

Binding site: 20, 21, 22, 23, 50, 54, 55, 113, 143, 165, 193, 194, 195, 196, 197, 198, 199, 200, 201, 207, 224, 240, 241, 242, 244, 248, 251
Most likely ligands at each site (Type): NAP
Centroid ligands at each site (TypeID): NAP294
All ligands in clusters (Type-Frequency): NAP-6, GLU-1, NAD-1, TES-2
Likely+centroid ligands at each site: NAP294
EC numbers: 1.1.1.50;1.1.1.213;1.1.1.188;1.1.1.-;1.1.1.274;
GO terms:
0005737;0008202;0016229;0016491;0021766;0047023;0047026;0055114;0004032;0004958;0006629;0006693;0007186;0007586;0008284;

JSmol view of ligand binding residues prediction for T0807

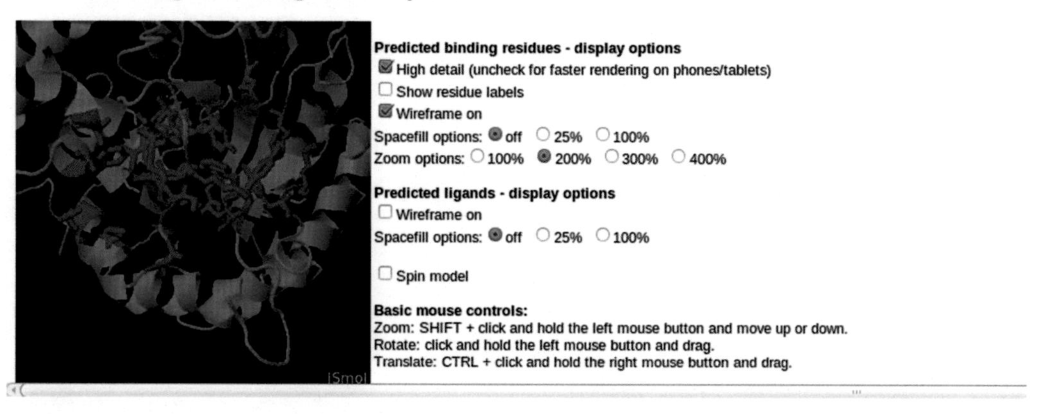

Fig. 2 The IntFOLD3-FN (FunFOLD3) server results page for CASP11 target T0807 (PDB ID 4wgh)

8. The results page contains graphical results for the target sequence, in addition to downloadable machine readable results in CASP format. Firstly, a graphical representation of the ligand binding site, showing putative binding site residues, rendered using PyMOL (www.pymol.org) is shown. The backbone of the protein is shown as a green ribbon, while the putative ligand binding site residues are labeled and shown as blue sticks. Secondly, a link is also available to download a PDB file containing the putative ligand binding site cluster within the top

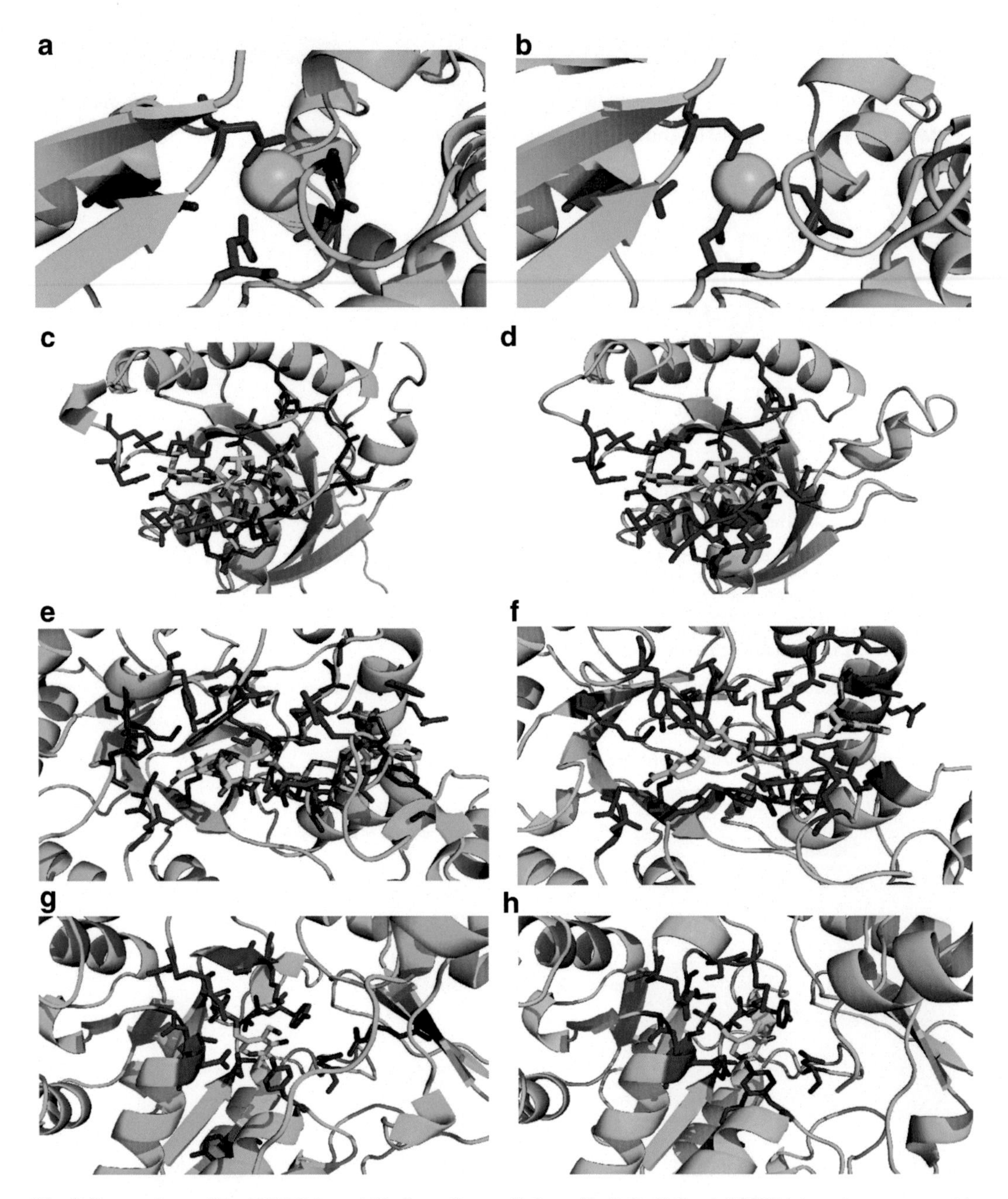

Fig. 3 Comparison of FunFOLD3 ligand binding site predictions (A, C, E, G) for 4 CASP11 targets, compared to the observed ligand binding sites (B, D, F, H). (**a**) Predicted ligand binding site for T0854 (PDB ID 4rn3), with correctly predicted binding site residues in *blue* and under- and over-predictions in *red*, the MG ligand is colored by element. BDT score of 0.845 and MCC score of 0.745. (**b**) The observed ligand binding site for T0854 (PDB ID 4rn3), with binding site residues colored in *blue* and the ligand MG colored by element. (**c**) Predicted ligand binding site for T0798 (PDB ID 4ojk), with correctly predicted binding site residues in *blue* and under- and over-predictions in *red*, the GDP ligand is colored by element. BDT score of 0.797 and MCC score of 0.754. (**d**) The observed ligand binding site for T0798 (PDB ID 4ojk), with binding site residues colored in *blue* and the ligand GDP colored by element. (**e**) Predicted ligand binding site for T0807 (PDB ID 4wgh), with correctly predicted binding site residues in *blue* and under- and over-predictions in *red*, the NAP ligand is colored by element.

IntFOLD [52] model. Thirdly, the CASP FN format results are shown. This includes a list of putative ligand binding site residues. The list also includes, the most likely ligand, which is the most likely ligand to be bound to the target protein according to the FunFOLD prediction. This is followed by the centroid ligand and a list of all ligands within the putative ligand cluster is also included. The centroid and most likely ligand have an associated residue number that corresponds to their residue number in the downloadable PDB file, the residue number can be easily used to locate the ligand in the PDB file for a more detailed examination of the results.

9. The final section of the results page is a JSmol view of the ligand binding site within the target protein, which can be easily used to examine the prediction in 3D space. There are a number of options to rotate the protein, show and hide the ligands as well as alter the way the ligands are represented.
10. Moreover, for the version of FunFOLD (FunFOLD3) integrated into the IntFOLD pipeline [44], putative EC [42] and GO [43] codes, derived from templates used in the prediction from the BioLip [39] database are included (*See* **Note 3** for details on the IntFOLD server [44, 53]).
11. In addition, predicted quality scores from FunFOLDQA [1] are also provided: BDTalign, Identity, Rescaled BLOSUM62 score, Equivalent Residue Ligand Distance Score, and Model Quality along with the predicted MCC and BDT scores (*See* Subheading 1.4 for a description of these scores). Furthermore, the propensity that the target protein binds to each ligand type (I, Ion; O, Organic; N, Nucleotide; P, Peptide) is also provided in CAMEO-LB format [33] (*See* **Note 2** for potential errors that may be encountered and **Note 4** for current method limitations).

3.2 The FunFOLD3 Executable

1. For large-scale analysis or to integrate the FunFOLD3 method into a structure prediction pipeline or web server (*See* **Notes 2** and **5**) a downloadable executable JAR file, which has been developed to run on linux-based operating systems is available (http://www.reading.ac.uk/bioinf/downloads/). This version

Fig. 3 (continued) BDT score of 0.849 and MCC score of 0.771. (**f**) The observed ligand binding site for T0807 (PDB ID 4wgh), with binding site residues colored in *blue* and the ligand NAP colored by element. (**g**) Predicted ligand binding site for T0819 (PDB ID 4wbt), with correctly predicted binding site residues in *blue* and under- and over-predictions in *red*, the PLP ligand is colored by element. BDT score of 0.753 and MCC score of 0.877. (**h**). The observed ligand binding site for T0819 (PDB ID 4wbt), with binding site residues colored in *blue* and the ligand PLP colored by element. All images were rendered using PyMOL (http://www.pymol.org/)

of the program has been tested on recent versions of Ubuntu, but it should work on all linux-based systems that have bash installed and meet the system requirements (*See* Subheading 2.2 and **item 1**).

2. To run the program you can simply edit the shell script (FunFOLD3.sh) or you can follow the steps below.
3. The user can optionally set the bash environment variable for Java, TM-align, and PyMOL if they have not installed it system wide, along with the location of the databases and database files, e.g.

```
export LC_ALL=en_US.utf-8
export PYMOL_HOME=/usr/bin/
export TMALIGN_HOME=/home/roche/bin/
export JAVA_HOME=/usr/bin/
export BIOLIP_Directory=/home/roche/bin/BioLip/FunFOLD
BioLip/
export BIOLIP_LIGAND=/home/roche/bin/BioLip/FunFOLD-
BioLip/ligand/
export  BIOLIP_RECEPTOR=/home/roche/bin/BioLip/Fun-
FOLDBioLip/receptor/
export     BIOLIP_TXT=/home/roche/bin/BioLip/FunFOLD
BioLip/BioLiP.txt
export CIF=/home/roche/bin/BioLip/FunFOLDBioLip/com-
ponents.cif
$BIOLIP_Directory = BioLip directory location
$BIOLIP_TXT = BioLip database text file including the
full directory path
$BIOLIP_LIGAND = BioLip ligand directory
$BIOLIP_RECEPTOR = BioLip receptor directory
$CIF = CIF file including the full directory path
```

4. For example, if the path of your model was "/home/roche/bin/FunFOLD3/MUProt_TS3", your list of templates was "/home/roche/bin/FunFOLD3/T0470_PARENTNew.dat" (all templates should be listed on a single line separated by a space), your FASTA sequence file was "/home/roche/bin/FunFOLD3/T0470.fasta", your output directory was "/home/roche/bin/FunFOLD3/" and your target was called T0470:

```
    $JAVA_HOME/java  -jar  FunFOLD3.jar  /home/roche/
bin/FunFOLD3/MUProt_TS3  T0470  /home/roche/bin/Fun-
FOLD3/ /home/roche/bin/FunFOLD3/T0470_PARENTNew.dat /
home/roche/bin/FunFOLD3/T0470.fasta $BIOLIP_TXT $BIOLIP_
LIGAND $BIOLIP_RECEPTOR $CIF
    Or, using the shell script provided:
```

```
./FunFOLD3.sh /home/roche/bin/FunFOLD3/MUProt_TS3
T0470 /home/roche/bin/FunFOLD3/ /home/roche/bin/Fun-
FOLD3/T0470_PARENTNew.dat /home/roche/bin/FunFOLD3/
T0470.fasta
```

5. Basically, the user requires a model generated for their target protein, this can be achieved using a homology modeling method either in-house or via a web server such as IntFOLD [37] (*see* **Note 3**). Additionally, the user needs a list of structurally similar templates. Again this list of templates can be generated from the list of templates used to generate the target protein model. The program utilizes the templates that have the same fold and contain biologically relevant ligands in the prediction process. Furthermore, it is important to download and install the BioLip databases [39] and CIF chemical components library file [51]. Additionally, it is important that the full paths for all input files are used, the output directory should also end with a "/" and must contain the input model, template list, and FASTA sequence file.
6. Additionally, a shell script is available called downloadBioLipdata.sh, which can be used to download and update the BioLip and CIF libraries. The shell script and the required perl script can be found on the downloads page, in a compressed directory: FunFOLD3Package.tar.gz. To run the shell script simply edit the file paths for the location of the BioLip databases and the executable directory.
7. A number of output files are produced in the output directory (e.g. "/home/roche/bin/FunFOLD3/") and a log of the prediction process is output to screen as standard output. A description of the output files are as follows:
   (a) The final ligand binding site prediction file "T0470_FN.txt" is supplied, conforming to CASP FN format. This file contains a list of predicted binding site residues, ligands, along with associated EC and GO terms.
   (b) The final binding site prediction file "T0470_FN2_CAMEO-LB.txt" is additionally supplied in CAMEO-LB format. This file contains the predicted propensity that each ligand type is in contact with the predicted binding site residues.
   (c) A PDB file "T0470_lig.pdb", which contains superpositions of all templates, having the same fold and containing biologically relevant ligands, onto the model is produced.
   (d) A reduced version of the PDB file "T0470_lig2.pdb", which contains only the target model with all possible ligands is also produced.

(e) Another reduced version of the PDB file "T0470_lig3.pdb", which contains only the target model with the predicted centroid ligand, is additionally output.

(f) A graphical representation of the protein–ligand interaction prediction "T0470_binding_site.png" is automatically generated using PyMOL.

(g) Finally, the PyMOL script "pymol.script" that was used to generate the image file is also output.

8. An example of output produced by FunFOLD3 for target T0470 can be found in the compressed directory: "T0470_Results.tar.gz" along with an example of the required input: "T0470_Input.tar.gz". These example directories can be found on the downloads page: http://www.reading.ac.uk/bioinf/downloads/, as part of the FunFOLD3 package - FunFOLD3Package.tar.gz.

3.3 Server Fair Usage Policy

To enable timely throughput and wide use of the server, a fair usage policy is implemented. Users are allowed to submit one prediction per IP address. Once the first job is complete, a notification is sent to the user via email, if an email address has been provided. If a user does not provide an email address, then a link to the results page is provided, which users are recommended to bookmark during the submission process. Once the job has been completed, the user's IP address is unlocked and the server is ready to receive the next submission. The results for each complete job is saved for 30 days. It is recommended for large-scale analysis of a large number of proteins (proteome level) to download the executable version of FunFOLD3 (See Subheading 3.2 and **Notes 2** and **5**).

3.4 Case Studies

The FunFOLD3 method and its previous implementation have been used in a number of studies [45–48], which have led to biologically significant findings, here we discuss one such study. Furthermore, in-house analysis of the CASP11 FN predictions produced by the FunFOLD3 algorithm, via the IntFOLD server are evaluated (CASP11 group ID: TS133).

3.5 Analysis of the Barley Powdery Mildew Proteome

The first study combined proteogenomic and in silico structural and functional annotations (prediction of protein–ligand interactions), to enable the investigation of the pathogen proteome of barley powdery mildew [45, 46]. Basically, genomic scale structure prediction was carried out using IntFOLD [53]. Both the global and per-residue model quality were assessed utilizing ModFOLD3 [52, 54] and putative protein–ligand interactions were additionally predicted using FunFOLD [3]. The results lead to interesting conclusions about the structural and functional diversity of the proteomes. Firstly, only six proteins could be modeled with a model quality score above 0.4, leading to a conclusion that the genome is very structurally diverse and may have many novel folds. Secondly, for the six predicted structures, FunFOLD [3] was able to predict that the proteins were carbohydrate binding, and using the models and other additional data it was concluded that they were probably glycosyl hydrolases. Furthermore, the putative functionality was experimentally verified. In conclusion the FunFOLD method was crucial in the putative functionality assignment of these enzymes, which were subsequently experimentally verified.

3.6 CASP11 Functional Prediction

The second case study focuses on the analysis of FunFOLD3 blind predictions from the CASP11 competition. Briefly, all CASP11 targets with associated PDB IDs were analyzed. Firstly, targets were analyzed using the BioLip [39] database to determine if they contained biologically relevant ligands. Secondly, targets deemed to contain biologically relevant ligands were further investigated to determine ligand binding site residues, using the standard CASP distance cut-off; the Van der Waal radius of the contacting atom of a residue and the contacting ligand atom plus 0.5 Å. This resulted in a set of 11 proteins containing biologically relevant ligands and binding site residues.

In CASP11, the FunFOLD3 method was integrated into the IntFOLD-TS predictions (TS133). Protein–ligand interactions were predicted for 8 out of the 11 FN targets (described above), with a mean MCC score of 0.554 and a mean BDT score of 0.478. Four of the top predictions are subsequently discussed in detail. Fig. 3 highlights the four assessed predictions, compared to the observed binding sites, with BDT scores ranging from 0.753 to 0.849. Figure 3a shows the predicted ligand binding site for a HAD-superfamily hydrolase, subfamily IA, variant 1 from *Geobacter sulfurreducens* (CASP ID T0854 and PDB ID 4rn3), with correctly predicted binding site residues in blue (16,18 and 173) and under (177) and over-predictions [18] in red, the MG ligand is colored by element. The prediction resulted in a BDT score of 0.845 and an MCC score of 0.745. Figure 3b shows the observed binding site for T0854 (PDB ID 4rn3), with binding site residues colored in blue and the ligand MG colored by element. A minority of residues were either under or over-predicted for this target as a result of the centroid ligand and the ligand cluster not being well superposed. The binding sites of the templates were not well superposed onto the model binding site, thus, the ligand cluster was not optimally located in the binding site.

The second CASP11 target is a cGMP-dependent protein kinase II from *Rattus norvegicus* (CASP ID T0798 and PDB ID 4ojk). Figure 3c shows the predicted ligand binding site, with correctly predicted binding site residues (14, 15, 16, 17, 18, 19, 29, 30, 31, 117, 118, 120, 121, 147, 148, 149) in blue and under [11, 31] and over-predictions (13, 33, 35, 36, 61, 62) in red, the GDP ligand is colored by element. This prediction has a BDT score of 0.797 and an MCC score of 0.754. The observed ligand binding site for T0798 (PDB ID 4ojk), with binding site residues colored in blue and the ligand GDP colored by element can be seen in Fig. 3d. Again, the minority of under- and over-predictions are caused by firstly having a very large ligand binding site, which did not have the ligands cluster in the correct location within the large binding site, in part due to a number of templates having larger cofactor ligands and others having an additional MG ion bound with the cofactor.

The third example is of an aldo/keto reductase from *Klebsiella pneumoniae* (CASP ID T0807 and PDB ID 4wgh). Figure 3e shows the predicted ligand binding site, with correctly predicted binding site residues (20, 21, 22, 50, 55, 143, 165, 193, 194, 195, 196, 198, 199, 201, 224, 240, 241, 242, 244, 248, 251) in blue and under- (80, 142, 243, 245, 252) and over-predictions (23, 54, 113, 197, 200, 207) in red, the NAP ligand is colored by element. This prediction resulted in a BDT score of 0.849 and an MCC score of 0.771. In addition, the observed ligand binding site can be seen in Fig. 3f, with binding site residues colored in blue and the ligand NAP colored by element. Furthermore, the over- and under-predictions seem to be a direct result of a number of templates having an additional ligand bound along with the cofactor, resulting in an extended ligand binding site.

The final CASP11 target that we will analyze is a histidinol-phosphate aminotransferase from *Sinorhizobium meliloti* (CASP ID T0819 and PDB ID 4wbt). Figure 3g shows the predicted ligand binding site, with correctly predicted binding site residues (93, 94, 95, 119, 167, 194, 197, 223, 225, 226, 234) in blue and under- (161, 196) and over-predictions (347) in red, the PLP ligand is colored by element. The prediction results in a BDT score of 0.753 and an MCC score of 0.877. In addition, Fig. 3h shows the observed ligand binding site for T0819 (PDB ID 4wbt), with binding site residues colored in blue and the ligand PLP colored by element. Here, the under- and over-predictions are a result of the incorrect orientation of residues in one case away from the binding site (TYR 161), in the other cases the under-predicted residue (ALA 196) and the over-predicted residue (ARG 347) are located on flexible loops.

These four CASP11 examples and the results [30–32] from previous CASP assessments, along with in-house evaluations [1, 3], highlight the usefulness of the FunFOLD methods for the accurate prediction of protein–ligand interactions, for a wide range of proteins and ligand binding sites. *See* **Note 4** for current method limitations.

4 Notes

1. When using the FunFOLD server [1–3], several problems may be encountered. These mainly include, but are not limited to, providing the incorrect data to the server. It is important to input a sequence in plain text and single-letter code format, into the text box labeled "Input sequence of target protein". Additionally, it is recommended not to submit sequences longer than 500 amino acids. Firstly, these sequences usually contain multiple domains, thus it may not be possible to find a good template to model multiple domains, resulting in one or more domains not being modeled well. Secondly, if both domains contain ligand binding sites only one will be predicted and displayed in the results page. Hence, it is advisable to partition the sequence into domains and submit each domain sequence as a separate job.

   The next place where errors can occur is the next submission box "Short name for protein target"; inputting a short name for your protein sequence is useful to keep track of your prediction by providing a meaningful description. The short descriptor is limited to a set of characters: letters A–Z (either case), the numbers 0–9, and the following characters: .~_-. The protein descriptor supplied by the user is subsequently utilized in the subject line of the email sent to the user, which contains a link to the FunFOLD results for their target protein.

   The final text box to be completed is the "E-mail address". This will enable a link of the graphical and machine readable results to be sent to the user, upon job completion. Here errors can occur if the user incorrectly inputs their email address.

2. For the downloadable Java application FunFOLD3, errors can occur but are not limited to the following reasons: Firstly, errors can occur if the dependencies—Java, TM-align [38], BioLip [39], and PyMOL—are not installed or not installed correctly; secondly, if the full paths to the input files, BioLip database, CIF database, and output directory are not included; thirdly, if the target model to be analyzed is not in the output directory; fourthly, if the list of templates used in the prediction contains non-existent PDB IDs or the PDB IDs (including chain identifiers) are not all on the same line of the text file, the program will not run; fifthly, if the input sequence file is not in FASTA format; finally, it is recommended to limit the template list to 40 template structures, for efficient prediction and this is near the limit of the number of structure files PyMOL can handle (*See* Subheading 3.2 and the README file downloaded with the executable).

Moreover, downloading the BioLip database may be time-consuming and is an area where problems may occur if the instructions available on the BioLip website and contained in the README are not followed. Alternatively, if the user has the I-TASSER [55] pipeline installed on their system, the BioLip databases [39] will have been installed as part of the I-TASSER installation process.

3. The IntFOLD server [44, 53] is a novel independent server, which gives users easy access to a number of cutting-edge methods, for the prediction of structure and function from sequence. The idea behind the IntFOLD server is to provide easy access to our methods from a single location, producing easily understandable integrated output of results, enabling ease of access for the non-expert user. The IntFOLD server provides output in graphical form, enabling users to interpret results at a glance as well as CASP formatted text files, allowing a more in-depth analysis of the prediction results. The IntFOLD pipeline integrates a number of methods, to enable users to simply input a target sequence and produce a set of models (IntFOLD3-TS [37]), with associated global and per-residue model quality (ModFOLD5 [54]), disorder prediction (DISOclust3 [56]), domain partitioning (DomFOLD3), and function prediction results utilizing FunFOLD3 [1–3]. The component methods of the IntFOLD server have been ranked amongst the top methods in their respective categories at recent CASP and CAMEO competitions.
4. Predicting protein–ligand interactions is a difficult task, which results in a number of limitations to current prediction methods. The following is a non-exhaustive list of the most common limitations currently encountered in the field: (1) If the server or prediction algorithm is unable to build a model for the target sequence, then no protein–ligand interactions are predicted. The solution to this problem is to utilize sequence-based methods (*see* Subheading 1.1 for suggestions of sequence-based prediction methods), which are less accurate. (2) If structurally similar templates to the target, which containing biologically relevant ligands cannot be found, then no prediction can be made. (3) The FunFOLD server currently outputs predictions based on the top IntFOLD model, which has the highest global model quality score. This model may not have the best per-residue model quality around the binding site location, resulting in under- or over-predicted ligand binding site residues.
5. The user has the option of using the server version of FunFOLD, IntFOLD, or the downloadable java application. The user has to leverage the option most appropriate to meet their needs. The server only permits users to submit one job at a time due to server load balancing. If the user would like to carry out large-

scale analysis, for example predicting protein–ligand interactions for a proteome, it is then recommended to download and use the executable java application for FunFOLD3. This allows the user the freedom in the number of structures they can analyze, provided they have adequate CPU capacity.

For light use (several predictions a week), server prediction is adequate for the user, whereas for heavy users (greater than 5–10 predictions a week) the downloadable application would be the most useful. Extensive help pages are available for the FunFOLD server. Furthermore, at least 30 GB of disc space is required to download the complete BioLip libraries. In addition, an extensive README file, example input and output files are available to aid the user in the installation and running of the FunFOLD3 downloadable java application.

Acknowledgements

Daniel Barry Roche is a recipient of a Young Investigator Fellowship from the Institut de Biologie Computationnelle, Université de Montpellier (ANR Investissements D'Avenir Bio-informatique: projet IBC).

References

1. Roche DB, Buenavista MT, Mcguffin LJ (2012) FunFOLDQA: a quality assessment tool for protein-ligand binding site residue predictions. PLoS One 7:e38219
2. Roche DB, Buenavista MT, Mcguffin LJ (2013) The FunFOLD2 server for the prediction of protein-ligand interactions. Nucleic Acids Res 41:W303–W307
3. Roche DB, Tetchner SJ, Mcguffin LJ (2011) FunFOLD: an improved automated method for the prediction of ligand binding residues using 3D models of proteins. BMC Bioinformatics 12:160
4. Oh M, Joo K, Lee J (2009) Protein-binding site prediction based on three-dimensional protein modeling. Proteins 77(Suppl 9): 152–156
5. Lopez G, Maietta P, Rodriguez JM et al (2011) Firestar--advances in the prediction of functionally important residues. Nucleic Acids Res 39:W235–W241
6. Lopez G, Valencia A, Tress ML (2007) Firestar--prediction of functionally important residues using structural templates and alignment reliability. Nucleic Acids Res 35:W573–W577
7. Talavera D, Laskowski RA, Thornton JM (2009) WSsas: a web service for the annotation of functional residues through structural homologues. Bioinformatics 25:1192–1194
8. Sankararaman S, Kolaczkowski B, Sjolander K (2009) INTREPID: a web server for prediction of functionally important residues by evolutionary analysis. Nucleic Acids Res 37: W390–W395
9. Ye K, Feenstra KA, Heringa J et al (2008) Multi-RELIEF: a method to recognize specificity determining residues from multiple sequence alignments using a Machine-Learning approach for feature weighting. Bioinformatics 24:18–25
10. Ashkenazy H, Erez E, Martz E et al (2010) ConSurf 2010: calculating evolutionary conservation in sequence and structure of proteins and nucleic acids. Nucleic Acids Res 38(Suppl):W529–W533
11. Wass MN, Sternberg MJ (2008) ConFunc--functional annotation in the twilight zone. Bioinformatics 24:798–806
12. Sankararaman S, Sha F, Kirsch JF et al (2010) Active site prediction using evolutionary and structural information. Bioinformatics 26:617–624
13. Dong-Jun Y, Jun H, Jing Y et al (2013) Designing template-free predictor for targeting

protein-ligand binding sites with classifier ensemble and spatial clustering. IEEE/ACM Trans Comput Biol Bioinform 10:994–1008
14. Chen P, Huang JHZ, Gao X (2014) LigandRFs: random forest ensemble to identify ligand-binding residues from sequence information alone. BMC Bioinformatics 15:S4
15. Brylinski M, Skolnick J (2008) A threading-based method (FINDSITE) for ligand-binding site prediction and functional annotation. Proc Natl Acad Sci U S A 105:129–134
16. Spitzer R, Cleves AE, Jain AN (2011) Surface-based protein binding pocket similarity. Proteins 79:2746–2763
17. Xie ZR, Liu CK, Hsiao FC et al (2013) LISE: a server using ligand-interacting and site-enriched protein triangles for prediction of ligand-binding sites. Nucleic Acids Res 41:W292–W296
18. Zhu X, Xiong Y, Kihara D (2015) Large-scale binding ligand prediction by improved patch-based method Patch-Surfer2.0. Bioinformatics 31:707–713
19. Cao Y, Li L (2014) Improved protein-ligand binding affinity prediction by using a curvature-dependent surface-area model. Bioinformatics 30:1674–1680
20. Fuller JC, Martinez M, Henrich S et al (2014) LigDig: a web server for querying ligand-protein interactions. Bioinformatics 31:1147–1149
21. Erdin S, Ward RM, Venner E et al (2010) Evolutionary trace annotation of protein function in the structural proteome. J Mol Biol 396:1451–1473
22. Madabushi S, Yao H, Marsh M et al (2002) Structural clusters of evolutionary trace residues are statistically significant and common in proteins. J Mol Biol 316:139–154
23. Hernandez M, Ghersi D, Sanchez R (2009) SITEHOUND-web: a server for ligand binding site identification in protein structures. Nucleic Acids Res 37:W413–W416
24. Yang J, Roy A, Zhang Y (2013) Protein-ligand binding site recognition using complementary binding-specific substructure comparison and sequence profile alignment. Bioinformatics 29:2588–2595
25. Roy A, Yang J, Zhang Y (2012) COFACTOR: an accurate comparative algorithm for structure-based protein function annotation. Nucleic Acids Res 40:W471–W477
26. Heo L, Shin WH, Lee MS et al (2014) GalaxySite: ligand-binding-site prediction by using molecular docking. Nucleic Acids Res 42:W210–W214
27. Izidoro SC, De Melo-Minardi RC, Pappa GL (2014) GASS: identifying enzyme active sites with genetic algorithms. Bioinformatics 31:864–870
28. Huang B, Schroeder M (2006) LIGSITEcsc: predicting ligand binding sites using the Connolly surface and degree of conservation. BMC Struct Biol 6:19
29. Andersson CD, Chen BY, Linusson A (2010) Mapping of ligand-binding cavities in proteins. Proteins 78:1408–1422
30. Lopez G, Ezkurdia I, Tress ML (2009) Assessment of ligand binding residue predictions in CASP8. Proteins 77(Suppl 9):138–146
31. Schmidt T, Haas J, Cassarino TG et al (2011) Assessment of ligand binding residue predictions in CASP9. Proteins: Structure, Function, and Bioinformatics 79 Suppl 10:126–136
32. Gallo Cassarino T, Bordoli L, Schwede T (2014) Assessment of ligand binding site predictions in CASP10. Proteins 82(Suppl 2):154–163
33. Haas J, Roth S, Arnold K et al (2013) The Protein Model Portal--a comprehensive resource for protein structure and model information. Database (Oxford) 2013:bat031
34. Wass MN, Sternberg MJ (2009) Prediction of ligand binding sites using homologous structures and conservation at CASP8. Proteins 77(Suppl 9):147–151
35. Matthews BW (1975) Comparison of the predicted and observed secondary structure of T4 phage lysozyme. Biochim Biophys Acta 405:442–451
36. Roche DB, Tetchner SJ, Mcguffin LJ (2010) The binding site distance test score: a robust method for the assessment of predicted protein binding sites. Bioinformatics 26:2920–2921
37. Buenavista MT, Roche DB, Mcguffin LJ (2012) Improvement of 3D protein models using multiple templates guided by single-template model quality assessment. Bioinformatics 28:1851–1857
38. Zhang Y, Skolnick J (2005) TM-align: a protein structure alignment algorithm based on the TM-score. Nucleic Acids Res 33:2302–2309
39. Yang J, Roy A, Zhang Y (2013) BioLiP: a semi-manually curated database for biologically relevant ligand-protein interactions. Nucleic Acids Res 41:D1096–D1103
40. Xu J, Zhang Y (2010) How significant is a protein structure similarity with TM-score = 0.5? Bioinformatics 26:889–895
41. Mcguffin LJ, Roche DB (2010) Rapid model quality assessment for protein structure predictions using the comparison of multiple models without structural alignments. Bioinformatics 26:182–188
42. Webb EC (1989) Nomenclature Committee of the International-Union-of-Biochemistry (Nc-Iub) - Enzyme Nomenclature - Recommendations 1984 - Supplement-2 -

Corrections and Additions. Eur J Biochem 179:489–533
43. Ashburner M, Ball CA, Blake JA et al (2000) Gene ontology: tool for the unification of biology. Nat Genet 25:25–29
44. Mcguffin LJ, Atkins JD, Salehe BR et al (2015) IntFOLD: an integrated server for modelling protein structures and functions from amino acid sequences. Nucleic Acids Research 43:W169–W173
45. Bindschedler LV, Mcguffin LJ, Burgis TA et al (2011) Proteogenomics and in silico structural and functional annotation of the barley powdery mildew Blumeria graminis f. sp. hordei. Methods 54:432–441
46. Pedersen C, Ver Loren Van Themaat E, Mcguffin LJ et al (2012) Structure and evolution of barley powdery mildew effector candidates. BMC Genomics 13:694
47. Zhou Y, Xue S, Yang JJ (2013) Calciomics: integrative studies of Ca2+-binding proteins and their interactomes in biological systems. Metallomics 5:29–42
48. Don CG, Riniker S (2014) Scents and sense: in silico perspectives on olfactory receptors. J Comput Chem 35:2279–2287
49. Finn RD, Bateman A, Clements J et al (2014) Pfam: the protein families database. Nucleic Acids Res 42:D222–D230
50. Letunic I, Doerks T, Bork P (2015) SMART: recent updates, new developments and status in 2015. Nucleic Acids Res 43:D257–D260
51. Feng Z, Chen L, Maddula H et al (2004) Ligand Depot: a data warehouse for ligands bound to macromolecules. Bioinformatics 20:2153–2155
52. Roche DB, Buenavista MT, Mcguffin LJ (2014) Assessing the quality of modelled 3D protein structures using the ModFOLD server. Methods Mol Biol 1137:83–103
53. Roche DB, Buenavista MT, Tetchner SJ et al (2011) The IntFOLD server: an integrated web resource for protein fold recognition, 3D model quality assessment, intrinsic disorder prediction, domain prediction and ligand binding site prediction. Nucleic Acids Res 39:W171–W176
54. Mcguffin LJ, Buenavista MT, Roche DB (2013) The ModFOLD4 server for the quality assessment of 3D protein models. Nucleic Acids Res 41:W368–W372
55. Roy A, Kucukural A, Zhang Y (2010) I-TASSER: a unified platform for automated protein structure and function prediction. Nat Protoc 5:725–738
56. Mcguffin LJ (2008) Intrinsic disorder prediction from the analysis of multiple protein fold recognition models. Bioinformatics 24:1798–1804

Chapter 2

Computational Modeling of Small Molecule Ligand Binding Interactions and Affinities

Marino Convertino and Nikolay V. Dokholyan

Abstract

Understanding and controlling biological phenomena via structure-based drug screening efforts often critically rely on accurate description of protein–ligand interactions. However, most of the currently available computational techniques are affected by severe deficiencies in both protein and ligand conformational sampling as well as in the scoring of the obtained docking solutions. To overcome these limitations, we have recently developed MedusaDock, a novel docking methodology, which simultaneously models ligand and receptor flexibility. Coupled with MedusaScore, a physical force field-based scoring function that accounts for the protein–ligand interaction energy, MedusaDock, has reported the highest success rate in the CSAR 2011 exercise. Here, we present a standard computational protocol to evaluate the binding properties of the two enantiomers of the non-selective β-blocker propanolol in the β2 adrenergic receptor's binding site. We describe details of our protocol, which have been successfully applied to several other targets.

Key words Flexible docking, MedusaDock, MedusaScore, Induced Fit, Gaia, Chiron, Protein–ligand interactions, Protein structure refinement

1 Introduction

The interactions between small molecules or small peptides and protein targets are at the basis of many biological processes; therefore, the scientific community has been very prolific in developing algorithms, protocols, and methodologies to describe, understand, and control the process of recognition and formation of protein–ligand and protein–peptide complexes [1–5]. The ability to elucidate the pharmacodynamical properties of low molecular weight compounds or small peptides, along with the possibility of rationally designing novel drugs, relies on the accurate prediction of atomic interactions between ligands and target proteins. However, the ligands' large number of degrees of freedom and proteins' backbone and side chains flexibility present a critical challenge for an effective computational description of the ligand–receptor

Barry L. Stoddard (ed.), *Computational Design of Ligand Binding Proteins*, Methods in Molecular Biology, vol. 1414,
DOI 10.1007/978-1-4939-3569-7_2,

interaction (i.e., docking calculations) [6–8]. Modeling the induced fit phenomenon, whereby both the target and the ligand undergo mutually adaptive conformational changes upon binding, is particularly demanding due to significant conformational sampling required for computational optimization of such interactions [8–10]. In order to properly account for this effect, experimentally (via X-ray crystallography or NMR spectroscopy) and/or computationally (via molecular dynamics or normal mode analysis) determined protein conformations have been included in current docking calculations [11–15]. However, multiple conformations of the protein may not be available, or be biased toward the protein–ligand complex conformations, and, thus not able to capture new rearrangements of protein binding sites upon binding of novel compounds.

To overcome these limitations, we have recently developed a new docking algorithm, namely MedusaDock [16], which accounts for ligand and receptor flexibility at the same time. In MedusaDock, we build a stochastic rotamer library for each ligand, and simultaneously model the protein sidechain conformation using a rotamer library for all natural amino acids. The efficient sampling of our docking is associated with the use of MedusaScore [17], a physical force field-based scoring function accounting for the protein–ligand interaction energy. The adoption of MedusaScore circumvents the problem of low transferability among different targets and ligands, which is typical of empirical scoring functions classically used in docking calculations [18, 19]. MedusaDock and MedusaScore have been successfully adopted in the evaluation of the binding properties of both peptides [5] and small molecules [16, 20, 21].

Our docking approach has successfully predicted the native conformations of 28 out of the 35 study cases proposed in the recent CSAR-2011 competition [20], more than any other group in the exercise (H. Carlson, personal communications). In this chapter, we present a standard protocol to perform the docking of the propanolol enantiomers in the binding site of the β2 adrenergic receptor (β2AR). We (1) assess the structural quality of this G protein-coupled receptor's structure using our in-house developed software Gaia, which compares the intrinsic properties of protein structural models to high-resolution crystal structures (http://chiron.dokhlab.org [22]); (2) generate the optimized starting structures of ligands using widely used molecular modeling tools; and finally (3) calibrate and run docking calculations using MedusaDock [16], which will eliminate any possible bias originated from the starting conformations of the amino acids in β2AR binding pockets.

2 Materials

To implement the reported docking calculation procedure, it is necessary to have access to an internet-connected computer running a Linux operative system and mount a licensed copy of the Schröedinger Suite (Schröedinger, LLC), as well as a licensed copy of the MedusaDock software (Molecules in Action, LLC).

3 Methods

3.1 Protein Preparation

1. Navigate through the Protein Data Bank (PDB) website [23] to download the crystallographic coordinates of the human β2AR at 2.8 Å resolution (PDB-ID: 3NY8 [24]). From the downloaded file, remove the coordinates of (1) the co-crystallized inverse agonist ICI 118,551; (2) water molecules not mediating the binding of ICI 118,551 to β2AR; and (3) molecules used for technical purposes and present in the final crystal structure.
2. In order to estimate the quality of the resulting β2AR protein structure, run the in-house developed software Gaia [22]. Navigate to the following address http://chiron.dokhlab.org. Click on the Submit Task button in the starting page (Fig. 1a). In the **step 1** section, enter a Job Title in the dedicated window, and upload the file containing the β2AR crystallographic structure in pdb format. You can choose to receive an e-mail notification when the submitted job is completed. In the **step 2** section, choose the task Gaia to validate the submitted protein structure. The status of the calculation can be monitored via the panel Gaia, which is accessible by clicking the Home/Overview button in the starting page (Fig. 1a). Upon completion of the job (indicated by a green mark in the Status), a short report of some protein features will be presented on the web page (Fig. 1b). The user can download a detailed report on the structural features of the protein clicking on the eye icon in the table (Fig. 1b, *see* **Note 1**).

3.2 Ligand Preparation

1. Several applications can be used to prepare the structure of ligands to be used in docking calculations. In this specific case, we will use a number of applications available via the Schrödinger Suite. Starting from the Maestro interface (v. 9.3.5), use the 2D Sketcher tool to draw the chemical structures of the inverse agonist ICI 118,551, co-crystallized with the β2AR protein, as well as the two propanolol enantiomers, whose binding modes will be investigated through docking.

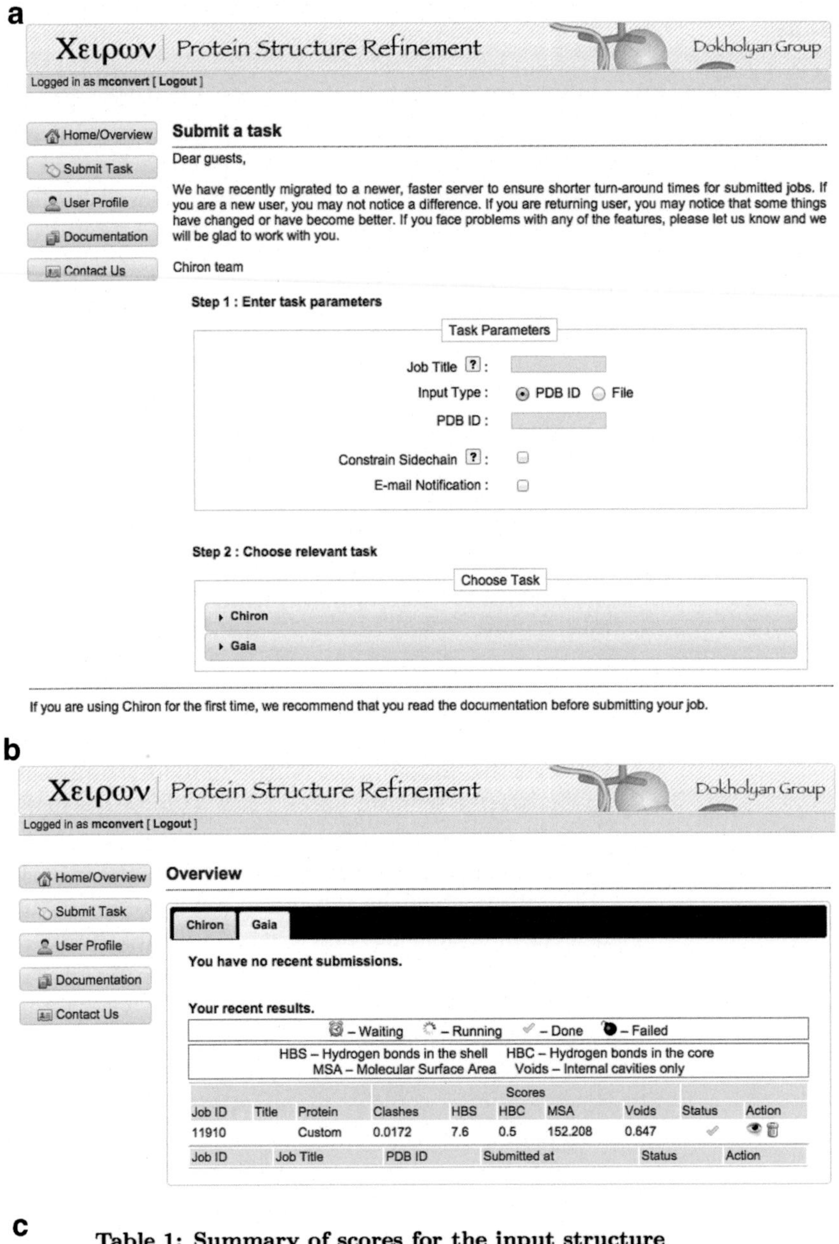

Table 1: Summary of scores for the input structure

Criterion		Observed	Target
Steric clashes		0.02	0.02
Hydrogen bonds	%Unsatisfied in shell	7.60	9.56
	%Unsatisfied in core	0.50	1.45
Solvent accessible surface area		243.36	221.64
Void volume		0.65	0.097

Fig. 1 (**a**) Home page of Chiron/Gaia server for protein structure refinement, which is available at the following link: http://chiron.dokhlab.org. (**b**) Short report of protein's structural features from the Chiron/Gaia server.

2. The ligand structures need to be further optimized using the LigPrep application. The user can choose the appropriate force field (in this case MMFFs [25]) for the optimization of atom distances, angles, and dihedral angles, along with the most appropriate pH for the determination of the formal charges of titratable groups (*see* **Note 2**). Several options are available for the determination of the ligands' stereochemistry. Since we have manually drawn the ligand structures, we determine the appropriate chiralities from the generated 3D structures without constructing any tautomers. The optimized structures of ligands are saved in mol2 format for docking calculations, and in Structure Data Format (i.e., SDF format by MDL Information Systems) for storage.

3.3 Docking Calibration

1. Docking calculations are executed via our Monte Carlo-based algorithm MedusaDock [16], which simultaneously accounts for ligands' and receptors' (side chains) flexibility. We calibrate docking calculations to the target protein by performing a self-docking of any co-crystallized binder as retrieved from the PDB to assess both the convergence of docking calculations, and the ability of reproducing the native pose of the co-crystallized ligand (i.e., ICI 118,551) in the β2AR binding site.
2. In order to test the convergence of docking results, submit several independent docking calculations of ICI 118,551 in the β2AR binding site (e.g., 100, 200, 500) using MedusaDock [16] (*see* **Note 3**), and plot the distributions of the binding energies as estimated by MedusaScore [17] (Fig. 2a). The number of calculations by which there is no more variation of the poses' binding energy distributions will be the minimal number of docking runs normally submitted to explore the binding modes of compounds (with similar molecular weight and rotatable bonds to ICI 118,551) in the β2AR binding site.
3. The estimated binding energies for all of the docking poses of ICI 118,551 (as for any docked compound) show a normal distribution (Fig. 2b). Therefore, according to the central limit theorem [26], it is possible to retrieve as statistical significant solutions from only those docking poses for which the

Fig. 1 (continued) The green mark below the Status column indicates the completion of the job; the eye icon in the table gives access to a detailed report, which can be downloaded in pdf format. (**c**) Initial summary about protein's structural features as downloaded from the Chiron/Gaia server. Values highlighted in red usually need the user attention in order to further refine the submitted protein structure (*see* **Note 1**). A detailed report about steric clashes, hydrogen bonds in the shell and in the core of the protein, solvent accessible surface area, and void volume is also available to the user

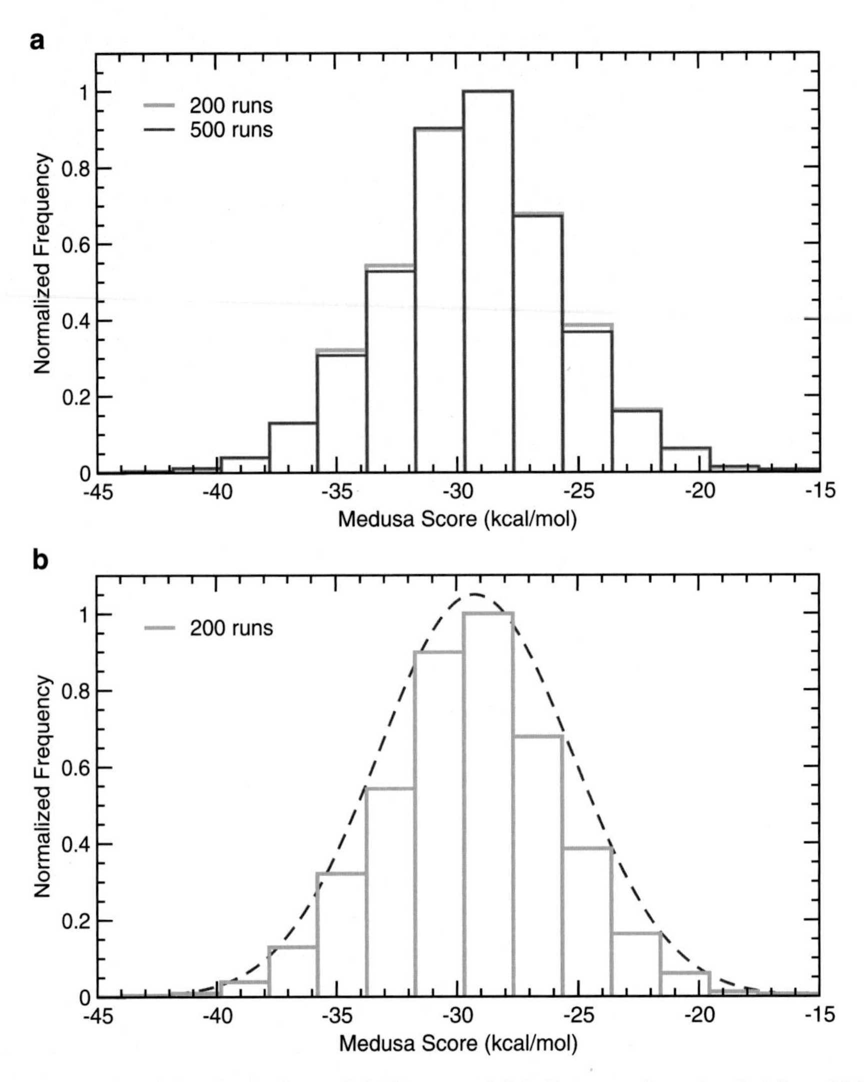

Fig. 2 (**a**) Convergence of the distributions of docking pose's binding energies extracted from 200 and 500 independent MedusaDock calculations are reported in *green* and *blue*, respectively. (**b**) Normal distribution (*red dashed curve*) of docking pose's binding energies extracted from 200 independent MedusaDock calculations (*green bars*)

Z-score is lower than −2 (i.e., less than 5 % probability that the specific docking pose is extracted by chance). In this case, *Z* is defined as:

$$Z = \frac{x - \mu}{\sigma}$$

where x is the estimated binding energy of a specific docking poses, and μ and σ are the mean and the standard deviation of the binding energies in the population of binding poses, respectively.

4. On the subset of extracted docking poses (i.e., poses with *Z*-score lower than −2), perform a cluster analysis to retrieve the most representative docking pose (i.e., centroid of the most populated cluster of poses). Cluster the ensemble of docking solutions according to the root mean square deviation (RMSD) computed over the ligand's heavy atoms. The optimal number of highly populated clusters can be identified by applying the average linkage method [27] and the Kelley penalty index [28] in order to minimize the number of clusters and the spread of internal values in each cluster. The clustering level with the lowest Kelley penalty represents a condition where the clusters are highly populated and concurrently maintain the smallest internal spread of RMSD values (*see* **Note 4**). The centroid of the most populated cluster is chosen as the representative conformation of the ICI 118,551 bound to β2AR.
5. Calculate the RMSD of the extracted solution of ICI 118,551 with respect to the original co-crystallized conformation of the ligand in β2AR. The RMSD computed over the ligand's heavy atoms (1.4 Å) is below the X-ray resolution (2.8 Å). Therefore, the applied strategy is successful in reproducing the native pose of ICI 118,551 as also demonstrated by the consistency with the electron-density map of the crystal as downloaded from the Uppsala Electron Density Server [29] (Fig. 3a).

3.4 Docking Calculations for Propanolol Enantiomers

1. Using MedusaDock submit the number of independent docking calculations determined in the **step 2** of docking calibration (*see* **Note 5**).
2. Isolate, cluster, and retrieve the obtained docking poses of propanolol enantiomers (Fig. 3b) as described in the **steps 3–5** of docking calibration.

4 Notes

1. Starting from Gaia panel in the Home/Overview page (Fig. 1b), the user can download a detailed report of the structural properties of the submitted protein in comparison with what observed in high-resolution crystal structures. The initial summary is reported in Fig. 1c. Values highlighted in red usually need the user attention in order to further refine the submitted protein structure. Such operation can be performed using the software Chiron [30], which minimizes the number of non-physical atom interactions (clashes) in the given protein structure.
2. The user can choose several options for the ligands' optimization. Available force fields are MMFFs [25] or OPLS_2005

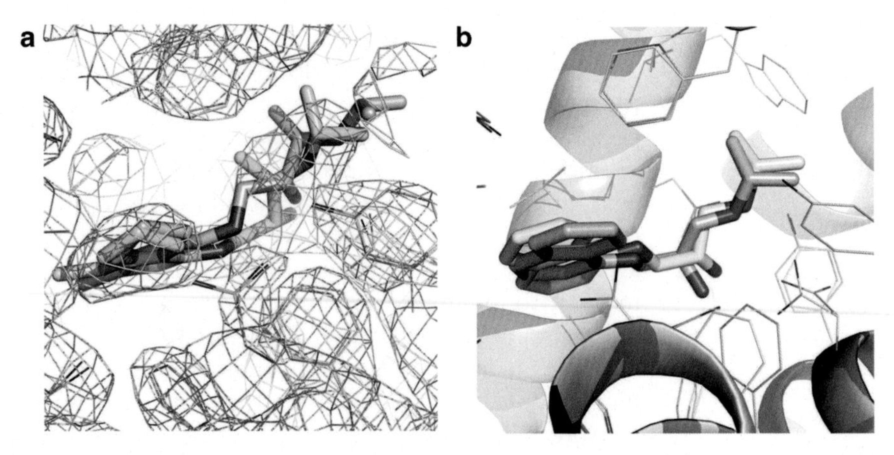

Fig. 3 (**a**) Superimposition of MedusaDock docking solution of ICI 118,551 to its crystallographic conformation in the β2AR binding site (PDB-ID: 3NY8). The described docking procedure demonstrates high reliability as it reproduces the binding pose of the original co-crystallized molecule with a RMSD computed over the ligand's heavy atoms of 1.4 Å, which is below the X-ray resolution (2.8 Å). The binding energy as estimated by MedusaDock is −39.4 kcal/mol and −37.9 kcal/mol for ICI 118,551 in its docked and crystallized conformation, respectively. Carbon atoms are represented in blue and green for ICI 118,551 in its docked and crystallized conformation, respectively. β2AR electron density map available from the Electron Density Server is reported as white mesh. (**b**) R/S propanolol bound conformations obtained by combining the MedusaScore values with a hierarchical cluster analysis of statistically significant docking solutions (i.e., poses with *Z*-score lower than −2, main text). The binding energy as estimated by MedusaScore is −38.1 kcal/mol and −38.8 kcal/mol for R- and S-propanolol, respectively. The reported solutions represent the centroids of the most populated clusters of statistically significant docking poses of R- and S-propanolol (i.e., 61.5 % and 57.7 % of the conformational ensembles, respectively). Carbon atoms are represented in pink and cyan for R- and S-enantiomers, respectively. The same color code is adopted to indicate the sidechains of β2AR amino acids when in complex with the two enantiomers

[31, 32]. The ionization state of titratable groups can be refined at the appropriate pH (the user should retrieve any available information about the pH value at the protein binding site) using either the Epik or the Ionizer application. The user can also decide to generate tautomers or all possible combinations of stereoisomers for each optimized ligand.

3. MedusaDock command can be submitted in a machine running a Linux operating system using the following command:

```
$> ./medusaDock.linux -i TARGET_PROTEIN -m MOLECULE_TO_DOCK -o DOCKING_SOLUTION -p ./ MEDUSADOCK_PARAMETERS/ -M BINDING_SITE_CENTER -r BINDING_SITE_RADIUS -S SEED_NUMBER -R
```

In this specific case `TARGET_PROTEIN` is β2AR; `MOLECULE_TO_DOCK` is ICI 118,551; `DOCKING_SOLUTION` is the output name for the calculation; `MEDUSADOCK_PARAMETERS` is the directory where parameters for docking calculations are stored; `BINDING_SITE_CENTER` is the centroid of the ICI 118,551's crystallographic coordinates as retrieved from the PDB (PDB ID: 3NY8), which has been chosen as center of the β2AR binding site; `BINDING_SITE_RADIUS` is 8 Å; `SEED_NUMBER` is a random number to be used to define a new independent Monte Carlo cycle; and `-R` is the flag which specify the initialization of a docking calculation in MedusaDock. The command is customizable for running multiple independent docking calculations as in the following *bash* script:

```
$> for i in $(seq -w 1 200 )
$> do
$>     rng = \$RANDOM     #random number generation
$>     ./medusaDock.linux  -i  TARGET_PROTEIN  -m
       MOLECULE_TO_DOCK -o DOCKING_SOLUTION -p ./
       MEDUSADOCK_PARAMETERS/  -M  BINDING_SITE_
       CENTER -r BINDING_SITE_RADIUS -S ${rng} -R
$> done
```

In this case, we perform 200 independent docking calculations of ICI 118,551 in β2AR. Even though MedusaDock can perform on a single 8-core CPU, each docking calculation requires on average 8 min to be completed, therefore the user should consider the use of supercomputer for the docking of small libraries of compounds.

4. We perform the cluster analysis using an *ad hoc* developed program. The less experienced user is advised to refer to the Conformer Cluster script available in the Resources of the Schrödinger Suite.

5. Perform MedusaDock calculations for propanolol enantiomers by adapting the command reported in **Note 3** to the new compounds.

Acknowledgments

This work was supported by the National Institute of Health grant 2R01GM080742. The authors are grateful to Dr. J. Das and B. Williams for critical reading of the manuscript. Calculations are performed on KillDevil high-performance computing cluster at the University of North Carolina at Chapel Hill.

References

1. Guedes IA, de Magalhães CS, Dardenne LE (2014) Receptor–ligand molecular docking. Biophys Rev 6:75–87
2. Grinter S, Zou X (2014) Challenges, applications, and recent advances of protein-ligand docking in structure-based drug design. Molecules 19:10150–10176
3. Audie J, Swanson J (2012) Recent work in the development and application of protein-peptide docking. Future Med Chem 4:1619–1644
4. Bhattacherjee A, Wallin S (2013) Exploring protein-peptide binding specificity through computational peptide screening. PLoS Comput Biol 9:e1003277

5. Dagliyan O, Proctor EA, D'Auria KM et al (2011) Structural and dynamic determinants of protein-peptide recognition. Structure 19:1837–1845
6. Leach AR, Shoichet BK, Peishoff CE (2006) Prediction of protein-ligand interactions. Docking and scoring: successes and gaps. J Med Chem 49:5851–5855
7. Sousa SF, Fernandes PA, Ramos MJ (2006) Protein-ligand docking: current status and future challenges. Proteins 65:15–26
8. Teague SJ (2003) Implications of protein flexibility for drug discovery. Nat Rev Drug Discov 2:527–541
9. Carlson HA, McCammon JA (2000) Accommodating protein flexibility in computational drug design. Mol Pharmacol 57:213–218
10. Teodoro ML, Kavraki LE (2003) Conformational flexibility models for the receptor in structure based drug design. Curr Pharm Des 9:1635–1648
11. Barril X, Morley SD (2005) Unveiling the full potential of flexible receptor docking using multiple crystallographic structures. J Med Chem 48:4432–4443
12. Damm KL, Carlson HA (2007) Exploring experimental sources of multiple protein conformations in structure-based drug design. J Am Chem Soc 129:8225–8235
13. Karplus M (2003) Molecular dynamics of biological macromolecules: a brief history and perspective. Biopolymers 68:350–358
14. Karplus M, Kuriyan J (2005) Molecular dynamics and protein function. Proc Natl Acad Sci U S A 102:6679–6685
15. Rueda M, Bottegoni G, Abagyan R (2009) Consistent improvement of cross-docking results using binding site ensembles generated with elastic network normal modes. J Chem Inf Model 49:716–725
16. Ding F, Yin SY, Dokholyan NV (2010) Rapid flexible docking using a stochastic rotamer library of ligands. J Chem Inf Model 50:1623–1632
17. Yin S, Biedermannova L, Vondrasek J, Dokholyan NV (2008) MedusaScore: an accurate force field-based scoring function for virtual drug screening. J Chem Inf Model 48:1656–1662
18. Gohlke H, Klebe G (2001) Statistical potentials and scoring functions applied to protein-ligand binding. Curr Opin Struct Biol 11:231–235
19. Golbraikh A, Tropsha A (2002) Beware of q(2)! J Mol Graph Model 20:269–276
20. Ding F, Dokholyan NV (2012) Incorporating backbone flexibility in medusadock improves ligand-binding pose prediction in the csar2011 docking benchmark. J Chem Inf Model 53:1871–1879
21. Serohijos AWR, Yin SY, Ding F et al (2011) Structural basis for mu-opioid receptor binding and activation. Structure 19:1683–1690
22. Kota P, Ding F, Ramachandran S, Dokholyan NV (2011) Gaia: automated quality assessment of protein structure models. Bioinformatics 27:2209–2215
23. Berman HM, Westbrook J, Feng Z et al (2000) The protein data bank. Nucleic Acids Res 28:235–242
24. Wacker D, Fenalti G, Brown MA et al (2010) Conserved binding mode of human beta2 adrenergic receptor inverse agonists and antagonist revealed by X-ray crystallography. J Am Chem Soc 132:11443–11445
25. Halgren TA (1995) The Merck molecular force field. I. basis, form, scope, parameterization, and performance of MMFF94. J Comp Chem 17:490–519
26. Central limit theorem. Encyclopedia of Mathematics. http://www.encyclopediaofmath.org/index.php?title=Central_limit_theorem&oldid=18508
27. Legendre P, Legendre L (1998) Numerical Ecology. Second English Edition. Developments in Environmental Modelling 20:302–305. Elsevier, Amsterdam
28. Kelley LA, Gardner SP, Sutcliffe MJ (1996) An automated approach for clustering an ensemble of NMR-derived protein structures into conformationally related subfamilies. Protein Eng 9:1063–1065
29. Kleywegt GJ, Harris MR, Zou J et al (2004) The Uppsala electron-density server. Acta Crystallogr D Biol Crystallogr 60:2240–2249
30. Ramachandran S, Kota P, Ding F, Dokholyan NV (2011) Automated minimization of steric clashes in protein structures. Proteins 79:261–270
31. Jorgensen WL, Tirado-Rives J (1988) The OPLS potential functions for proteins. Energy minimizations for crystals of cyclic peptides and crambin. J Am Chem Soc 110(6): 1657–1666
32. Jorgensen WL, Maxwell DS, TiradoRives J (1996) Development and testing of the OPLS all-atom force field on conformational energetics and properties of organic liquids. J Am Chem Soc 118:11225–11236

Chapter 3

Binding Site Prediction of Proteins with Organic Compounds or Peptides Using GALAXY Web Servers

Lim Heo, Hasup Lee, Minkyung Baek, and Chaok Seok

Abstract

We introduce two GALAXY web servers called GalaxySite and GalaxyPepDock that predict protein complex structures with small organic compounds and peptides, respectively. GalaxySite predicts ligands that may bind the input protein and generates complex structures of the protein with the predicted ligands from the protein structure given as input or predicted from the input sequence. GalaxyPepDock takes a protein structure and a peptide sequence as input and predicts structures for the protein–peptide complex. Both GalaxySite and GalaxyPepDock rely on available experimentally resolved structures of protein–ligand complexes evolutionarily related to the target. With the continuously increasing size of the protein structure database, the probability of finding related proteins in the database is increasing. The servers further relax the complex structures to refine the structural aspects that are missing in the available structures or that are not compatible with the given protein by optimizing physicochemical interactions. GalaxyPepDock allows conformational change of the protein receptor induced by peptide binding. The atomistic interactions with ligands predicted by the GALAXY servers may offer important clues for designing new molecules or proteins with desired binding properties.

Key words GALAXY, Binding site prediction, Peptide docking, Ligand docking, Ligand design

1 Introduction

Proteins are involved in numerous biological processes such as enzymatic activities and signal transductions [1–3]. The biological functions of proteins result from their molecular interactions with other molecules such as metal ions, small organic compounds, lipids, peptides, nucleic acids, or other proteins. Typically, proteins interact with other molecules by binding them at specific sites. Therefore, identification of the binding sites on the three-dimensional protein surfaces can be an important step for inferring protein functions [4, 5] and for designing novel molecules that control protein functions [6, 7] or designing new proteins with desired interaction properties [8, 9]. Various methods have been developed to predict ligand binding sites of proteins from protein sequences or structures. Those methods are based on geometry,

Barry L. Stoddard (ed.), *Computational Design of Ligand Binding Proteins*, Methods in Molecular Biology, vol. 1414,
DOI 10.1007/978-1-4939-3569-7_3, © Springer Science+Business Media New York 2016

energy, evolutionary information, or combinations of them [10]. Methods utilizing available experimentally resolved structures of homologous protein–ligand complexes were proven to be successful in predicting binding sites in the community-wide blind prediction experiments [11–13]. Those methods predict binding sites by transferring the available binding information for homologs, assuming that binding sites are conserved among homologs. However, methods based on evolutionary information alone may not be sufficient to predict interactions at the binding sites in atomic detail, and physicochemical interactions may have to be considered in addition.

In this chapter, we introduce two methods that predict binding sites of small organic compounds and peptides that are available on the GALAXY web server called GalaxyWEB [14]. These methods effectively search the protein structure database to find available experimental structures of related proteins complexed with ligands, build three-dimensional protein–ligand complex structures from the available information, and further refine the complex structure to go beyond the available information by optimizing physicochemical energy. The GalaxySite server predicts binding sites of small organic compounds from input protein structure or sequence [15]. Binding ligands are first predicted and the predicted ligands are then docked to the given protein structure or a predicted protein structure if sequence is given. The predicted complex structures are optimized by protein–ligand docking simulations which take into account the binding information derived from related proteins and additional physicochemical energy that do not rely on evolutionary information. GalaxySite was ranked among top methods in the recent critical assessment techniques for protein structure prediction (CASP) experiments when evaluated in terms of predicted binding site residues [16, 17]. GalaxyPepDock predicts protein–peptide complex structures from input protein structure and peptide sequence [18]. It also combines information on interactions found in homologous complexes in the protein structure database and additional physicochemical energy to optimize the protein–peptide complex structures. The protein structure is allowed to change flexibly according to its interaction with the peptide ligand during optimization.

The method proved its usefulness in the recent critical assessment of prediction of interactions (CAPRI) experiments ([19], http://www.ebi.ac.uk/msd-srv/capri/round28/round28.html). Both GalaxySite ligand binding site prediction server and the GalaxyPepDock peptide binding site prediction server rely on similarity to the protein–ligand complexes of known structures and provide detailed protein–ligand atomic interactions by sophisticated energy optimization.

2 Materials

1. A personal computer or device and a web browser are required to access the GalaxyWEB server through the Internet. A JavaScript enabled web browser is highly recommended to see the results on the web browser: The server compatibility was tested on Google Chrome, Firefox, Safari, and Internet Explorer.
2. The following input materials are required to use GalaxySite and GalaxyPepDock on GalaxyWEB.
   (a) To run GalaxySite for ligand binding site prediction, a sequence in FASTA format or a structure file in standard PDB format for the protein of interest is required. The input target protein sequence/structure file must contain 20 standard amino acids in one/three-letter codes. The input should be a single-chain protein, and the number of amino acids should be greater than 30 and less than 500. The user may judiciously delete irrelevant protein chains or termini before job submission to meet this requirement and/or to save computational cost. An example input sequence (Fig. 1,

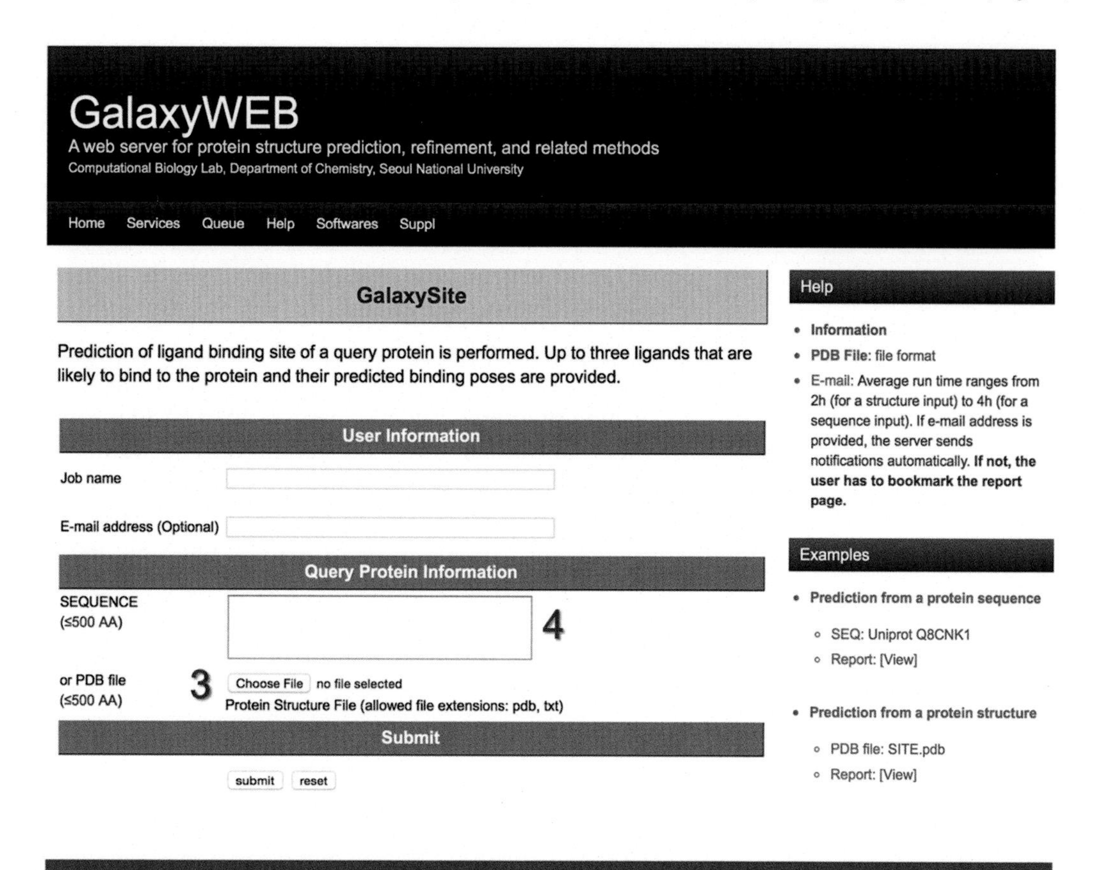

Fig. 1 The GalaxySite input page

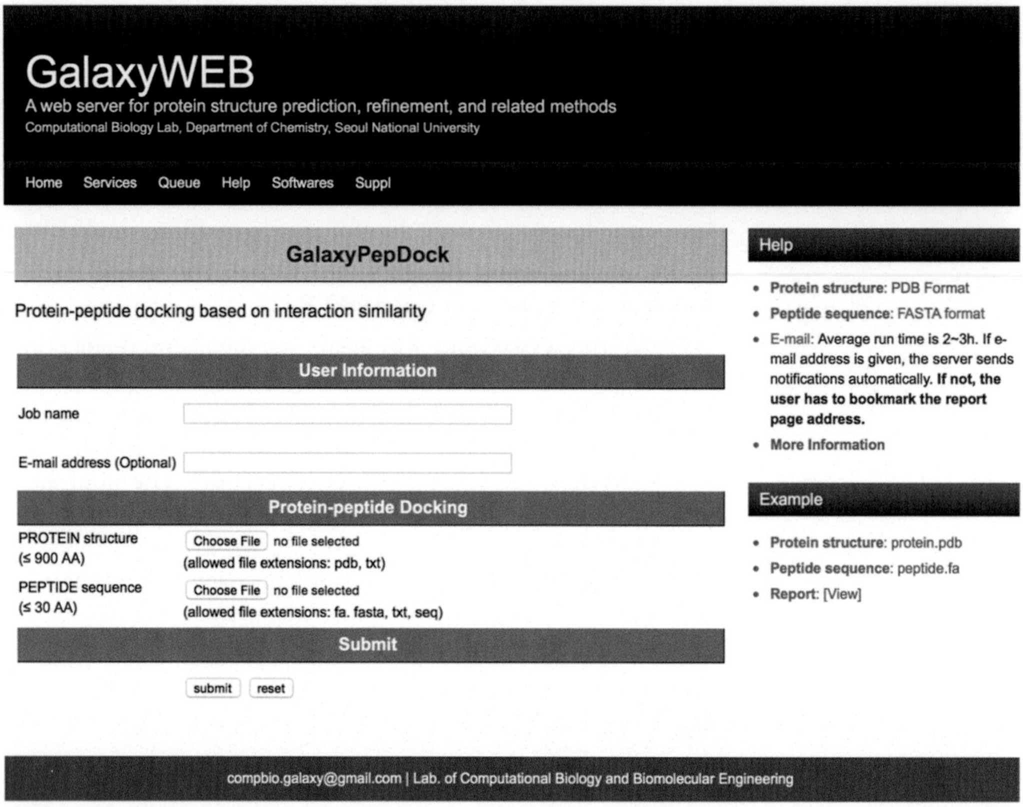

Fig. 2 The GalaxyPepDock input page

Label 1) and structure file (Fig. 1, *Label 2*) can be obtained from the GalaxySite web page.

(b) To run GalaxyPepDock for peptide binding site prediction, a structure file in standard PDB format for the receptor protein of interest and a sequence file in FASTA format for the peptide of interest are required. The number of amino acids of the receptor protein should be less than 900 and that of the peptide less than 30. The input peptide sequence file must contain 20 standard amino acids in one-letter codes. Example input files (Fig. 2, *Label 1*) can be obtained from the GalaxyPepDock web page.

3 Methods

3.1 Ligand Binding Site Prediction Using GalaxySite

1. Go to GalaxyWEB, http://galaxy.seoklab.org. Click "Site" in the "Services" tab at the top of the page.
2. In the "User Information" section, enter job name (defaults to "None"). The user can provide e-mail address so that the server sends progress reports of the submitted job automatically.

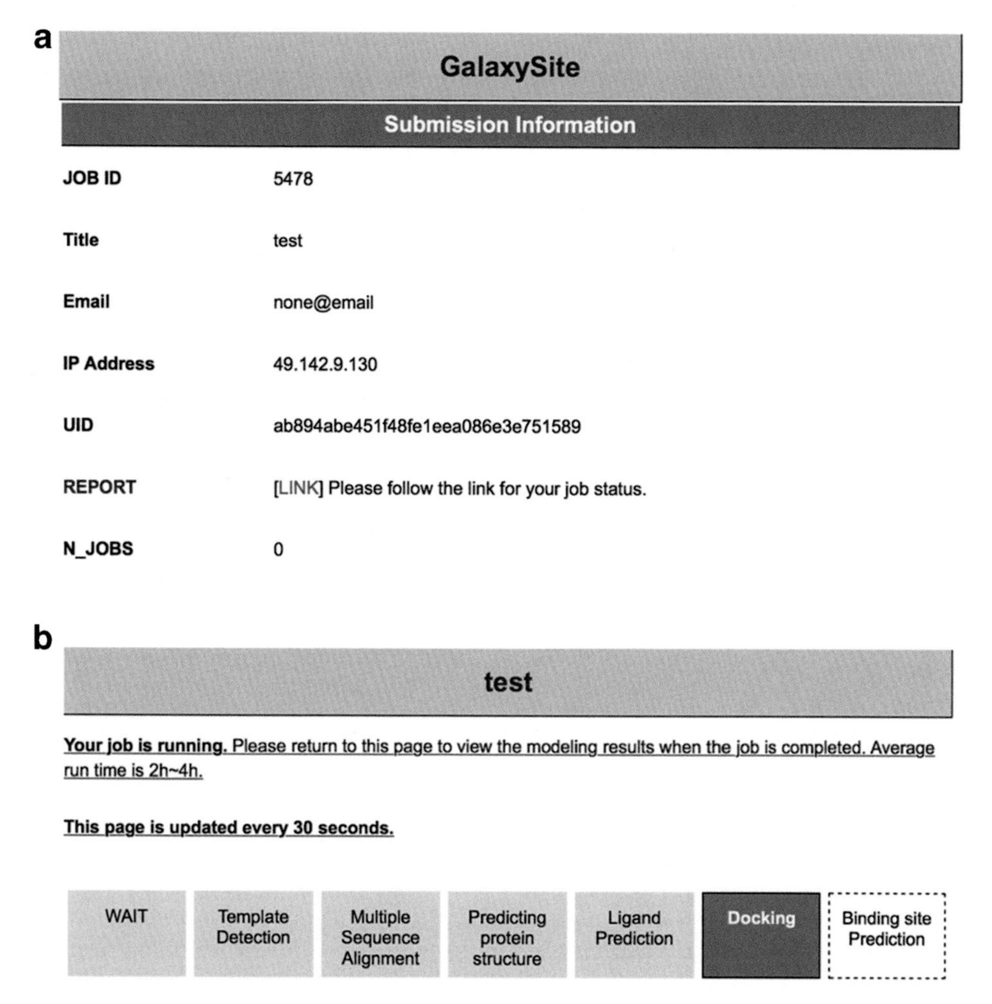

Fig. 3 (**a**) A summary page showing the submission information of a GalaxySite job. (**b**) An example report page showing the status of the GalaxySite job

Otherwise, the user should bookmark the report page (Fig. 3b) after submitting the job.

3. In the "Query Protein Information" section, provide a FASTA-formatted protein sequence or a standard PDB-formatted protein structure file. If the structure of query protein has been already determined or predicted, the user may simply upload the protein structure file in PDB format (Fig. 1, *Label 3*). If only the sequence of the query protein is known, the user may provide a FASTA-formatted protein sequence by copying the sequence and pasting it into the text box (Fig. 1, *Label 4*). When sequence information is provided, the GalaxySite server predicts its protein structure by using a simplified version of GalaxyTBM [20], a template-based protein structure prediction method (*see* **Note 1**).

Ligands predicted to bind

No	Ligand Name	Ligand Structure	Templates for protein-ligand complex
1	FMN 1	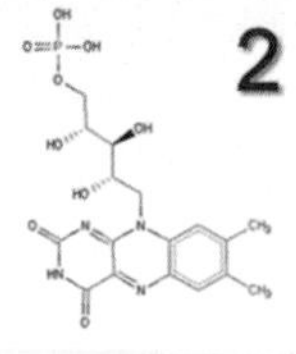 2	3 3qe2_A, 1tll_A, 2bpo_A, 3hr4_A, 1bvy_F, 1ykg_A, 3f6r_A, 1f4p_A, 2wc1_A, 2fcr_A, 1yob_A, 1obo_A, 1czn_A, 1ag9_A, 5nul_A, 2q9u_A, 1e5d_A, 1ycg_A, 2ohh_A, 2xod_A, 3n3a_C, 1rlj_A
2	FAD	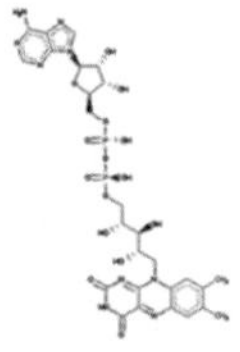	2bpo_A
3	BEN	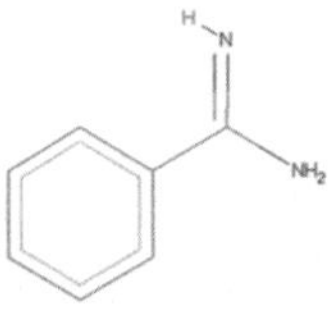	2bmv_A

Fig. 4 An example of the "Ligands predicted to bind" section on the GalaxySite report page

4. Press the submit button to queue the job. If any errors occur with the provided input, the user will get a notice about the errors that need to be corrected. If the submission is successful, the user will be directed to the summary page of the submission information which has a link to the report page (Fig. 3a). The number of jobs in the "WAIT" or "RUN" status allowed per user is limited to three.
5. Click "LINK" in the submission information page to access to the report page. The user can track the status of the submitted job in the report page which will be refreshed every 30 s (Fig. 3b). When the job is completed, predicted results will be automatically presented. Average run time of GalaxySite is 2–4 h.
6. Ligands predicted to bind: GalaxySite predicts up to three ligands that are likely to bind to the target protein (*see* **Note 2**). The predicted ligands are presented in the descending order of the estimated likelihood of binding (Fig. 4). For each ligand, ligand name in a three-letter code (Fig. 4, *Label 1*) and two-dimensional chemical structure (Fig. 4, *Label 2*) are shown. Ligand name is hyperlinked to the ligand summary page of RCSB PDB (http://www.rcsb.org) [21] for detailed information on the molecule. PDB IDs for protein–ligand complexes used for the prediction are also provided and

a Predicted ligand-binding residues

No	Ligand Name	Binding Residues	Interaction Analysis
1	FMN	1 26S 27Q 28T 30T 31A 78A 79T 80Y 81G 113L 114G 115N 118Y 120H 121F 122N 148D 152L	2 LINK
2	FAD	28T 30T 80Y 115N 118Y 148D	LINK
3	BEN	39S 40K 42A 43H 48R 49G	LINK

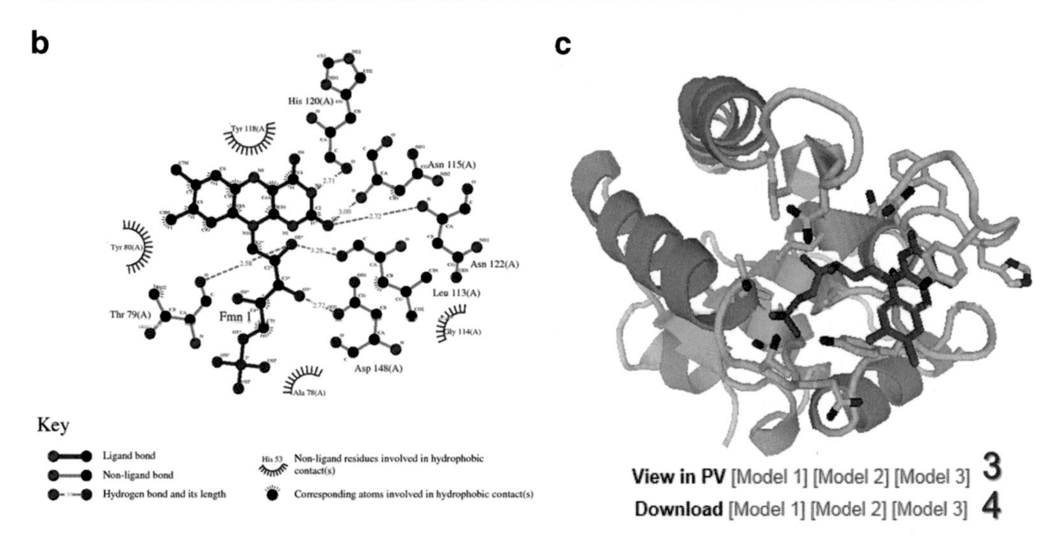

Fig. 5 (**a**) An example of the "Predicted ligand binding residue" section on the GalaxySite report page. (**b**) An example of interaction analysis between ligand and ligand binding residues made by LIGPLOT. (**c**) An example of the "Predicted binding poses" section on the GalaxySite report page

hyperlinked to the structure summary page of RCSB PDB (Fig. 4, *Label 3*).

7. Predicted ligand binding residues: For each predicted ligand, information on the predicted ligand binding residues is provided (Fig. 5a). Ligand binding residues are defined from the protein–ligand complex structure obtained by molecular docking in GalaxySite (Fig 4a, *Label 1*). If the distance of any amino acid residue from any ligand atom is less than the sum of van der Waals radii of the two atoms + 0.5 Å, the residue is considered to bind the ligand. In addition, detailed atomic interactions between ligand and ligand binding residues are analyzed by using LIGPLOT [22] and can be seen through LINK (Fig 4a, *Label 2*). On the LIGPLOT page (Fig. 5b), the ligand molecule and the protein amino acid residues are depicted in violet and brown, respectively. Hydrogen bonds are shown in green dashed lines with their lengths, and hydrophobic contacts are shown in red spikes. Ideas for designing ligands or ligand binding site residues may be gained from this interaction analysis.

Re-submission with other possible ligands

Ligand Name	Ligand Structure	Templates for protein-ligand complex	Re-run
BTB		1ag9_A	Submit 1
FNR		2fz5_A	Submit

Fig. 6 An example of the "Re-submission with other possible ligands" section on the GalaxySite report page

8. Predicted binding poses: For each predicted ligand, a predicted protein–ligand complex structure can be seen on the page using PV (http://biasmv.github.io/pv/), a JavaScript protein viewer, if the web browser supports JavaScript (Fig. 5c). Users can zoom in and out by scrolling mouse wheel and change the focusing center by double clicking. Different predicted protein–ligand complex structures are shown by clicking the model number in the "View in PV" line (Fig 4c, *Label 3*). Predicted protein–ligand complex structures can be downloaded in PDB-formatted file for further analyses (Fig 4c, *Label 4*).
9. Re-submission with other ligands: Other ligands that are likely to bind to the query protein are listed in another table (Fig. 6). Similarly to the top three ligands with the highest estimated likelihood of binding (*see* **step 6**), ligand names, two-dimensional chemical structures, and PDB IDs for the corresponding protein–ligand complexes are shown in the table. By clicking the "Submit" button (Fig. 6, *Label 1*), the user can re-submit a new ligand binding site prediction job with a selected ligand.
10. Detailed explanations on the GalaxySite web server are also provided on the GalaxySite help page; click "Help" tab at the top of the page, and then click "GalaxySite" on the right of the help page. The prediction method used for the GalaxySite program is described in the original paper [15].

3.2 Peptide Binding Site Prediction Using GalaxyPepDock

1. Go to GalaxyWEB, http://galaxy.seoklab.org. Click "PepDock" in the "Services" tab at the top of the page.
2. In the "User Information" section, enter job name (defaults to "None"). The user can provide e-mail address so that the server sends progress reports of submitted job automatically.

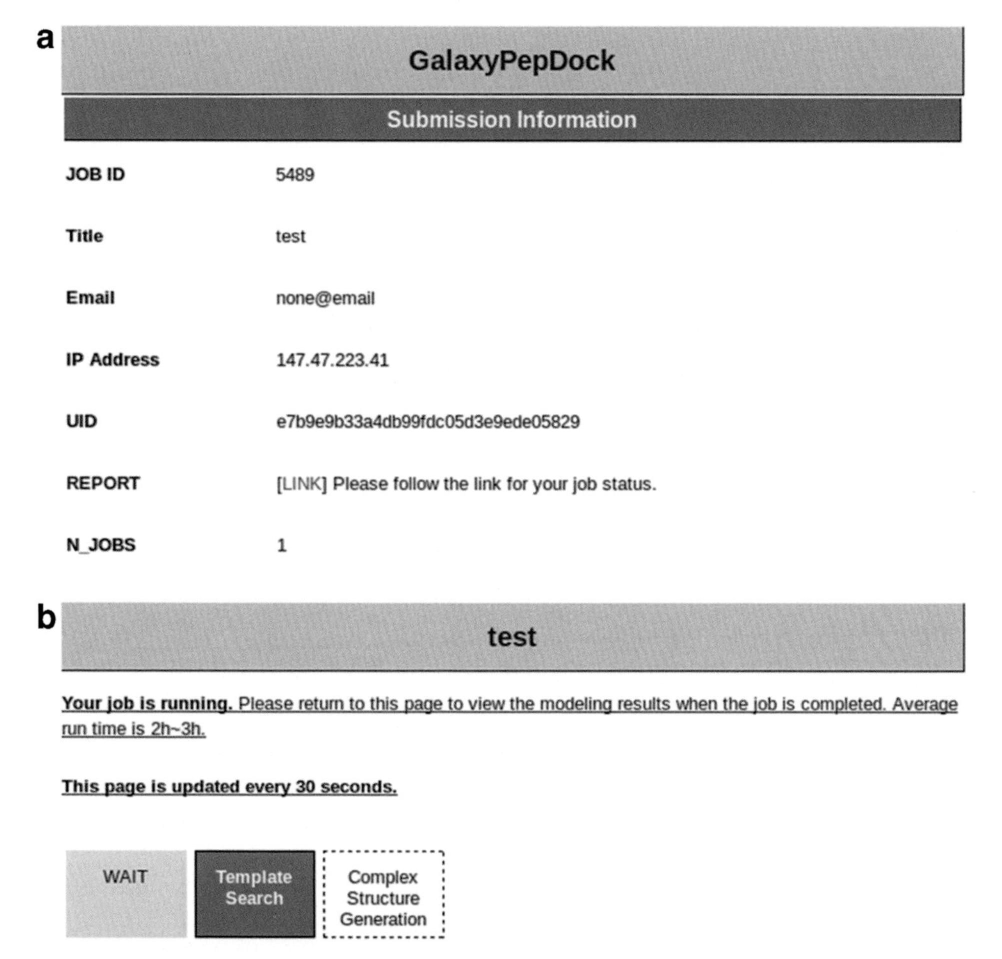

Fig. 7 (**a**) A summary page showing the submission information of a GalaxyPepDock job. (**b**) An example report page showing the status of the GalaxyPepDock job

Otherwise, the user should bookmark the report page after submitting job.

3. In the "Protein–peptide Docking" section, provide a standard PDB-formatted protein structure file (Fig. 2, *Label 2*) and a FASTA-formatted peptide sequence file (Fig. 2, *Label 3*).
4. Press the submit button to queue the job. If the submission is successful, a "Submission Information" page will appear (Fig. 7a).
5. Click "LINK" of the submission information page to access the report page. The report page will be refreshed every 30 s, updating the status of the submitted job. When the job is completed, the predicted results will be presented. Average run time of GalaxyPepDock is 2–3 h (Fig. 7b).
6. Predicted protein–peptide complex structures: Predicted structures of the query protein–peptide complex can be visualized on the report page using PV (http://biasmv.github.io/pv/), a JavaScript protein viewer, if the web browser supports JavaScript

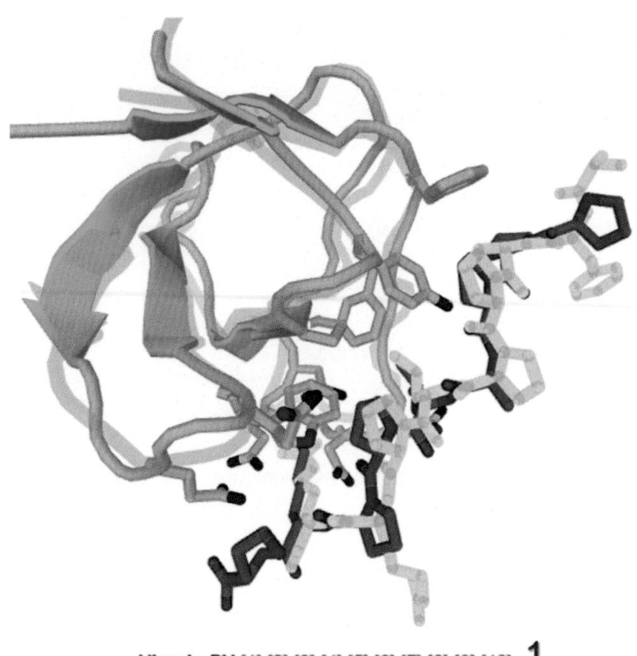

Fig. 8 An example of the "Predicted protein–peptide complex structures" section on the GalaxyPepDock report page

(Fig. 8). Users can zoom in and out by scrolling mouse wheel and change the focusing center by double clicking. Template structures selected from the database of protein–peptide complex structures to be used in the prediction are shown in light colors; protein and peptide structures are in light red and blue, respectively. Different protein–peptide complex model structures can be seen by clicking the model number in the "View in PV" line (Fig. 8, *Label 1*). Predicted protein–peptide complex structures can also be downloaded in PDB-formatted files for further analyses (Fig. 8, *Label 2*).

7. Additional information: Additional information on predicted models and intermediate results generated during the GalaxyPepDock run is provided in a table (Fig. 9a). Structures of protein template and peptide template are given as PDB IDs and can also be downloaded (Fig. 9a, *Labels 1 and 2*, respectively). Sequences and alignments of the query and the template used for the prediction are provided (Fig. 9a, *Label 3*) for both protein and peptide (Fig. 9b). Structure similarity between the predicted protein structure and the protein template structure is presented in terms of TM-score [23] and RMSD (Fig. 9a, *Label 4*). A score called interaction similarity score [18] that was designed to describe the similarity of the amino acids of the query complex aligned to the interacting

a

Model	Protein template	Peptide template	Sequences& alignments	Protein structure similarity (TM-score/RMSD)	Interaction similarity score	Estimated accuracy	Predicted binding site residues in protein
1	1PRM_C 1	1PRM_A 2	LINK 3	0.805 / 1.71 4	139 5	0.746 6	LINK 7
2	1PRM_C	1PRM_A	LINK	0.805 / 1.72	139	0.746	LINK
3	1PRM_C	1PRM_A	LINK	0.805 / 1.67	139	0.746	LINK
4	1RLQ_C	1RLQ_R	LINK	0.807 / 1.56	124	0.714	LINK
5	1RLQ_C	1RLQ_R	LINK	0.807 / 1.62	124	0.714	LINK
6	1RLQ_C	1RLQ_R	LINK	0.807 / 1.64	124	0.714	LINK
7	1RLP_C	1RLP_R	LINK	0.779 / 1.82	128	0.696	LINK
8	1RLP_C	1RLP_R	LINK	0.779 / 1.85	128	0.696	LINK
9	1GBR_A	1GBR_B	LINK	0.818 / 1.34	78	0.622	LINK
10	1GBR_A	1GBR_B	LINK	0.818 / 1.35	78	0.622	LINK

b

Query protein : AEYVRALFDFNGNDEEDLPFKKGDILRIFDKPEEQWWNAEDSE-GKRGMIPVPYVEKY
Templ protein : —TFVALYDYESRTETDLSFKKGEPLQIVNNTEGDWWLAHSLTTGQTGYIPSNYVAPS
Query peptide : –PPPALPPKK
Templ peptide : AFAPPLPRR–

c

8 PHE
9 ASP
10 PHE
12 GLY
13 ASN
14 ASP
16 GLU
17 ASP
33 GLU
35 GLN
36 TRP
48 MET
50 PRO
52 PRO
53 TYR

Fig. 9 An example of the "Additional information" section on the GalaxyPepDock report page. (**a**) A summary table showing the results of the protein–peptide complex structure predictions. (**b**) An example of structure/sequence alignments between the query protein/peptide and the template protein/peptide. (**c**) An example of the list of predicted binding residues of protein

residues of the template complex is reported for each prediction. This is to give an idea on the degree of the relative differences in similarity to the selected templates among different models (Fig. 9a, *Label 5*).

8. Predicted binding site residues: Binding site residues of the protein taken from the predicted complex structure (Fig. 9a, *Label 7* and 9c) and the estimated prediction accuracy of the binding site (Fig. 9a, *Label 6*) are provided (*see* **Note 3**). Those residues with any heavy atom within 5 Å from any peptide heavy atom in the predicted structure are reported as binding residues.
9. GalaxyPepDock help page is also available; click the "Help" tab at the top of the page, and click "GalaxyPepDock" on the right of the help page. More detailed description of the prediction method of GalaxyPepDock can be found in the original paper [18].

4 Notes

1. When a protein sequence is provided as input, GalaxySite predicts its protein structure first by using a simplified version of the GalaxyTBM template-based protein structure prediction program. Protein structure is required because ligand binding

sites are predicted by structure-based protein–ligand docking with additional information from available protein–ligand complex structures in the database. For computational efficiency, loop/termini modeling and further refinement step employed in the original GalaxyTBM are skipped during the GalaxySite runs. If the user desires to use a protein structure predicted by the full components of GalaxyTBM, he/she can run the GalaxyTBM program on GalaxyWEB. Select "TBM" in the "Services" tab at the top of the GalaxyWEB page. The same FASTA-formatted protein sequence described in the Materials section is sufficient to run GalaxyTBM.

2. Because GalaxySite predicts ligand binding sites using available protein–ligand complex structures, it cannot predict ligand binding sites if no structures for similar protein–ligand complexes are identified. In such cases, GalaxySite generates the message, "No template for binding site prediction has been found".
3. The estimated prediction accuracy in GalaxyPepDock means the estimated fraction of correctly predicted binding site residues. This value is obtained by using the linear regression data obtained from the prediction and experimental results on the PeptiDB test set [24]. A low value of estimated prediction accuracy implies that proper templates were not able to be selected, and the current similarity-based method may not provide reliable results for the query. When a very low value of estimated accuracy is returned, the user is recommended to try an ab initio protein–peptide docking method such as PEP-SiteFinder [25] that does not rely on similarity to the known structures.

Acknowledgement

This work was supported by the National Research Foundation of Korea grants funded by the Ministry of Science, ICT & Future Planning (No. 2013R1A2A1A09012229).

References

1. Kristiansen K (2004) Molecular mechanisms of ligand binding, signaling, and regulation within the superfamily of G-protein-coupled receptors: molecular modeling and mutagenesis approaches to receptor structure and function. Pharmacol Ther 103(1):21–80. doi:10.1016/j.pharmthera.2004.05.002
2. Negri A, Rodriguez-Larrea D, Marco E, Jimenez-Ruiz A, Sanchez-Ruiz JM, Gago F (2010) Protein-protein interactions at an enzyme-substrate interface: characterization of transient reaction intermediates throughout a full catalytic cycle of Escherichia coli thioredoxin reductase. Proteins 78(1):36–51. doi:10.1002/prot.22490
3. Pawson T, Nash P (2000) Protein-protein interactions define specificity in signal transduction. Genes Dev 14(9):1027–1047

4. Campbell SJ, Gold ND, Jackson RM, Westhead DR (2003) Ligand binding: functional site location, similarity and docking. Curr Opin Struct Biol 13(3):389–395
5. Kinoshita K, Nakamura H (2003) Protein informatics towards function identification. Curr Opin Struct Biol 13(3):396–400
6. Laurie AT, Jackson RM (2006) Methods for the prediction of protein-ligand binding sites for structure-based drug design and virtual ligand screening. Curr Protein Pept Sci 7(5):395–406
7. Sotriffer C, Klebe G (2002) Identification and mapping of small-molecule binding sites in proteins: computational tools for structure-based drug design. Farmaco 57(3):243–251
8. Damborsky J, Brezovsky J (2014) Computational tools for designing and engineering enzymes. Curr Opin Chem Biol 19:8–16. doi:10.1016/j.cbpa.2013.12.003
9. Feldmeier K, Hocker B (2013) Computational protein design of ligand binding and catalysis. Curr Opin Chem Biol 17(6):929–933
10. Tripathi A, Kellogg GE (2010) A novel and efficient tool for locating and characterizing protein cavities and binding sites. Proteins 78(4):825–842. doi:10.1002/prot.22608
11. Lopez G, Ezkurdia I, Tress ML (2009) Assessment of ligand binding residue predictions in CASP8. Proteins 77(Suppl 9):138–146. doi:10.1002/prot.22557
12. Lopez G, Rojas A, Tress M, Valencia A (2007) Assessment of predictions submitted for the CASP7 function prediction category. Proteins 69(Suppl 8):165–174. doi:10.1002/prot.21651
13. Oh M, Joo K, Lee J (2009) Protein-binding site prediction based on three-dimensional protein modeling. Proteins 77(Suppl 9):152–156. doi:10.1002/prot.22572
14. Ko J, Park H, Heo L, Seok C (2012) GalaxyWEB server for protein structure prediction and refinement. Nucleic Acids Res 40(Web Server Issue):W294–W297. doi:10.1093/nar/gks493
15. Heo L, Shin WH, Lee MS, Seok C (2014) GalaxySite: ligand-binding-site prediction by using molecular docking. Nucleic Acids Res 42(Web Server Issue):W210–W214. doi:10.1093/nar/gku321
16. Gallo Cassarino T, Bordoli L, Schwede T (2014) Assessment of ligand binding site predictions in CASP10. Proteins 82(Suppl 2):154–163. doi:10.1002/prot.24495
17. Schmidt T, Haas J, Gallo Cassarino T, Schwede T (2011) Assessment of ligand-binding residue predictions in CASP9. Proteins 79(Suppl 10):126–136. doi:10.1002/prot.23174
18. Lee H, Heo L, Lee MS, Seok C (2015) GalaxyPepDock: a protein-peptide docking tool based on interaction similarity and energy optimization. Nucleic Acids Res. doi:10.1093/nar/gkv495
19. Lensink MF, Wodak SJ (2013) Docking, scoring, and affinity prediction in CAPRI. Proteins 81(12):2082–2095. doi:10.1002/prot.24428
20. Ko J, Park H, Seok C (2012) GalaxyTBM: template-based modeling by building a reliable core and refining unreliable local regions. BMC Bioinformatics 13:198. doi:10.1186/1471-2105-13-198
21. Bernstein FC, Koetzle TF, Williams GJ, Meyer EF Jr, Brice MD, Rodgers JR, Kennard O, Shimanouchi T, Tasumi M (1977) The Protein Data Bank: a computer-based archival file for macromolecular structures. J Mol Biol 112(3):535–542
22. Wallace AC, Laskowski RA, Thornton JM (1995) LIGPLOT: a program to generate schematic diagrams of protein-ligand interactions. Protein Eng 8(2):127–134
23. Zhang Y, Skolnick J (2005) TM-align: a protein structure alignment algorithm based on the TM-score. Nucleic Acids Res 33(7):2302–2309. doi:10.1093/nar/gki524
24. London N, Movshovitz-Attias D, Schueler-Furman O (2010) The structural basis of peptide-protein binding strategies. Structure 18(2):188–199. doi:10.1016/j.str.2009.11.012
25. Saladin A, Rey J, Thevenet P, Zacharias M, Moroy G, Tuffery P (2014) PEP-SiteFinder: a tool for the blind identification of peptide binding sites on protein surfaces. Nucleic Acids Res 42(Web Server issue):W221–226. doi:10.1093/nar/gku404

Chapter 4

Rosetta and the Design of Ligand Binding Sites

Rocco Moretti, Brian J. Bender, Brittany Allison, and Jens Meiler

Abstract

Proteins that bind small molecules (ligands) can be used as biosensors, signal modulators, and sequestering agents. When naturally occurring proteins for a particular target ligand are not available, artificial proteins can be computationally designed. We present a protocol based on RosettaLigand to redesign an existing protein pocket to bind a target ligand. Starting with a protein structure and the structure of the ligand, Rosetta can optimize both the placement of the ligand in the pocket and the identity and conformation of the surrounding sidechains, yielding proteins that bind the target compound.

Key words Computational design, Protein/small molecule interaction, Sequence optimization, Protein design, Ligand docking

1 Introduction

Proteins which bind to small molecules (i.e. ligands) are involved in many biological processes such as enzyme catalysis, receptor signaling, and metabolite transport. Designing these interactions can produce reagents which can serve as biosensors, in vivo diagnostics, signal modulators, molecular delivery devices, and sequestering agents [1–5]. Additionally, the computational design of proteins which bind small molecules serves as a critical test of our understanding of the principles that drive protein/ligand interactions.

While in vitro techniques for the optimization of protein/ligand interactions have shown success [6], these are limited in the number of sequence variants which can be screened, and often require at least a modest starting affinity which to further optimize [7]. Computational techniques allow searching larger regions of sequence space and permit design in protein scaffolds with no detectable intrinsic affinity for the target ligand. Computational and in vitro techniques are often complementary and starting activity achieved via computational design can often be improved via in vitro techniques ([8] and Chapter 9 of this volume).

Barry L. Stoddard (ed.), *Computational Design of Ligand Binding Proteins*, Methods in Molecular Biology, vol. 1414,
DOI 10.1007/978-1-4939-3569-7_4,

Although challenges remain, computational design of small molecule interactions have yielded success on a number of occasions [5, 9], and further attempts will refine our predictive ability to generate novel ligand binders.

The Rosetta macromolecular modeling software suite [10, 11] has proven to be a robust platform for protein design, having produced novel protein folds [12, 13], protein/DNA interactions [14], protein/peptide interactions [15], protein/protein interactions [16], and novel enzymes [17–19]. Technologies for designing protein/ligand interactions have also been developed and applied [4, 8, 20]. Design of ligand binding proteins using Rosetta approaches the problem in one of two ways. One method derives from enzyme design, where predefined key interactions to the ligand are emplaced onto a protein scaffold and the surrounding context is subsequently optimized around them [8]. The other derives from ligand docking, in which the interactions with a movable ligand are optimized comprehensively [4, 20]. Both approaches have proven successful in protein redesign, and features from both can be combined using the RosettaScripts system [21], tailoring the design protocol to particular design needs.

Here we present a protocol derived from RosettaLigand ligand docking [22–25], which designs a protein binding site around a given small molecule ligand (Fig. 1). After preparing the protein and ligand structures, the placement of the ligand in the binding pocket is optimized, followed by optimization of sidechain identity and conformation. This process is repeated iteratively, and the proposed designs are sorted and filtered by a number of relevant structural metrics, such as predicted affinity and hydrogen bonding. This design process should be considered as part of the integrated program of computational and experimental work, where proteins designed computationally are tested experimentally and the experimental results are used to inform subsequent rounds of computational design.

2 Materials

1. A computer running a Unix-like operating system such as Linux or MacOS. Use of a multi-processor computational cluster is recommended for productions runs, although test runs and small production runs can be performed on conventional laptop and desktop systems.
2. Rosetta. The Rosetta modeling package can be obtained from the RosettaCommons website (https://www.rosettacommons.org/software/license-and-download). Rosetta licenses are available free to academic users. Rosetta is provided as source code and must be compiled before use. See the Rosetta

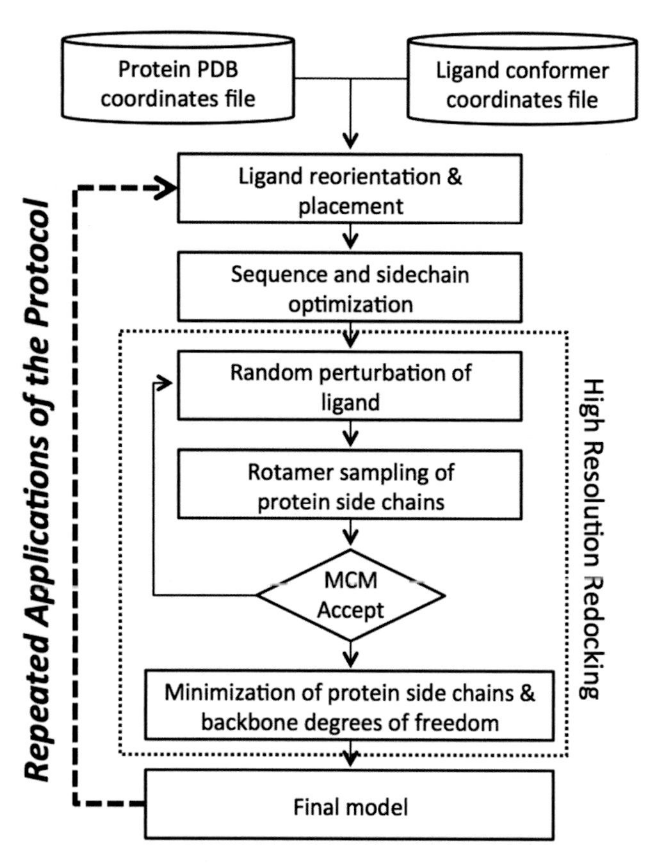

Fig. 1 Flowchart of RosettaLigand design protocol. From the combined input coordinates of the protein and ligand, the position of the ligand is optimized. Next, residues in the protein/ligand interface are optimized for both identity and position. After several cycles of small molecule perturbation, sidechain rotamer sampling, Monte Carlo minimization with Metropolis (MCM) criterion, and a final gradient-based minimization of the protein to resolve any clashes ("high resolution redocking"), the final model is the output. Further optimization can occur by using the final models of one round of design as the input models of the next round. Most variables in this protocol are user-defined, and will be varied to best fit the protein–ligand complex under study

Documentation (https://www.rosettacommons.org/docs/latest/) for instructions on how to compile Rosetta. The protocol in this paper has been tested with Rosetta weekly release version 2015.12.57698.

3. A program to manipulate small molecules. OpenBabel [26] is a free software package which allows manipulation of many small molecule file formats. See http://openbabel.org/ for download and installation information. The protocol in this paper has been tested with OpenBabel version 2.3.1. Other small molecule manipulation programs can also be used.

4. A ligand conformer generation program. We recommend the BCL [27] which is freely available from http://meilerlab.org/index.php/bclcommons for academic use but does require an additional license to the Cambridge Structural Database [28] for conformer generation. The protocol in this paper has been tested with BCL version 3.2. Other conformer generation programs such as Omega [29], MOE [30], or RDKit [31] can also be used.
5. The structure of the target small molecule in a standard format such as SDF or SMILES (*see* **Note 1**).
6. The structure of the protein to be redesigned, in PDB format (*see* **Notes 2** and **3**).

3 Methods

Throughout the protocol ${ROSETTA} represents the directory in which Rosetta has been installed. File contents and commands to be run in the terminal are in *italics*. The use of a bash shell is assumed—users of other shells may need to modify the syntax of command lines.

3.1 Pre-relax the Protein Structure into the Rosetta Scoring Function [32]

Structure from non-Rosetta sources or structures from other Rosetta protocols can have minor structural variations resulting in energetic penalties which adversely affect the design process (*see* **Notes 4** and **5**).

${ROSETTA}/main/source/bin/relax.linuxgccrelease -ignore_unrecognized_res -ignore_zero_occupancy_false -use_input_sc -flip_HNQ -no_optH false -relax:constrain_relax_to_start_coords -relax:coord_constrain_sidechains -relax:ramp_constraints false -s PDB.pdb

For convenience, rename the output structure.

mv PDB_0001.pdb PDB_relaxed.pdb

3.2 Prepare the Ligand

1. Convert the small molecule to SDF format, including adding hydrogens as needed (*see* **Note 6**).
   obabel LIG.smi --gen3D -O LIG_3D.sdf
   obabel LIG_3D.sdf -p 7.4 -O LIG.sdf
2. Generate a library of ligand conformers (*see* **Notes 7** and **8**).
   bcl.exe molecule: ConformerGenerator -top_models 100 -ensemble_filenames LIG.sdf -conformers_single_file LIG_conf.sdf
3. Convert the conformer library into a Rosetta-formatted "params file" (*see* **Notes 9 and 10**).
   ${ROSETTA}/main/source/src/python/apps/public/molfile_to_params.py -n LIG -p LIG --conformers-in-one-file LIG_conf.sdf
   This will produce three files: "LIG.params", a Rosetta-readable description of the ligand; "LIG.pdb", a selected ligand conformer; and "LIG_conformers.pdb", the set of all conformers (*see* **Note 11**).

3.3 Place the Ligand into the Protein (See **Notes 12** *and* **13***)*

1. Identify the location of desired interaction pockets. Visual inspection using programs like PyMol or Chimera [33] is normally the easiest method (*see* **Note 14**). Use the structure editing mode of PyMol to move the LIG.pdb file from step 3.2.3 into the starting conformation. Save the repositioned molecule with its new coordinates as a new file (LIG_positioned.pdb) (*see* **Note 15**).
2. If necessary, use a text editor to make the ligand be residue 1 on chain X (*see* **Note 16**).
3. Using a structure viewing program, inspect and validate the placement of the ligand (LIG_positioned.pdb) in the binding pocket of the protein (PDB_relaxed.pdb) (*see* **Note 17**).

3.4 Run Rosetta Design

1. Prepare a residue specification file. A Rosetta resfile allows specification of which residues should be designed and which should not. A good default is a resfile which permits design at all residues at the auto-detected interface (*see* **Note 18**).

```
ALLAA
AUTO
start
1 X NATAA
```

2. Prepare a docking and design script ("design.xml"). The suggested protocol is based off of RosettaLigand docking using the RosettaScripts framework [22–25]. It will optimize the location of ligand in the binding pocket (low_res_dock), redesign the surrounding sidechains (design_interface), and refine the interactions in the designed context (high_res_dock). To avoid spurious mutations, a slight energetic bonus is given to the input residue at each position (favor_native).

```
<ROSETTASCRIPTS>
    <SCOREFXNS>
        <ligand_soft_rep weights=ligand_soft_rep />
        <hard_rep weights=ligandprime />
    </SCOREFXNS>
    <TASKOPERATIONS>
        <DetectProteinLigandInterface name=design_
        interface cut1=6.0 cut2=8.0 cut3=10.0 cut4=12.0
        design=1 resfile="PDB.resfile"/> # see Note 19
    </TASKOPERATIONS>
    <LIGAND_AREAS>
        <docking_sidechain chain=X cutoff=6.0 add_
        nbr_radius=true all_atom_mode=true minimize_
        ligand=10/>
        <final_sidechain chain=X cutoff=6.0 add_nbr_
        radius=true all_atom_mode=true/>
        <final_backbone chain=X cutoff=7.0 add_
        nbr_radius=false all_atom_mode=true Calpha_
        restraints=0.3/>
    </LIGAND_AREAS>
    <INTERFACE_BUILDERS>
```

```
            <side_chain_for_docking ligand_areas=docking_
            sidechain/>
            <side_chain_for_final     ligand_areas=final_
            sidechain/>
            <backbone ligand_areas=final_backbone extension_
            window=3/>
        </INTERFACE_BUILDERS>
        <MOVEMAP_BUILDERS>
            <docking sc_interface=side_chain_for_docking
            minimize_water=true/>
            <final  sc_interface=side_chain_for_final  bb_
            interface=backbone minimize_water=true/>
        </MOVEMAP_BUILDERS>
        <SCORINGGRIDS ligand_chain=X width=15> # see Note 20
            <vdw grid_type=ClassicGrid weight=1.0/>
        </SCORINGGRIDS>
        <MOVERS>
            <FavorNativeResidue name=favor_native bonus=
            1.00 /> # see Notes 21 and 22
            <Transform name=transform chain=X box_size=
            5.0  move_distance=0.1  angle=5  cycles=500
            repeats=1  temperature=5  rmsd=4.0  />  #  see
            Note 23
            <HighResDocker name=high_res_docker cycles=6
            repack_every_Nth=3 scorefxn=ligand_soft_rep
            movemap_builder=docking/>
            <PackRotamersMover name=designinterface score-
            fxn=hard_rep  task_operations=design_inter-
            face/>
            <FinalMinimizer  name=final  scorefxn=hard_rep
            movemap_builder=final/>
            <InterfaceScoreCalculator   name=add_scores
            chains=X scorefxn=hard_rep />
            <ParsedProtocol name=low_res_dock>
                <Add mover_name=transform/>
            </ParsedProtocol>
            <ParsedProtocol name=high_res_dock>
                <Add mover_name=high_res_docker/>
                <Add mover_name=final/>
            </ParsedProtocol>
            </MOVERS>
            <PROTOCOLS>
                <Add mover_name=favor_native/>
                <Add mover_name=low_res_dock/>
                <Add mover_name=design_interface/> # see
                 Note 24
                <Add mover_name=high_res_dock/>
                <Add mover_name=add_scores/>
        </PROTOCOLS>
    </ROSETTASCRIPTS>
```

3. Prepare an options file ("design.options"). Rosetta options can be specified either on the command line or in a file. It is convenient to put options which do not change run-to-run (such as

those controlling packing and scoring) into an options file rather than the command line.

```
-ex1
-ex2
-linmem_ig 10
-restore_pre_talaris_2013_behavior # see Note 25
```

4. Run the design application (*see* **Notes 26** and **27**). This will produce a number of output PDB files (named according to the input file names, *see* **Note 28**) and a summary score file ("design_results.sc").

```
    ${ROSETTA}/main/source/bin/rosetta_scripts.linuxgccre-
lease @design.options -parser:protocol design.xml -extra_
res_fa LIG.params -s "PDB_relaxed.pdb LIG_positioned.pdb"
-nstruct <number of output models> -out:file:scorefile
design_results.sc
```

3.5 Filter Designs

1. Most Rosetta protocols are stochastic in nature. The output structures produced will contain a mixture of good and bad structures. The large number of structures produced need to be filtered to a smaller number of structures taken on to the next step.

   A rule of thumb is that filtering should remove unlikely solutions, rather than selecting the single "best" result. Successful designs are typically good across a range of relevant metrics, rather than being the best structure on a single metric (*see* **Note 29**).

   The metrics to use can vary based on the desired properties of the final design. Good standard metrics include the predicted interaction energy of the ligand, the stability score of the complex as a whole, the presence of any clashes [34], shape complementarity of the protein/ligand interface [35], the interface area, the energy density of the interface (binding energy per unit of interface area), and the number of unsatisfied hydrogen bonds formed on binding.

2. Prepare a file ("metric_thresholds.txt") specifying thresholds to use in filtering the outputs of the design runs. IMPORTANT: The exact values of the thresholds need to be tuned for your particular system (*see* **Note 30**).

```
    req total_score value < -1010 # measure of protein
stability
    req if_X_fa_rep value < 1.0# measure of ligand
clashes
    req ligand_is_touching_X value > 0.5# 1.0 if ligand
is in pocket
    output sortmin interface_delta_X# binding energy
```

3. Filter on initial metrics from the docking run. This will produce a file ("filtered_pdbs.txt") containing a list of output PDBs which pass the metric cutoffs.

```
perl ${ROSETTA}/main/source/src/apps/public/enzdes/DesignSelect.pl -d <(grep SCORE design_results.sc) -c metric_thresholds.txt -tag_column last > filtered_designs.sc
awk '{print $NF ".pdb"}' filtered_designs.sc> filtered_pdbs.txt
```

4. Calculate additional metrics (*see* **Note 31**). Rosetta's InterfaceAnalyzer [36] calculates a number of additional metrics. These can take time to evaluate, though, so are best run on only a pre-filtered set of structures. After the metrics are generated, the structures can be filtered as in **steps 3.5.1** and 3.5.2. This will produce a score file ("design_interfaces.sc") containing the calculated metric values for the selected PDBs.

```
${ROSETTA}/main/source/bin/InterfaceAnalyzer.linuxgccrelease -interface A_X -compute_packstat -pack_separated -score:weights ligandprime -no_nstruct_label -out:file:score_only design_interfaces.sc -l filtered_pdbs.txt -extra_res_fa LIG.params
```

5. Filter on additional metrics. The commands are similar to those used in step 3.5.2, but against the design_interfaces.sc score file, and with a new threshold file.

```
perl ${ROSETTA}/main/source/src/apps/public/enzdes/DesignSelect.pl -d <(grep SCORE design_results.sc) -c metric_thresholds.txt -tag_column last > filtered_designs.sc
awk '{print $NF ".pdb"}' filtered_designs.sc> filtered_pdbs.txt
```

Example contents of metric_thresholds2.txt:

```
req packstat value > 0.55 # packing metric; 0-1 higher better
req sc_value value > 0.45# shape complementarity; 0-1 higher better
req delta_unsatHbonds value < 1.5# unsatisfied hydrogen bonds on binding
req dG_separated/dSASAx100 value < -0.5 # binding energy per contact area
output sortmin dG_separated# binding energy
```

3.6 Manually Inspect Selected Sequences

While automated procedures are continually improving and can substitute to a limited extent [37], there is still no substitute for expert human knowledge in evaluating designs. Visual inspection of interfaces by a domain expert can capture system-specific requirements that are difficult to encode into an automated filter (*see* **Note 32**).

3.7 Reapply the Design Protocol, Starting at Step 3.4

Improved results can be obtained by repeating the design protocol on the output structures from previous rounds of design. The number of design rounds depends on your system and how quickly

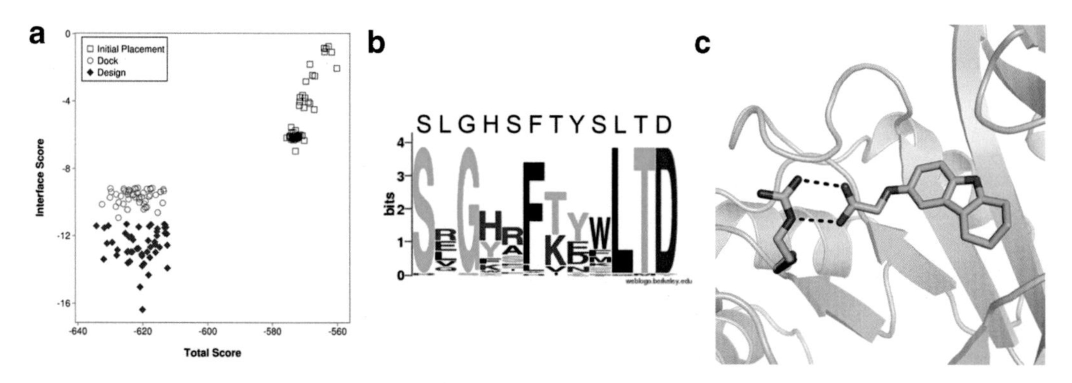

Fig. 2 Protein/ligand interface design with RosettaLigand. (**a**) Comparison in improvements in Interface Score and Total Score for top models from an initial placement, docking without sequence design, and docking with design. (**b**) Sequence logo of mutation sites among the top models from a round of interface design [43]. For most positions, the consensus sequence resembles the native sequence. Amino acids with sidechains that directly interact with the ligand show a high prevalence to mutation as seen in the positions with decreased consensus. (**c**) Example of a typical mutation introduced by RosettaLigand. The protein structure is represented in cartoon (*cyan*). The native alanine (*pink*) is mutated to an arginine residue (*green*) to match ionic interactions with the negatively charged ligand (*green*). Image generated in PyMol [44]

it converges, but 3–5 rounds of design, each starting from the filtered structures of the previous one, is typical (*see* **Note 33**).

3.8 Extract Protein Sequences from the Final Selected Designs into FASTA Format

```
${ROSETTA}/main/source/src/python/apps/public/pdb2fasta.py $(cat final_filtered_pdbs.txt) > selected_sequences.fasta
```

3.9 Iteration of Design

Only rarely will the initial design from a computational protocol give exactly the desired results. Often it is necessary to perform iterative cycles of design and experiment, using information learned from experiment to alter the design process (Fig. 2).

4 Notes

1. While Rosetta can ignore chain breaks and missing loops far from the binding site, the structure of the protein should be complete in the region of ligand binding. If the binding pocket is missing residues, remodel these with a comparative modeling protocol, using the starting structure as a template.
2. Acceptable formats depend on the capabilities of your small molecule handling program. OpenBabel can be used to convert most small molecule representations, including SMILES and InChI, into the sdf format needed by Rosetta.

3. High resolution experimental structures determined in complex with a closely related ligand are most desirable, but not required. Experimental structures of the unliganded protein and even homology models can be used [38, 39].
4. The option "-relax:coord_constrain_sidechains" should be omitted if the starting conformation of the sidechains are from modeling rather than experimental results.
5. Rosetta applications encode the compilation conditions in their filename. Applications may have names which end with *.linuxgccrelease, *.macosclangrelease, *.linuxiccrelease, etc. Use whichever ending is produced for your system. Applications ending in "debug" have additional error checking which slows down production runs.
6. It is important to add hydrogens for the physiological conditions under which you wish to design. At neutral pH, for example, amines should be protonated and carboxylates deprotonated. The "-p" option of OpenBabel uses heuristic rules to reprotonate molecules for a given pH value. Apolar hydrogens should also be present.
7. Visually examine the produced conformers and manually remove any which are folded back on themselves or are otherwise unsuitable for being the target design conformation.
8. It is unnecessary to sample hydrogen positions during rotamer generation, although any ring flip or relevant heavy atom isomeric changes should be sampled.
9. molfile_to_params.py can take a number of options—run with the "-h" option for details. The most important ones are: "-n", which allows you to specify a three letter code to use with the PDB file reading and writing, permitting you to mix multiple ligands; "-p", which specifies output file naming; "--recharge", which is used to specify the net charge on the ligand if not correctly autodetected; and "--nbr_atom", which allows you to specify a neighbor atom (*see* **Note 10**)
10. Specifying the neighbor atom is important for ligands with offset "cores". The neighbor atom is the atom which is superimposed when conformers are exchanged. By default the neighbor atom is the "most central" atom. If you have a ligand with a core that should be stable when changing conformers, you should specify an atom in that core as the neighbor atom.
11. LIG.params expects LIG_conformers.pdb to be in the same directory, so keep them together when moving files to a new directory. If you change the name of the files, you will need to adjust the value of the PDB_ROTAMERS line in the LIG. params file.
12. Rosetta expects the atom names to match those generated in the molfile_to_params.py step. Even if you have a starting

structure with the ligand correctly placed, you should align the molfile_to_params.py generated structure into the pocket so that atom naming is correct.

13. Other methods of placing the ligand in the pocket are also possible. Notably, Tinberg et al. [8] used RosettaMatch [40] both to place the ligand in an appropriate scaffold and to place key interactions in the scaffold.
14. Other pocket detection algorithms can also be used (see Chapter 1 of this volume and [41] for a review).
15. If you have a particularly large pocket, or multiple potential pockets, save separate ligand structures at different positions and perform multiple design runs. For a large number of locations, the StartFrom mover in RosettaScripts can be used to randomly place the ligand at multiple specified locations in a single run.
16. Being chain X residue 1 should be the default for molfile_to_params.py produced structures. Chain identity is important as the protocol can be used to design for ligand binding in the presence of cofactors or multiple ligands. For fixed-location cofactors, simply change the PDB chain of the cofactor to something other than X, add the cofactor to the input protein structure, and add the cofactors' params file to the -extra_res_fa command line option. For designing to multiple movable ligands, including explicit waters, see Lemmon et al. [42].
17. To refine the initial starting position of the ligand in the protein, you can do a few "design" runs as in step 3.4, but with design turned off. Change the value of the design option in the DetectProteinLigandInterface tag to zero. A good starting structure will likely have good total scores and good interface energy from these runs, but will unlikely result in ideal interactions. Pay more attention to the position and orientation of the ligand than to the energetics of this initial placement docking run.
18. The exact resfile to use will depend on system-specific knowledge of the protein structure and desired interactions. Relevant commands are ALLAA (allow design to all amino acids), PIKAA (allow design to only specified amino acids) NATAA (disallow design but permit sidechain movement), and NATRO (disallow sidechain movement). The AUTO specification allows the DetectProteinLigandInterface task operation to remove design and sidechain movement from residues which are "too far" from the ligand.
19. Change the name of the resfile in the XML script to match the full path and filename of the resfile you are using. The cut values decide how to treat residues with the AUTO specification. All AUTO residues with a C-beta atom within cut1 Angstroms

of the ligand will be designed, as will all residues within cut2 which are pointing toward the ligand. The logic in selecting sidechains is similar for cut3 and cut4, respectively, but with sidechain flexibility rather than design. Anything outside of the cut shells will be ignored during the design phase, but may be moved during other phases.

20. The grid width must be large enough to accommodate the ligand. For longer ligands, increase the value to at least the maximum extended length of the ligand plus twice the value of box_size in the Transform mover.
21. Allison et al. [20] found that a value of 1.0 for the FavorNativeSequence bonus worked best over their benchmark set. Depending on your particular requirements, though, you may wish to adjust this value. Do a few test runs with different values of the bonus and examine the number of mutations which result. If there are more mutations than desired, increase the bonus. If fewer than expected, decrease the bonus.
22. More complicated native favoring schemes can be devised by using FavorSequenceProfile instead of FavorNativeSequence. For example, you can add weights according to BLOSUM62 relatedness scores, or even use a BLAST-formatted position-specific scoring matrix (PSSM) to weight the bonus based on the distribution of sequences seen in homologous proteins.
23. The value of box_size sets the maximum rigid body displacement of the ligand from the starting position. The value of rmsd sets the maximum allowed root mean squared deviation from the starting position. Set these to smaller values if you wish to keep the designed ligand closer to the starting conformation, and to larger values if you want to permit more movement. These are limits for the active sampling stage of the protocol only. Additional movement may occur during other stages of the protocol.
24. The provided protocol only does one round of design and minimization. Additional rounds may be desired for further refinement. Simply replicate the low_res_dock, design_interface, and high_res_dock lines in the PROTOCOLS section to add additional rounds of design and optimization. Alternatively, the EnzRepackMinimize mover may be used for finer control of cycles of design and minimization (although it does not incorporate any rigid body sampling).
25. Refinement of the Rosetta scorefunction for design of protein/ligand interfaces is an area of current active research. The provided protocol uses the standard ligand docking scorefunction which was optimized prior to the scorefunction changes in 2013, and thus requires an option to revert certain changes. Decent design performance has also been seen with the "enzdes" scorefunction (which also requires the -restore_

pre_talaris_2013 option) and the standard "talaris2013" scorefunction.

26. Use of a computational cluster is recommended for large production runs. Talk to your local cluster administrator for instructions on how to launch jobs on your particular cluster system. The design runs are "trivially parallel" and can either be manually split or run with an MPI-compiled version. If splitting manually, change the value of the -nstruct option to reduce the number of structures produced by each job, and use the options -out:file:prefix or -out:file:suffix to uniquely label each run. The MPI version of rosetta_scripts can automatically handle distributing structures to multiple CPUs, but requires Rosetta to be compiled and launched in cluster-specific ways. See the Rosetta documentation for details.
27. The Rosetta option "-s" takes a list of PDBs to use as input for the run. The residues from multiple PDBs can be combined into a single structure by enclosing the filenames in quotes on the command line. Multiple filenames not enclosed in quotes will be treated as independent starting structures.
28. The number of output models needed (the value passed to -nstruct) will depend on the size of the protein pocket and the extent of remodeling needed. Normally, 1000–5000 models is a good sized run for a single starting structure and a single protocol variant. At a certain point, you will reach "convergence" and the additional models will not show appreciable metric improvement or sequence differences. If you have additional computational resources, it is often better to run multiple smaller runs (100–1000 models) with slightly varying protocols (different starting location, number of rounds, extent of optimization, native bonus, etc.), rather than have a larger number of structures from the identical protocol.
29. Relevant metrics can be determined by using "positive controls". That is, run the design protocol on known protein–ligand interactions which resemble your desired interactions. By examining how the known ligand–protein complexes behave under the Rosetta protocol, you can identify features which are useful for distinguishing native-like interactions from non-native interactions. Likewise, "negative controls", where the design protocol is run without design (*see* **Note 17**) can be useful for establishing baseline metric values and cutoffs.
30. The thresholds to use are system-specific. A good rule of thumb is to discard at least a tenth to a quarter by each relevant metric. More important metrics can receive stricter thresholds. You may wish to plot the distribution of scores to see if there is a natural threshold to set the cut at. You will likely need to do several test runs to adjust the thresholds to levels which give

the reasonable numbers of output sequences. "Negative controls" (the protocol run with design disabled, *see* **Note 17**) can also be used to determine thresholds.

31. Other system-specific metric values are available through the RosettaScripts interface as "Filters". Adding "confidence = 0" in the filter definition tag will turn off the filtering behavior and will instead just report the calculated metric for the final structure in the final score file. Many custom metrics, such as specific atom–atom distances, can be constructed in this fashion. See the Rosetta documentation for details.
32. Certain automated protocol can ease this post-analysis. For example, Rosetta can sometimes produce mutations which have only a minor influence on binding energy. While the native bonus (*see* **Notes 21** and **22**) mitigates this somewhat, explicitly considering mutation-by-mutation reversions can further reduce the number of such "spurious" mutations seen. Nivon et al. [37] presents such a protocol.
33. In subsequent rounds, you will likely want to decrease the aggressiveness of the low resolution sampling stage (the box_size and rmsd values of the Transform mover in step 3.4.2) as the ligand settles into a preferred binding orientation. As the output structure contains both the protein and ligand, the quotes on the values passed to the "-s" option (*see* step 3.4.4 and **Note 27**) are no longer needed. Instead, you may wish to use the "-l" option, which takes the name of a text file containing one input PDB per line. Each input PDB will each produce "-nstruct" models. Reduce this value such that the total number of unfiltered output structures in each round is approximately the same.

Acknowledgements

This work was supported through NIH (R01 GM099842, R01 DK097376, R01 GM073151) and NSF (CHE 1305874). RM is further partially supported by grant from the RosettaCommons.

References

1. Leader B, Baca QJ, Golan DE (2008) Protein therapeutics: a summary and pharmacological classification. Nat Rev Drug Discov 7(1):21–39. doi:10.1038/nrd2399
2. Knudsen KE, Scher HI (2009) Starving the addiction: new opportunities for durable suppression of AR signaling in prostate cancer. Clin Cancer Res 15(15):4792–4798. doi:10.1158/1078-0432.CCR-08-2660
3. Baeumner AJ (2003) Biosensors for environmental pollutants and food contaminants. Anal Bioanal Chem 377(3):434–445. doi:10.1007/s00216-003-2158-9
4. Morin A, Kaufmann KW, Fortenberry C, Harp JM, Mizoue LS, Meiler J (2011) Computational design of an endo-1,4-beta-xylanase ligand binding site. Protein Eng Des Sel 24(6):503–516. doi:10.1093/protein/gzr006

5. Morin A, Meiler J, Mizoue LS (2011) Computational design of protein-ligand interfaces: potential in therapeutic development. Trends Biotechnol 29(4):159–166. doi:10.1016/j.tibtech.2011.01.002
6. Jackel C, Kast P, Hilvert D (2008) Protein design by directed evolution. Annu Rev Biophys 37:153–173. doi:10.1146/annurev.biophys.37.032807.125832
7. Nannemann DP, Birmingham WR, Scism RA, Bachmann BO (2011) Assessing directed evolution methods for the generation of biosynthetic enzymes with potential in drug biosynthesis. Future Med Chem 3(7):809–819. doi:10.4155/fmc.11.48
8. Tinberg CE, Khare SD, Dou J, Doyle L, Nelson JW, Schena A, Jankowski W, Kalodimos CG, Johnsson K, Stoddard BL, Baker D (2013) Computational design of ligand-binding proteins with high affinity and selectivity. Nature 501(7466):212–216. doi:10.1038/nature12443
9. Feldmeier K, Hocker B (2013) Computational protein design of ligand binding and catalysis. Curr Opin Chem Biol 17(6):929–933 doi: 10.1016/j.cbpa.2013.10.002
10. Schueler-Furman O, Wang C, Bradley P, Misura K, Baker D (2005) Progress in modeling of protein structures and interactions. Science 310(5748):638–642. doi:10.1126/science.1112160
11. Leaver-Fay A, Tyka M, Lewis SM, Lange OF, Thompson J, Jacak R, Kaufman K, Renfrew PD, Smith CA, Sheffler W, Davis IW, Cooper S, Treuille A, Mandell DJ, Richter F, Ban YE, Fleishman SJ, Corn JE, Kim DE, Lyskov S, Berrondo M, Mentzer S, Popovic Z, Havranek JJ, Karanicolas J, Das R, Meiler J, Kortemme T, Gray JJ, Kuhlman B, Baker D, Bradley P (2011) ROSETTA3: an object-oriented software suite for the simulation and design of macromolecules. Methods Enzymol 487:545–574. doi:10.1016/B978-0-12-381270-4.00019-6
12. Kuhlman B, Dantas G, Ireton GC, Varani G, Stoddard BL, Baker D (2003) Design of a novel globular protein fold with atomic level accuracy. Science 302(5649):1364–1368 doi: 10.1126/science.1089427
13. Koga N, Tatsumi-Koga R, Liu G, Xiao R, Acton TB, Montelione GT, Baker D (2012) Principles for designing ideal protein structures. Nature 491(7423):222–227. doi:10.1038/nature11600
14. Ashworth J, Taylor GK, Havranek JJ, Quadri SA, Stoddard BL, Baker D (2010) Computational reprogramming of homing endonuclease specificity at multiple adjacent base pairs. Nucleic Acids Res 38(16):5601–5608 doi: 10.1093/nar/gkq283
15. Sammond DW, Bosch DE, Butterfoss GL, Purbeck C, Machius M, Siderovski DP, Kuhlman B (2011) Computational design of the sequence and structure of a protein-binding peptide. J Am Chem Soc 133(12):4190–4192. doi:10.1021/ja110296z
16. Fleishman SJ, Whitehead TA, Ekiert DC, Dreyfus C, Corn JE, Strauch EM, Wilson IA, Baker D (2011) Computational design of proteins targeting the conserved stem region of influenza hemagglutinin. Science 332(6031):816–821. doi:10.1126/science.1202617
17. Jiang L, Althoff EA, Clemente FR, Doyle L, Rothlisberger D, Zanghellini A, Gallaher JL, Betker JL, Tanaka F, Barbas CF 3rd, Hilvert D, Houk KN, Stoddard BL, Baker D (2008) De novo computational design of retro-aldol enzymes. Science 319(5868):1387–1391. doi:10.1126/science.1152692
18. Rothlisberger D, Khersonsky O, Wollacott AM, Jiang L, DeChancie J, Betker J, Gallaher JL, Althoff EA, Zanghellini A, Dym O, Albeck S, Houk KN, Tawfik DS, Baker D (2008) Kemp elimination catalysts by computational enzyme design. Nature 453(7192):190–195. doi:10.1038/nature06879
19. Siegel JB, Zanghellini A, Lovick HM, Kiss G, Lambert AR, St Clair JL, Gallaher JL, Hilvert D, Gelb MH, Stoddard BL, Houk KN, Michael FE, Baker D (2010) Computational design of an enzyme catalyst for a stereoselective bimolecular Diels-Alder reaction. Science 329(5989):309–313. doi:10.1126/science.1190239
20. Allison B, Combs S, DeLuca S, Lemmon G, Mizoue L, Meiler J (2014) Computational design of protein-small molecule interfaces. J Struct Biol 185(2):193–202. doi:10.1016/j.jsb.2013.08.003
21. Fleishman SJ, Leaver-Fay A, Corn JE, Strauch EM, Khare SD, Koga N, Ashworth J, Murphy P, Richter F, Lemmon G, Meiler J, Baker D (2011) RosettaScripts: a scripting language interface to the rosetta macromolecular modeling suite. PLoS One 6(6):20161. doi:10.1371/journal.pone.0020161
22. Meiler J, Baker D (2006) ROSETTALIGAND: protein-small molecule docking with full side-chain flexibility. Proteins 65(3):538–548. doi:10.1002/prot.21086
23. Davis IW, Baker D (2009) RosettaLigand docking with full ligand and receptor flexibility. J Mol Biol 385(2):381–392. doi:10.1016/j.jmb.2008.11.010
24. Lemmon G, Meiler J (2012) Rosetta Ligand docking with flexible XML protocols. Methods Mol Biol 819:143–155. doi:10.1007/978-1-61779-465-0_10
25. DeLuca S, Khar K, Meiler J (2015) Fully Flexible Docking of Medium Sized Ligand Libraries with RosettaLigand. PLoS One

10(7):e0132508. doi: 10.1371/journal.pone.0132508
26. O'Boyle NM, Banck M, James CA, Morley C, Vandermeersch T, Hutchison GR (2011) Open Babel: an open chemical toolbox. J Cheminform 3:33. doi:10.1186/1758-2946-3-33
27. Kothiwale S, Mendenhall JL, Meiler J (2015) BCL::Conf: small molecule conformational sampling using a knowledge based rotamer library. J Cheminform 7:47. doi: 10.1186/s13321-015-0095-1
28. Allen FH (2002) The Cambridge Structural Database: a quarter of a million crystal structures and rising. Acta Crystallogr B 58(Pt 3 Pt 1):380–388 doi: 10.1107/S0108768102003890
29. Hawkins PC, Skillman AG, Warren GL, Ellingson BA, Stahl MT (2010) Conformer generation with OMEGA: algorithm and validation using high quality structures from the Protein Databank and Cambridge Structural Database. J Chem Inf Model 50(4):572–584. doi:10.1021/ci100031x
30. Labute P (2010) LowModeMD--implicit low-mode velocity filtering applied to conformational search of macrocycles and protein loops. J Chem Inf Model 50(5):792–800. doi:10.1021/ci900508k
31. Ebejer JP, Morris GM, Deane CM (2012) Freely available conformer generation methods: how good are they? J Chem Inf Model 52(5):1146–1158. doi:10.1021/ci2004658
32. Nivon LG, Moretti R, Baker D (2013) A Pareto-optimal refinement method for protein design scaffolds. PLoS One 8(4), e59004. doi:10.1371/journal.pone.0059004
33. Pettersen EF, Goddard TD, Huang CC, Couch GS, Greenblatt DM, Meng EC, Ferrin TE (2004) UCSF Chimera--a visualization system for exploratory research and analysis. J Comput Chem 25(13):1605–1612. doi:10.1002/jcc.20084
34. Sheffler W, Baker D (2009) RosettaHoles: rapid assessment of protein core packing for structure prediction, refinement, design, and validation. Protein Sci 18(1):229–239. doi:10.1002/pro.8
35. Lawrence MC, Colman PM (1993) Shape complementarity at protein/protein interfaces. J Mol Biol 234(4):946–950. doi:10.1006/jmbi.1993.1648
36. Stranges PB, Kuhlman B (2013) A comparison of successful and failed protein interface designs highlights the challenges of designing buried hydrogen bonds. Protein Sci 22(1):74–82. doi:10.1002/pro.2187
37. Nivon LG, Bjelic S, King C, Baker D (2014) Automating human intuition for protein design. Proteins 82(5):858–866. doi:10.1002/prot.24463
38. Combs SA, Deluca SL, Deluca SH, Lemmon GH, Nannemann DP, Nguyen ED, Willis JR, Sheehan JH, Meiler J (2013) Small-molecule ligand docking into comparative models with Rosetta. Nat Protoc 8(7):1277–1298. doi:10.1038/nprot.2013.074
39. Song Y, DiMaio F, Wang RY, Kim D, Miles C, Brunette T, Thompson J, Baker D (2013) High-resolution comparative modeling with RosettaCM. Structure 21(10):1735–1742. doi:10.1016/j.str.2013.08.005
40. Zanghellini A, Jiang L, Wollacott AM, Cheng G, Meiler J, Althoff EA, Rothlisberger D, Baker D (2006) New algorithms and an in silico benchmark for computational enzyme design. Protein Sci 15(12):2785–2794. doi:10.1110/ps.062353106
41. Henrich S, Salo-Ahen OM, Huang B, Rippmann FF, Cruciani G, Wade RC (2010) Computational approaches to identifying and characterizing protein binding sites for ligand design. J Mol Recognit 23(2):209–219. doi:10.1002/jmr.984
42. Lemmon G, Meiler J (2013) Towards ligand docking including explicit interface water molecules. PLoS One 8(6), e67536. doi:10.1371/journal.pone.0067536
43. Crooks GE, Hon G, Chandonia JM, Brenner SE (2004) WebLogo: a sequence logo generator. Genome Res 14(6):1188–1190. doi:10.1101/gr.849004
44. DeLano WL (2007) The PyMOL Molecular Graphics System 1.0 edn. DeLano Scientific LLC, Palo Alto, CA, USA

Chapter 5

PocketOptimizer and the Design of Ligand Binding Sites

Andre C. Stiel, Mehdi Nellen, and Birte Höcker

Abstract

PocketOptimizer is a computational method to design protein binding pockets that has been recently developed. Starting from a protein structure an existing small molecule binding pocket is optimized for the recognition of a new ligand. The modular program predicts mutations that will improve the affinity of a target small molecule to the protein of interest using a receptor ligand scoring function to estimate the binding free energy. PocketOptimizer has been tested in a comprehensive benchmark and predicted mutations have also been used in experimental tests. In this chapter, we will provide general recommendations for usage as well as an in-depth description of all individual PocketOptimizer modules.

Key words Computational protein design, Protein–small molecule interaction, Ligand binding design, Enzyme engineering, PocketOptimizer

1 Introduction

Computational design of ligand binding pockets is related to the well-known field of molecular docking. It aims at identifying mutations in the binding pocket that establish or improve the affinity and specificity of a given ligand. From a search space point of view, it can be regarded as docking of a ligand against an ensemble containing all allowed permutations of the binding pocket.

In the last decade, the field of computational protein design has progressed considerably. However, the number of versatile and robust algorithms (beyond training-set optimized specialized cases) is still small. One reason might be that despite the large number of new and innovative tools for computational design, consistent benchmark sets and strategies for comparing algorithm performance are lacking. For example, in prior studies we compared the energy functions of CADD-Suite [1] and Autodock-Vina [2] and could already identify individual strengths and weaknesses [3]. Such data can be a first step toward building better energy functions. Another important topic is the implementation of backbone flexibility, especially with respect to ambitious design

Barry L. Stoddard (ed.), *Computational Design of Ligand Binding Proteins*, Methods in Molecular Biology, vol. 1414, DOI 10.1007/978-1-4939-3569-7_5, © Springer Science+Business Media New York 2016

tasks involving pronounced changes in the binding pocket. In these cases, different sources of backbone ensembles should be benchmarked including experimentally (e.g. native crystal structures, NMR structures) as well as computationally derived ones (e.g. snapshots from molecular dynamics simulation, geometric programs such as Backrub [4] or BRDEE [5]).

Consequently, we developed PocketOptimizer, a tool for computational binding pocket design [3]. The defining feature of this program is its modularity. All components: sampling of the ligand position and the binding pocket conformers, scoring of pairwise and self-energies as well as calculation of solutions are crafted as individual modules relying on human-readable input- and output-file formats. Within a single framework this allows the user to substitute sampling strategies, energy functions, or complete algorithms, e.g. to compare techniques or to benchmark own developments toward binding pocket design. Beyond that, a modular program easily allows the implementation of consensus scoring which is likely to provide a more robust result than the use of only one algorithm. PocketOptimizer has already been tested against a benchmark set of 12 proteins and proofed to perform similar to the design program Rosetta [6]. We hope that this addition to the family of programs provides a further step toward addressing the comparability issues raised above.

The present chapter consists of two parts: (1) An introduction and guide to PocketOptimizer covering general strategic questions (a complementary hands-on user guide is available in the manual and the tutorial provided with the program). (2) A detailed mechanistic description and tech-notes on all PocketOptimizer modules. Apart from providing necessary information to exploit the modularity of the program, this part will also aid the user in troubleshooting problems during general use.

2 Methods

2.1 PocketOptimizer General Strategies and Considerations

PocketOptimizer is comprised of seven main modules (Fig. 1) (*see* **Note 1**): **poseGenerator** and **createPocketSidechainConformers** provide the sampling capacity for the ligand and the binding pocket residues. Scoring of the self-energies of the ligand and the binding pocket residues is accomplished by **calculateLigandScaffoldScores*** and **sidechainScaffoldEnergyCalculator**. The pairwise energies in the binding pocket are computed via **calculateLigandSidechainScore*** and **calculateSidechainPairEnergies** for the interaction of the ligand with the binding pocket residues and for the residues among themselves, respec-

These programs are available in two versions utilizing CADD-Suite or Autodock-Vina as scoring algorithms: calculateLigandScaffoldBALLScores, calculateLigandScaffoldVinaScores, calculateLigandSidechainBALLScore, calculateLigandSidechainVinaScore

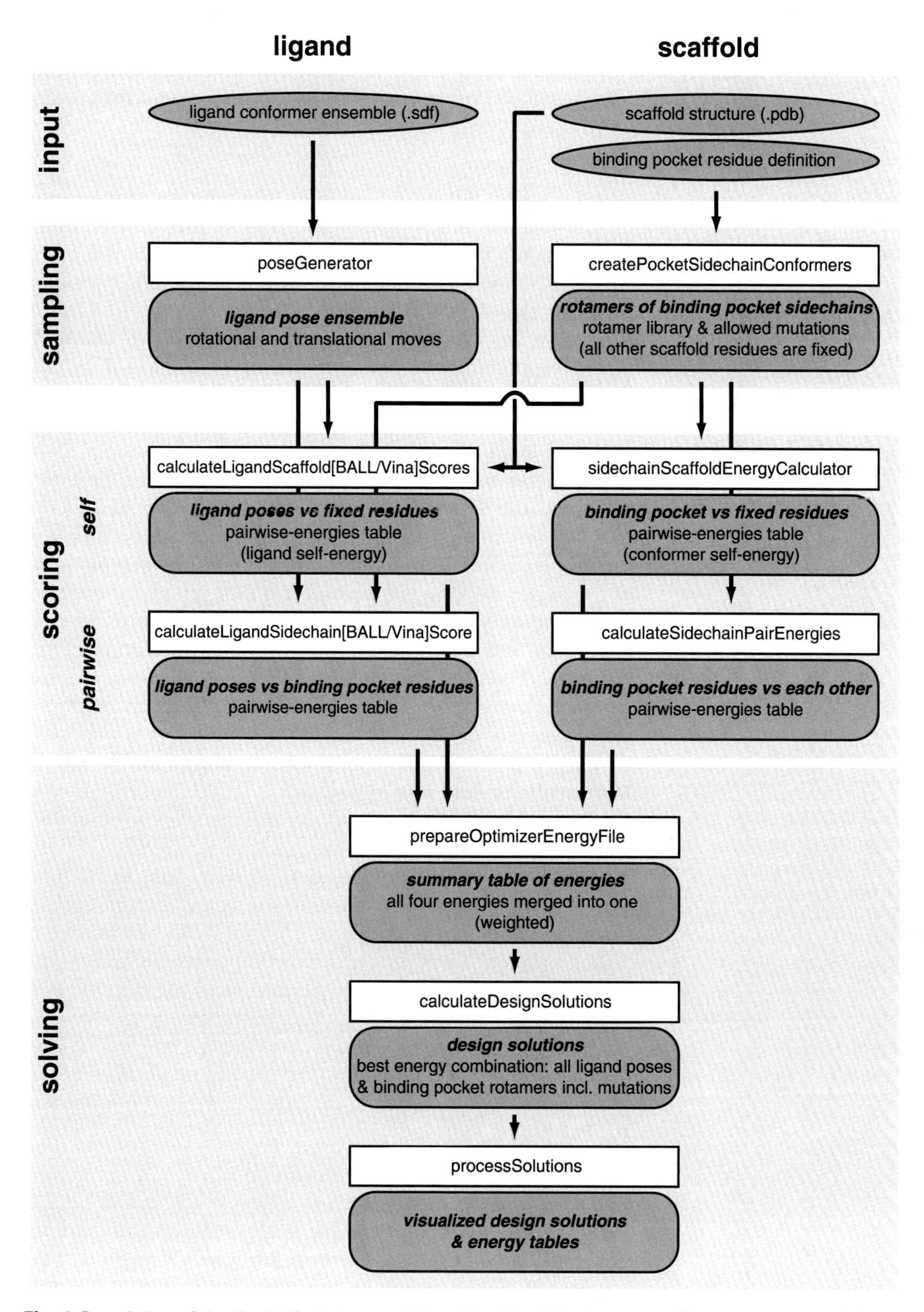

Fig. 1 Description of the PocketOptimizer workflow. User input is shown as *ellipses*, program modules as *boxes*, and module output as *rounded* and *filled boxes*. The elements are ordered by dependence on the input ligand or the scaffold as well as by belonging to the sampling, scoring, or solving group of program modules

tively. Finally, based on the calculated energies the module **calculateDesignSolutions** employs a linear programming algorithm to identify the best energy solution(s). Below we describe considerations regarding the different components.

1. Ligand
   Based on the input ligand, poseGenerator builds a ligand pose ensemble in the binding pocket employing user-defined translational and rotational movements with a subsequent filtering for clashes. Some considerations are important: (1) poseGenerator does not sample internal degrees of freedom. Thus, if the ligand has rotatable bonds an input conformer ensemble has to be provided by the user. It can be generated by, e.g. FROG [7] or confab [8]. (2) The coordinates of the input ligand ("–"conformers) have to match those of the receptor structure since the origin of the transformations applied by poseGenerator is the initial ligand position. Please be aware that manually created ligands or ligands derived from a chemical component library will have coordinates mostly centered at the origin of the space (0,0,0). (3) Naturally, the completeness of the pocket sampling is dependent on the initial placement of the ligand and extent of movement (provided as parameters maximum and step size). Thus, choosing the limits of the movements slightly beyond the pocket boundaries ensures complete coverage of the binding pocket and can rectify an initial misplacement of the ligand (see point 2). (4) Especially for larger pockets with unknown binding-mode it can be worthwhile to first scan the pocket relatively broadly (i.e. to run PocketOptimizer based on a coarse ligand pose ensemble) and, once a reasonable binding position is identified, to use finer sampling to identify the ideal binding pose and a most convincing energy. (5) In general, the ligand should contain hydrogens and proper charges: e.g. a ligand can be obtained as "ideal instance" from the chemical component dictionary [9], while the required charges can be calculated for example with antechamber (http://ambermd.org/antechamber/antechamber.html). For further details on the input formats *see* Subheading 2. A script for automatic ligand preparation is part of the PocketOptimizer program downloadable at our homepage [10].

2. Receptor and binding pocket definition
   The receptor structure is given as a file in PDB format (`.pdb`). It is recommended that the file is cleaned and standardized since this simplifies data analysis and troubleshooting: i.e. chain breaks and special amino acids should be avoided, waters or ions should be deleted or treated explicitly.

The binding pocket residues are defined in a plain text file (`.txt`). Additionally, this file is used to specify mutagenesis positions and the range of allowed amino acids at this position. The number of pocket residues or mutations is not limited, however, the calculation time exponentially scales with the number of binding pocket residues.

3. Waters, ions, and cofactors
   Heteroatoms other than the ligand (waters, ions, or other cofactors) play a significant role in binding events and can be an essential part of the calculation. There are multiple approaches to include these molecules in PocketOptimizer: (1) They can be treated as extensions to the ligand. In this case each member of the conformer ensemble has to be solvated (e.g. using the leap program from the Amber package) before being subjected to poseGenerator. (2) Alternatively, the additional heteroatoms can be treated in the same way as the main ligand. That is: sampling of accessible positions, self- as well as pairwise-energies to the binding pocket, the ligand and potentially other accessory molecules. This option is computationally very expensive since all combinations of poses and conformers between the different ligands need a pairwise energy calculation to be performed. (3) An intermediate solution is to only perform these calculations on manually identified positions for the additional heteroatoms and let PocketOptimizer evaluate if they "improve" the binding or not. To this extend PocketOptimizer defines these molecules as two "conformers": molecule is present at that position and molecule is not present.

Subheading 2.2 **item 7** explains how additional molecules can be invoked for the calculation of solutions. In principle there is no limit to the number of additional molecules other than the increasing computational complexity.

4. PocketOptimizer run
   All modules of PocketOptimizer can be run individually by calling them, together with their specific keywords (*see* **Note 2**), as arguments of the python script `PocketOptimizer.py` (*see* **Note 3**). This provides full flexibility in the usage of the modules. However, if the modules are not used in a consecutive manner, the input for each module must be otherwise produced by the user. Dependencies are visualized in Fig. 1 and explained explicitly (including file formats) in Subheading 2.2. As an alternative to the step-by-step module calling, there is a script, which runs through all modules automatically (*see* **Note 4**). This script is part of the PocketOptimizer distribution and is

explained briefly at the end of the chapter. Based on a parameter file, input ligands and receptor structure as well as binding pocket definition, the script runs all PocketOptimizer modules with intermittent output checks in one go.

5. Interpreting results
   As an output PocketOptimizer calculates binding-, packing- and total-energy of the solution(s) as well as the respective ligand pose and binding pocket conformers. The nature of the linear programming algorithm of CalculateDesignSolutions results in the single best solution; however, any arbitrary number of lower-value solutions can also be computed. This approach differs from the heuristic Monte-Carlo strategy of Rosetta, which provides a large number of design solutions that have to be ranked and analyzed subsequently to derive the "best" solutions.

For a given question it is advisable to run PocketOptimizer with multiple different parameters (ligand charge/protonation, sidechain-rotamer library, backbone positions, weight set, and so on) and compare the results numerically and structurally to retrieve a more comprehensive and insightful solution.

2.2 Module Description

In this section every module is described in detail together with the necessary input data and the output that is generated. Running the script with the `--help` flag provides a brief description of each module and its arguments.

1. poseGenerator (C++)

   description: The module creates a ligand pose ensemble employing translational (`--translation-step`, `--max-translation`) and rotational movements (`--rotation-step`, `--max-rotation`, `--axis-detail`) of the supplied ligand conformer library. The resulting ligand poses can be filtered for proximity to the binding site (`--max-pocket-distance`, `--min-pocket-fraction`) and potential clashes with the scaffolds' backbone (`--vdw-cutoff`). The module is build upon functionalities of the BALL library (http://www.ball-project.org/caddsuite).

   input: (1) a ligand rotamer library file (SDF format, `.sdf`) with the associated data field "`<AMBER TYPES>`" contains Amber type naming of all atoms, (2) the scaffold structure (`.pdb`), (3) plain text file indicating the binding pocket residues (`positions.txt`). The atom nomenclature needs to be provided in the field "`<AMBER TYPES>`". Amber conform atom types can be created using the `antechamber` program from the Amber-tools package [11] with the atom-types option set to amber ("-at amber").

output: ligand_poses.sdf contains all poses delimited by "$$$$", including the Amber type naming as described above.

2. createPocketSidechainConformers (python)

   Description: This module creates rotamers for every binding pocket residue position specified and, if mutagenesis is performed, for every possible amino acid type at the given position. The rotamers are minimized and filtered for clashes with scaffold residues based on the van der Waals (vdW) energy. Internally, all calculations are done by TINKER [12]. A TINKER key file (".key") that contains all minimization parameters will be generated in the directory specified with the --temp-dir option. It is possible to edit this file allowing maximal control. Based on the calculated energy rotamer, solutions are accepted or rejected (--energy-threshold). Possible rotamers are read from a rotamer library directory (--conformer-lib-dir) that contains the rotamers for each amino acid in single .pdb and single .sdf files. The employed force-field can be changed via --ff-param-file.

   input: (1) the scaffold structure (.pdb), (2) plain text file indicating the binding pocket residues (positions.txt).

   output: A folder (scaffold_rotamers) with sub-folders of all binding pocket positions (format: [chain]_[residue_number]). If mutations are performed, the sub-directory contains all allowed residues, otherwise only the wild type residue. For each residue there is one .pdb file with the rotamers separated by "TER" cards. Entries are numbered consecutively and not as in a multimodel pdb. Besides the .pdb file, there is a similar structured .sdf file. If no suitable rotamer was found this is indicated in the console output by "No suitable conformers found for [position]".

3. calculateLigandScaffoldBALLScores (python), calculateLigandScaffoldVinaScores (python)

   description: The module calculates pairwise energies between the ligand poses and the fixed residues of the scaffold (i.e. each residue of the scaffold that is not specified in positions.txt). This can be interpreted as the self-energy of the ligand pose in the binding pocket. The python script itself calls ReceptorDesignScorer.exe for every ligand pose–scaffold combination. This step can be parallelized using the option -s. Since the executable builds on BALL functionalities, several docking parameters (e.g. vdW forces and electrostatic cut-off) can be accessed by editing the file scoring_options.ini (in share/BALL_scorer/scoring_options.ini). Internally, the calculations are split into batches of 50 ligand poses per run. The program requires the poses as a single .sdf file together with the scaffold structure as a `.sdf`. The output of each run contains the structure as described below and is eventually combined with the other files to one final output table (*see* **Note 5**).

For the AutoDock-Vina implementation, the Autodock executable (`vina.exe`) is called internally in a similar fashion as described above. Autodock requires the presence of a pdbqt (description) file, which is generated internally by the function `prepareReceptorPDBQT` in `Autodock.py` (`py/Common`).

input: (1) a ligand pose ensemble file (`ligand_poses.sdf`) with the associated data field “<AMBER TYPES>” (c.f. 2.2.1), (2) the scaffold structure (`.pdb`), (3) an ascii-text file indicating the binding pocket residues (`positions.txt`)

output: A tab formatted text file (`ligand_energies/ligand/ligand.dat`) containing a matrix with the ligand poses as lines and the residue energy terms as columns. For every residue five columns exist, corresponding to the five energy terms: adv, vdW, solvation, HB, and rotamer. “adv” represents the electrostatics term in the BALL framework. For Autodock-Vina the output follows the same format but includes: gauss1, gauss2, repulsion, hydrophobic, and hydrogen (see Autodock for details). Since in the current implementation of PocketOptimizer (version 1.2.0) the individual energies are simply added up, the different composition of the score terms does not matter.

4. sidechainScaffoldEnergyCalculator (C++)

   description: The program module calculates the energies of the individual rotamers of all binding pocket residues with respect to the scaffold. Various parameters for treatment of electrostatics and vdW forces can be adjusted (`--es-scaling-factor`, `--es-distance-cutoff`, `--dist-dep-dielectric`, `--vdw-distance-cutoff`, `--vdw-softening-limit`, `--vdw-radius-scaling-factor`, `--vdw-method`). The module uses components of the BALL framework.

   input: (1) Binding pocket residue rotamers in the format described for the rotamer creation above. (2) The scaffold structure (`.pdb`)

   output: Directory structure as described for the rotamer creation above. For every residue there is a tab delimited text file (`.out`) that contains a matrix with rotamers for the residue as rows and columns for vdW and electrostatic energies.

5. calculateSidechainPairEnergies (C++)

   description: This part calculates pairwise energies between the respective rotamers of all binding pocket residues. The scaling factors for electrostatics and vdW forces can be adjusted (`--es-factor`, `--vdw-radius-factor`). Internally, for every combination `sidechainPairEnergyCalculator-static.exe` is called. The module uses components of the BALL framework.

input: Binding pocket residue rotamers in the format described for the rotamer creation above.

output: Files for every binding pocket residue combination (naming: [res1]-[res2] with each res: [chain]_[residue_number]_[3-letter amino acid-type]) stored in a directory called `scaffold_rotamers_pair_energies`. The files contain a matrix with rotamers of residue one as rows and two columns for each rotamer of residue two that contain the vdW and electrostatic energy, respectively.

6. calculateLigandSidechainBALLScore (python), calculateLigandSidechainVinaScore (python)

   description: This module calculates pairwise energies between the respective rotamers of every binding pocket residue and all ligand poses. For each combination of rotamer set and ligand poses an instance of `calcLigandSidechainScore.py` is called that executes `ReceptorDesignScorer.exe` in the same fashion as described above for calculateLigandScaffoldBALLScores. The batch size is 20 ligand poses per run. For the AutoDock-Vina implementation `computeVinaScores.py` is called that besides the pdbqt generation (see above for **calculateLigandScaffoldVinaScores**) calls `vina.exe`. The script `computeVinaScores.py` also accepts weighting terms that cannot be accessed in the main python module.

   input: (1) The binding pocket residue rotamers in the format described for the rotamer creation above. (2) The ligand poses in the format described for ligand pose ensemble creation.

   output: Files for every binding pocket residue and mutation (format: [ligand]_[ligand]-[chain]_[residue_number]_[3-letter amino acid-type]) stored in a directory called by default `Ligand_Scaffold_Pair_E`. The file is a tab delimited text file containing a matrix with the ligand poses as rows and the residue-rotamers energy terms as columns. The number of rows should be equal to the number of ligand poses, while for the columns there are five energy terms per rotamer (adv, vdW, solvation, HB, rotamer). For the AutoDock-Vina score terms see description in **calculateLigandScaffoldVinaScores**.

7. prepareOptimizerEnergyFile (python)

   description: The module creates one single energy table file from the individual output of the various modules described above. Furthermore, auxiliary files are created that define the order of the energy file corresponding to the residues and ligand. The readout of the previously generated energy files distinguishes between self and pairwise energies (`py/Optimizer/energyReader.py`). Energy tables with only two energy terms are treated in the same way as tables with five terms by summing up the energies. The columns for the respective energy array (two or five) are selected based on the recurring header of the type "`[3-letter-residue-type]_[number of e.g. rotamer]_`". All energies are negated, and the ligand energies can be scaled (`--ligand-factor`). Waters can be included using the water flag (`--water`) with their index as argument (e.g. "`water180`", multiple waters can be invoked by repeated calls of the water flag). The same works for metals (`--metal`, e.g. `mg_ion:mg`) and cofactors (`--cofactor`).

input: (1) The paths to all required energy files. (2) A text file indicating the binding pocket residues (`positions.txt`)

output: (1) The primary output is a space-separated text file containing the complete energy table (`lambdas.txt`). The file can be separated in four sections (although not visually): (a) First the pairwise energies of the respective rotamers of all binding pocket residues with each other. Each line in this section is dedicated to a residue combination and the line provides all energies of all possible rotamer combinations. (b) The second section stores the pairwise energies of the ligand with each binding pocket residue. Each line represents the energies for all combinations of ligand poses with rotamers at the respective residue position. (c) The third section contains all self-energies of the binding pocket residues. Each line provides the energies for all rotamers of the given residue. (d) The fourth section does the same for all ligand poses (i.e. only one line). (2) Two additional identical files (`regions.txt, intersects.txt`) describe the order of the energies in `lambdas.txt` based on the numbering (order) of binding pocket residues given in `positions.txt` and the ligand as last entry. Consequently, the files consist of two parts (a and b described in (1) above) with two columns providing the pairwise combinations and two parts (c and d) providing the numbering for the residues and ligand. (3) Another file (`region_intersects.txt`) contains the same information in another numbering scheme, starting with the last sorting number of the combinations. Additionally, a frontal third (or second for c and d) column provides a running number starting with one. (4) A text file (`var_sizes.txt`) contains just the number of rotamers of the respective residue in a line, with the number of ligand poses in the last line. (5) A summary file (`index.dat`) contains all this information in a more readable format. This file contains the ligand scaling factor ("[SCALING]") at the very bottom. It can be used to scale the energies regarding the ligand energies vs. packing energies.

8. calculateDesignSolutions (python)

   description: The module is used to call the MPLP solver[13] (`algo_triplet.exe`) with arguments for the algorithm (`--niter`, `--niter-later`, `--nclust_to_add`, `--obj-del-thr`, `--int-gap-thr`). The algorithm utilizes cluster-based linear programming with belief propagation to efficiently identify the best energy combination of rotamers of binding pocket residues and a ligand pose. If several amino acids at a binding pocket position are allowed (mutations), the solution helps to identify the energetically most favored mutations. The number of output solutions (2nd best ...) can be adjusted (`--number-of-solutions`).

   input: All files prepared by **prepareOptimizerEnergyFile**

   output: (1) A text file (`all_solutions.txt`) giving all solutions with one solution per line in decreasing order. Each solution is represented by space-separated numbers, with each number being the rotamer of the respective binding pocket residue and the last number being the ligand pose. The numbering follows the one given in `index.dat`. (2) Individual files for each solution with the same convention as for the all solutions file (`res00.txt` ...).

9. processSolutions (python)

   description: This module can be used to present the solution found by the **calculateDesignSolutions** as `.pdb` file with a detailed energy contribution description. Prior to processing the solutions, the solver file containing all solutions can be used to identify solutions sharing same mutants, rotamers, or ligand poses. This provides a better overview over the possible results, their energies and thus the confidence regarding the top-score solution.

   input: (1) The solutions file. (2) All necessary files and directories allowing to build the representative `.pdb` and to compile the solutions' energy report from the individual energies.

   output: A directory named with the solution number containing `.pdb` and `.pml` files of the solution as well as the energy report in `.txt` and `.html` format.

3 Notes

1. Binary packages of PocketOptimizer can be obtained from (https://webdav.tue.mpg.de/u/birtehoecker). For convenience we provide PocketOptimizer also as a completely set-up image-container which can be loaded using the Docker software [14].
2. PocketOptimizer is able to work with flag files; with one flag/argument per line. The only difference to the command line is

that equal signs are required between the flags and the arguments. Moreover, trailing spaces are not allowed (e.g. "--water water180" becomes "--water=water180").

3. Version 1.2.0 of PocketOptimizer (as of 2016) provides a wrapper script called `PocketOptimizer.py`. All functionalities can be called via this script (see manual and tutorial for details). The modules can be still used as stand-alones but the user then has take care of setting proper paths to libraries and additional content (like CADD-Suite).
4. Due to the modular nature of PocketOptimizer, we suggest to process multiple structures in an automated fashion, since otherwise all modules have to be started individually and their appropriate output has to be checked. An adequate script needs to check the proper termination of all modules by validating the individual output formats described above. For example, the number of computed pairwise ligand versus binding pocket residue energies has to match with the number of previously computed ligand poses and binding pocket rotamers. For Ubuntu systems, a modification of the PocketOptimizer start script performing an automated multi-structure processing is part of the PocketOptimizer distribution. The script has to be called with a command-file specifying the individual command line arguments for each module. The script is also able to do an automatic alignment of the prepared ligand to the binding pocket of the scaffold structure. For this, a special input file has to be provided (`static.txt`) that should contain three atoms that are linearly independent of each other (e.g. not lie in one straight line) and that do not move between the different conformers (e.g. an aromatic ring at the root of the structure). This will result in the atoms listed in the `static.txt` file to be aligned to each other and to the correct position in the binding pocket.
5. A temporary directory (`--temp-dir`) with enough free space is of importance especially when a large number of ligand poses or large scaffolds are used. Temporary directories are used in the following modules: calculateLigandScaffoldScores, createPocketSidechainConformers, calculateSidechainPairEnergies, calculateLigandSidechainScore, and calculateDesignSolutions.

Acknowledgments

Financial support from the German Research Foundation (DFG grant HO 4022/2-3) is acknowledged. M.N. was supported by the Erasmus+ mobility program. The authors like to thank Steffen Schmidt for comments on the manuscript.

References

1. Kohlbacher O (2012) CADDSuite – a workflow-enabled suite of open-source tools for drug discovery. J Cheminform 4:O2. doi:10.1186/1758-2946-4-S1-O2
2. Trott O, Olson AJ (2010) AutoDock Vina: improving the speed and accuracy of docking with a new scoring function, efficient optimization, and multithreading. J Comput Chem 31:455–61. doi:10.1002/jcc.21334
3. Malisi C, Schumann M, Toussaint NC et al (2012) Binding pocket optimization by computational protein design. PLoS One 7, e52505. doi:10.1371/journal.pone.0052505
4. Smith CA, Kortemme T (2008) Backrub-like backbone simulation recapitulates natural protein conformational variability and improves mutant side-chain prediction. J Mol Biol 380:742–56. doi:10.1016/j.jmb.2008.05.023
5. Georgiev I, Keedy D, Richardson JS et al (2008) Algorithm for backrub motions in protein design. Bioinformatics 24:i196–204. doi:10.1093/bioinformatics/btn169
6. Richter F, Leaver-Fay A, Khare SD et al (2011) De novo enzyme design using Rosetta3. PLoS One 6, e19230. doi:10.1371/journal.pone.0019230
7. Leite TB, Gomes D, Miteva MA et al (2007) Frog: a FRee Online druG 3D conformation generator. Nucleic Acids Res 35:W568–72. doi:10.1093/nar/gkm289
8. O'Boyle NM, Vandermeersch T, Flynn CJ et al (2011) Confab - Systematic generation of diverse low-energy conformers. J Cheminform 3:8. doi:10.1186/1758-2946-3-8
9. wwPDB (2008) Chemical Component Dictionary. http://www.wwpdb.org/ccd.html. Accessed 17 Feb 2016
10. Höcker Lab (2015) Algorithms and software. https://webdav.tue.mpg.de/u/birte-hoecker//. Accessed 17 Feb 2016
11. AMBER (2015) The amber molecular dynamics package. http://ambermd.org. Accessed 17 Feb 2016
12. Jay Ponder Lab (2015) TINKER molecular modeling package. http://dasher.wustl.edu/tinker/. Accessed 17 Feb 2016
13. Sontag D, Choe DK, Li Y (2012) Efficiently searching for frustrated cycles in MAP inference. arXiv preprint arXiv:1210.4902
14. DOCKER (2015) Docker software. http://www.docker.com. Accessed 17 Feb 2016

Chapter 6

Proteus and the Design of Ligand Binding Sites

Savvas Polydorides, Eleni Michael, David Mignon, Karen Druart, Georgios Archontis, and Thomas Simonson

Abstract

This chapter describes the organization and use of Proteus, a multitool computational suite for the optimization of protein and ligand conformations and sequences, and the calculation of pK_α shifts and relative binding affinities. The software offers the use of several molecular mechanics force fields and solvent models, including two generalized Born variants, and a large range of scoring functions, which can combine protein stability, ligand affinity, and ligand specificity terms, for positive and negative design. We present in detail the steps for structure preparation, system setup, construction of the interaction energy matrix, protein sequence and structure optimizations, pK_α calculations, and ligand titration calculations. We discuss illustrative examples, including the chemical/structural optimization of a complex between the MHC class II protein HLA-DQ8 and the vinculin epitope, and the chemical optimization of the compstatin analog Ac-Val4Trp/His9Ala, which regulates the function of protein C3 of the complement system.

Key words Protein design, Ligand design, Monte Carlo, Implicit solvent, Generalized Born model

1 Introduction

Computational protein design (CPD) is a set of methods to engineer proteins (and ligands) and optimize molecular properties such as stability, binding affinity, and binding specificity. Many successful CPD examples have been reported in recent years [1–15], and their impact will certainly increase with the continuous improvement in CPD tools and computational hardware.

We have developed the Proteus (v. 2.1) software package for computational protein and ligand design [16–18]. It consists of (1) a modified version of the XPLOR program [19], which performs the initial setup of the system under study, computes an energy matrix used in the design, and re-assesses the conformations and sequences suggested by the design; (2) a library of scripts in the XPLOR command language that control the calculations; (3) the proteus program (v. 30.4), which conducts the actual

Barry L. Stoddard (ed.), *Computational Design of Ligand Binding Proteins*, Methods in Molecular Biology, vol. 1414, DOI 10.1007/978-1-4939-3569-7_6, © Springer Science+Business Media New York 2016

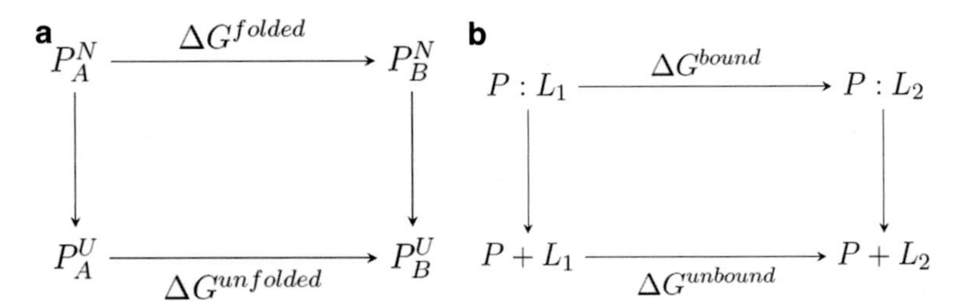

Fig. 1 Thermodynamic cycles employed in CPD of stability (**a**) and ligand specificity (**b**)

search in the protein and ligand's structure and sequence space; (4) a set of Perl scripts to help analyze the solutions provided by proteus. Shell scripts that automate the whole procedure are also available. For the sake of clarity, in this chapter we describe a detailed design protocol, so that new users can follow it step by step.

1.1 Thermodynamic Cycles

The concepts of stability or specificity design, as implemented in Proteus, are illustrated in the thermodynamic cycles of Fig. 1. The cycle on the left compares the stabilities of two sequences *A* and *B*. The folding processes are depicted by the vertical legs; the horizontal legs display the (unphysical) transformations from sequence *A* into *B*, in the folded (*N*) and unfolded (*U*) states. The difference between the free energy changes for the horizontal (or vertical) legs yields the difference in stability between the two sequences:

$$\Delta\Delta G_{\mathrm{f}} = \left[G\left(P_{\mathrm{B}}^{\mathrm{N}}\right) - G\left(P_{\mathrm{A}}^{\mathrm{N}}\right)\right] - \left[G\left(P_{\mathrm{B}}^{\mathrm{U}}\right) - G\left(P_{\mathrm{A}}^{\mathrm{U}}\right)\right] \quad (1)$$

Stability calculations seek to minimize the above free energy difference $\Delta\Delta G_{\mathrm{f}}$.

Specificity calculations are illustrated by the thermodynamic cycle on the right of Fig. 1. The vertical legs represent the binding of two ligands L_1 and L_2 to a protein *P*; the horizontal legs represent the (unphysical) chemical transformation between the two ligands, either in the protein complex (top leg) or in solution (bottom leg). If L_1 is a *reference* ligand and L_2 a modified analog, the calculations seek to minimize the relative binding free energy

$$\Delta\Delta G_{\mathrm{b}} = \left[G\left(P:L_2\right) - G\left(P:L_1\right)\right] - \left[G\left(L_2\right) - G\left(L_1\right)\right] \quad (2)$$

The above expression assumes that the protein relaxes to the same state (*P*) upon dissociation of the two complexes (unlike some MM-PBSA or MM-GBSA methods [20, 21]).

1.2 Energy Model

The free energies appearing in Eqs. 1–2 are computed via a physical energy function with the general form:

$$G = E_{\mathrm{bond}} + E_{\mathrm{angle}} + E_{\mathrm{dihe.}} + E_{\mathrm{impr.}} + E_{\mathrm{vdW}} + E_{\mathrm{coul.}} + E_{\mathrm{GB}} + E_{\mathrm{SA}} + E_{\mathrm{corr}} \quad (3)$$

The first six terms describe the internal and nonbonded contributions to the potential energy of the protein or ligand under study, and are borrowed from a molecular mechanics energy function. The parameterizations currently available in Proteus are the Charmm19 force field [22] and the Amber ff99SB force field [23]. The next two terms capture solvent effects via a generalized Born (GB) approximation and an accessible surface area (SA) term. Simpler energy functions that model solvent electrostatic screening via a homogeneous ("cdie") or distance-dependent ("rdie") dielectric constant are also available. The last term represents an optional "correction" energy, whose interpretation depends on the design criterion (*see below*).

1.3 Unfolded State

The above free energies are functions of the atomic coordinates. This poses a difficulty in the case of unfolded states, for which structural models are not readily available. In stability calculations, we make the assumption that the sidechains do not interact with each other in the unfolded state, but only with nearby backbone and solvent [24–26]. We implement this idea by considering any sidechain X as a part of a tripeptide Ala-X-Ala. We compute the average free energy for a large number of backbone conformations of the tripeptide, using Eq. 3, and assign this value to chemical type X. An empirical correction can be added to this value (*see* last term of Eq. 3), chosen so that the resulting amino acid compositions are reasonable during the design of whole protein sequences. The calculation of this term can be done ahead of time and is explained in Ref. 18. The total free energy of a given protein sequence in its unfolded state is the sum of the individual contributions of its constituent residue types.

1.4 Ligand Titration

In the case of binding calculations, the contribution of the free protein cancels out in relative binding free energies, as explained above. The free energies of the unbound ligands can be averaged over single or multiple structures, obtained from experiments or simulations; alternatively, it may be assumed that the ligands (and possibly the protein) maintain the same conformations in solution and in the complexes. A correction (*see* last term of Eq. 3) can be added to the energy of the unbound ligand L, to express the dependence of binding free energies on the ligand concentrations:

$$E_{\text{corr}}^{\text{L}} = +k_{\text{B}} T \ln[L] \tag{4}$$

with k_{B} the Boltzmann's constant, T the temperature, and $[L]$ the ligand concentration (set by the user). The ratio of concentrations of two complexes obeys the equation

$$\frac{[PL_2]}{[PL_1]} = \exp\left[-\beta\left(\Delta\Delta G_{\text{b}} - k_{\text{B}} T \ln\left(L_2 / L_1\right)\right)\right] \tag{5}$$

One can vary the ligand concentration ratio $[L_2]/[L_1]$ progressively during ligand design, and monitor the ratio of predicted concentrations $[PL_1]$, $[PL_2]$; the binding free energy difference $\Delta\Delta G_b$ is then obtained as $k_B T \ln([L_2]/[L_1])$, for the concentration ratio $([L_2]/[L_1])$ that yields equal concentrations $[PL_1]=[PL_2]$.

1.5 Proton Binding

The thermodynamic cycle on the right of Fig. 1 can also describe proton binding (or release) by titratable protein residues (e.g., Asp→AspH). This can be of use to determine sidechain protonation states and prepare a system for design or other simulations. Proton binding in the protein environment is described by the upper horizontal leg, and in solution by the lower leg. The solution state is a model compound—typically a single amino acid X with blocking terminal groups (ACE-X-NME). The free energy change upon protonation in the protein, relative to the model compound in solution, is:

$$\Delta\Delta G_\mathrm{p} = \left[G(P-XH)-G(P-X)\right]-\left[G(XH)-G(X)\right] \quad (6)$$

and corresponds to the pK_α difference between the sidechain in the protein and the model compound. In titration calculations, as in ligand optimization, we add a correction term to the free energy of the model compound in its protonated state to account for the proton concentration $\left[\ ^{+}\right]$:

$$E_\mathrm{corr}^\mathrm{X} = 2.303 k_\mathrm{B} T\left(\mathrm{pH}-\mathrm{p}K_\mathrm{a}^\mathrm{model}\right) \quad (7)$$

where p$K_\alpha^{\mathrm{model}}$ is the experimental pK_α value for model compound [27, 28]. The fraction f of protonated states at different pH values can usually be described by the following titration curve:

$$f = \frac{[XH]}{[X]+[XH]} = \frac{1}{1+10^{\mathrm{n}(\mathrm{pH}-\mathrm{p}K\_(\S\,))}} \quad (8)$$

To apply the above equation, titration calculations are conducted for different pH values. The pK_α of residue X is the pH for which the protonated and unprotonated states are equiprobable. The Hill coefficient n represents the maximum slope of the curve, which occurs at the titration mid-point.

1.6 Multi-Objective Optimization

As described above, Proteus is a multitool CPD suite, which is applicable to typical sequence/structure optimization calculations, but also to more refined pK_α and relative binding affinity calculations. Its physical scoring function, with the addition of appropriate correction terms, can be easily adjusted to describe different situations. Eqs. 1 and 2 can be decomposed into protein–ligand intramolecular and intermolecular energy contributions, which can be enhanced or diminished during energy minimization via appropriate weighting factors (positive, negative, or zero); and combined to produce more sophisticated, multi-objective energy, or cost functions, as follows:

$$\tilde{G} = w_1 \cdot G(P) + w_2 \cdot G(P:L) + w_3 \cdot G(L) + w_4 \cdot G_{\text{dc}}(P) + w_5 \cdot G_{\text{dc}}(L) \qquad (9)$$

The subscript "dc" denotes duplicate copies of the protein and ligand groups, which share the same amino acid sequence, but sample different conformations during exploration. Energy threshold values can also be included in Eq. 9 to refine the sequence optimization.

1.7 Energy Matrix

The design begins by separating the protein (and ligand, if present) into groups (residues), which can contain backbone and sidechain moieties. Part of the system, typically the backbone and selected sidechains, is classified as "frozen"; i.e., it retains its conformation and chemical composition during the calculation. Other parts can change both their chemical identity and conformation ("active"), or only their conformation ("inactive"). Sidechain conformations are taken from a rotamer library [29]. Multiple backbone conformations can also be specified (*see* Eq. 9). We then pre-compute and store in a matrix the interaction energies for all intra- and intermolecular residue pairs, taking into account all chemical types and conformations compatible with the classification of each residue (active or inactive). This calculation is done by XPLOR and a library of command scripts, using the energy function of Eq. 3. The GB and SA terms of the energy function are not rigorously pairwise-additive; i.e., even though they can be expressed as contributions from particular residue pairs, each contribution depends on the geometry of the entire molecule. To solve this problem, we employ a "Native Environment Approximation" (NEA) for the GB term, and a "sum over atom pairs" approximation for the SA term; more details are supplied below and in Ref. 30.

The entries of the resulting interaction matrix correspond to distinct rotamer orientations of the active and inactive parts, and to a given conformation of the "frozen" part. Often, it is desirable to take into account multiple conformations of the frozen part (e.g., several backbone conformations from an MD trajectory). Separate interaction matrices can be constructed for each of these conformations, and employed in the design.

1.8 Sequence/Structure Exploration

The interaction energy matrices are read by the C program proteus, which performs the exploration (or "optimization") in structure and sequence space. Three exploration methods are available in proteus; a heuristic protocol, first introduced by Wernisch et al. [26], a mean-field approach [31, 32], and a Monte Carlo (MC) method [33, 34]. The Monte Carlo method can use a single "walker", exploring a single trajectory. Alternatively, it can use multiple walkers, which have distinct temperatures, explore distinct

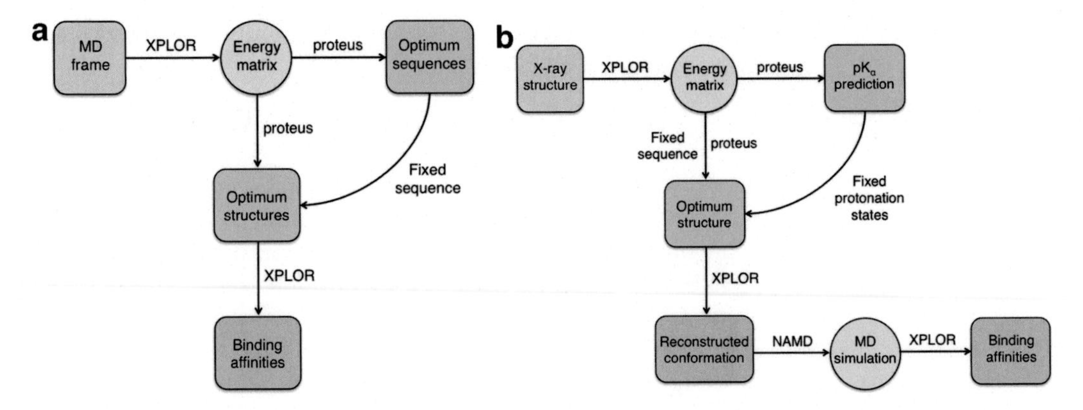

Fig. 2 Calculation flowchart diagrams for the test cases: (**a**) ligand redesign, and (**b**) preparation of a structure for MD simulations

trajectories, and occasionally exchange their temperatures. The multi-walker variant corresponds to a "replica exchange" Monte Carlo simulation, which we refer to as REMC.

All the exploration methods output multiple "solutions", sampled along the MC trajectory or the heuristic exploration. Each solution or time-step is described by a list of chemical types and rotamers for all the active and inactive positions. Subsequently, the corresponding conformations can be reconstructed and subjected to energy minimization and/or MD simulations with the same force field used in the design. Average binding free energies can be obtained from the resulting trajectories, and/or post-processed using a GBSA or PBSA approximation, as a further test of the design.

1.9 Flowcharts

The above calculations are summarized in the flowcharts of Fig. 2. The left flowchart portrays a structure/sequence optimization of a complex, which starts from an initial conformation taken from an MD trajectory. A related example, described in the Methods section, involves the redesign of the cyclic 13-residue peptide compstatin, which regulates the function of protein C3 of the complement system. Binding of this molecule and related analogs has been the subject of numerous experimental and computational studies in recent years [35–39]. The right flowchart describes the preparation of an X-ray structure for MD simulations. A related example in Methods describes the chemical and structural optimization of a complex between the MHC class II protein HLA-DQ8 and the vinculin epitope.

2 Materials: Software and Data Files

To carry out a complete protein design calculation with Proteus, the user needs the Proteus 2.1 CPD package. The appropriate files can be downloaded from *http://biology.polytechnique.fr/biocomputing/*

proteus.html. In what follows, we refer to specific files from this distribution. Furthermore, the user needs an initial structural model for the molecule (or complex) under study.

3 Methods

3.1 Structure Preparation

1. Split the PDB file into separate files for each protein segment (e.g., multiple chains), the ligand, and the crystallographic waters. Rename atoms and residues to match the Amber or Charmm force field. Renumber residues of each segment starting from 1000 for chain A, 2000 for chain B, etc., to ensure unique residue numbers; name the various segments "PROA", "PROB", "PROC" or "LIGA" and "XWAT" (*see* **Note 1**).
2. Use the XPLOR script *build.inp* to generate a protein structure file (*system.psf*) which describes the topology of the protein–ligand system and a coordinate file (*system.pdb*) in XPLOR pdb format (*see* **Note 2**).

3.2 System Setup

1. The XPLOR stream file *parameters.str* contains important information about the energy calculation setup. Edit the file to select between the Amber "ff99SB" [23] and Charmm "toph19" [22] force fields. These two force fields are consistent, respectively, with the GB/HCT [40] and GB/ACE [41] implicit solvent models. Add a surface area term to the energy function to account for the nonpolar contribution to the solvation energy. Include X-ray sidechain conformations ("native rotamers") in the rotamer library, and choose the number of minimization steps before the computation of pairwise interaction energies. Set the protein dielectric constant and define parameters employed by the solvation model and the corresponding nonbonded energy terms.
2. Modify the XPLOR stream file *sele.str* to define the sequence and conformation space. Select the modifiable residues (active), the flexible sidechains (inactive), the ligand (active or inactive), and the fixed part (backbone plus any glycines, prolines, cysteines in disulfide bonds, and crystallographic waters/ions).
3. The file *mutation_space.dat* lists the amino acid types available for each active position. The mutation space includes up to 26 amino acid types, including all natural amino acids (except glycine and proline), three histidine tautomers (protonated on N_δ, N_ε, or both), and the minor protonation states of titratable residues Lys, Asp, Glu, Tyr, Cys.
4. The system setup is done via two XPLOR scripts. The first one, *setup.inp*, prepares the system for residue pairwise energy calculations. The structure file *setup.psf* defines each active residue, including its crystallographic backbone and a set of

sidechains corresponding to all considered mutations (defined in *mutation_space.dat*). Entries of these amino acid sidechains at each modifiable position are included in the coordinate file *setup.pdb*, with arbitrary coordinates $(x = y = z = 9999.0)$. The B-factor column of the coordinate file labels the corresponding residue as active $(b = 2.00)$, inactive $(b = 1.00)$, or frozen $(b = 0.00)$. The Q-factor column labels buried $(q = 0.50)$ and exposed $(q = 1.00)$ residues, with $q = 0.00$ for hydrogens. At this point the GB solvation radii of the backbone atoms are computed and stored in the file *bsolv.pdb*.

5. The Perl script *make_position_list.pl* reads the file *setup.pdb*, and lists in *position_ list.dat* the active, inactive, and ligand positions, including the number of all possible pairwise interactions to be computed at each position.
6. The Shell script *make_mutation_space.sh* creates individual files for each active, inactive, and ligand position, listing the compatible amino acid types at each position. These files are stored locally and read later by the XPLOR scripts during the residue pairwise interaction calculations.
7. The second XPLOR script for system setup is *setupI.inp*. For each position *I*, we loop over its allowed amino acid types (depending on whether it is active, inactive, frozen, or part of the ligand). For each amino acid type we loop over rotamer states taken from a rotamer library [29]. We also include the native orientation as a separate rotamer. At this stage, we compute and store GB solvation radii for all residues, assuming the Native Environment Approximation (NEA). In a standard GB formulation, the GB energy function is not pairwise-additive, since the solvation radius of each atom depends on the position and chemical type of all other atoms in the molecule. To render the GB function pairwise-additive, we assume during the solvation radii calculation that each residue is surrounded by the native sequence and conformation. Thus, for each rotamer, we compute the GB solvation radii in the presence of residue *I*, the whole backbone (fixed part) and all remaining portions of the molecule, further than 3.0 Å away from sidechain *I*, considered in their native sequence and structure. The 3.0 Å cutoff distance excludes native sidechain atoms that might overlap with sidechain *I* in its new rotamer; this cutoff can be adjusted to a different value in *parameters.str*. Importantly, to alleviate possible clashes of a sidechain in a particular rotamer with the backbone, we do $N_{min} = 15$ steps of Powell energy minimization (*see* **Note 3**), keeping everything else (everything but sidechain *I*) fixed. If a resulting solvation radius is too large (e.g., due to overlap of the residue with the rest of the molecule), it is reset to a maximum value (999.0 Å). After the minimization, sidechain coordinates and solvation

radii are stored in a local PDB file (*matrix/local/Rota/1025.pdb*; 1025 is the residue number *I*) to be used in **step 3** from Subheading 3.3.

3.3 Interaction Energy Matrix

1. First, we compute the diagonal terms of the interaction energy matrix using the file *matrixI.inp*. This rather fast calculation is usually run sequentially over all nonfrozen positions; it is also possible to run the separate positions in parallel on multiple cores. For each position *I*, we reread the solvation radii and sidechain coordinates (*matrix/local/Rota/1025.pdb*). We loop over the allowed amino acid types (depending on whether position *I* is active, inactive, frozen, or part of the ligand) and the corresponding rotamer states. For each rotamer, we compute the energy due to interactions that sidechain *I* makes with itself and with the backbone. The energy function includes bond, angle, dihedral, improper, van der Waals, Coulomb, GB, and SASA energies. The results are printed in local files (*matrix/dat/matrix_I_1025.dat*), and can be displayed either in standard or enriched format. The basic information for each position is printed with the standard format: residue number (1025), amino acid type (ARG), one letter code (R), rotamer index number (5) followed by four energy values: the unfolded state (or unbound ligand) energy (*estimated by* Eq. 3), the bonded terms plus vdW, the electrostatic term, including GB, and the surface area term. A further decomposition of individual energy terms is displayed when the "enriched format" is requested in *parameters.str*.
2. Use the Shell script *make_rotamer_space.sh* to examine the rotamer van der Waals energies and exclude those exceeding a locally defined threshold value. Excluding "bad" rotamers for each amino acid type at each position reduces the conformational space.
3. The energy matrix calculation continues with the off-diagonal terms, using *matrixIJ.inp*, which computes the interaction between sidechains *I* and *J*. Only the lower triangle of the matrix $I < J$ is needed. The fastest approach for this part of the calculation evaluates single residue pairs $I - J$ simultaneously, on multiple cores. It is also possible to calculate all the residue pair interactions sequentially. For each residue pair, we loop over the sidechain type/rotamer space of residue *I*; we retrieve the coordinates and atomic solvation radii of the current sidechain from the rotamer PDB file (*matrix/local/Rota/1025.pdb*), created in **step 7** from Subheading 3.2. For each rotamer we loop over all residues $J < I$ and apply a first distance filter. Residues that are too far from I (e.g., $C_{\dagger} - C_{\dagger}$ distance $> 30^{-}$) are omitted. For each residue *J* within the first distance filter, we loop over the sidechain type/rotamer space of residue *J* and

read the coordinates and solvation radii from the corresponding rotamer PDB files. For both residues *I* and *J* we employ only the "good" rotamers, determined in the previous step. With the current sidechains in place, we apply a second distance filter, where interactions between sidechains are ignored if the minimum distance between the two sidechains exceeds 12 Å, say. The interaction energies of sidechain pairs that pass the second distance filter are computed. Recall that the final coordinates of two sidechains are produced via the independent minimization of each sidechain in the presence of the fixed backbone. Consequently, it is possible that the two sidechains overlap for some rotamer combinations. If the minimum sidechain–sidechain distance is smaller than a cutoff (3 Å), we perform N_{min} $(15-50)$ steps of Powell minimization (*see* **Note 3**) to improve the sidechain geometry and alleviate bad contacts. During this minimization, everything except the two sidechains is kept fixed, and the two sidechains interact with each other and the backbone. The results are stored in local files (*matrix/dat/matrix_IJ_1025_1022.dat*). The standard display format consists of a line indicating the residue numbers and names of a given pair (1025 ARG 1022 VAL), followed by a list of entries for each computed rotamer pair, for the given pair of amino acid types. Each entry reports the two rotamer numbers, the vdW interaction term, the sum of electrostatic and GB terms, and the surface area term. Similarly to **step 1**, an "enriched format" option is possible, which prints a more detailed output.

4. Finally, run the shell script *concat_matrix.sh* to join all the energy elements in a global matrix file *matrix.dat*, to be read by the proteus exploration program.

3.4 Protein Design

3.4.1 Sequence Optimization

The sequence exploration is done by the proteus program, controlled by setting various options in an input script, *proteus.conf*.

1. One may want to use a protein dielectric constant that is different from the one used in the energy matrix calculations (defined in *parameters.str*). To use a different value, first use the Perl script *modify_matrix.pl* to modify the original matrix accordingly (*see* **Note 4**).
2. During the energy matrix construction (*see* Subheading 3.3, **steps 1** and **3**), a large set of active and inactive positions can be defined. During sequence exploration, we may want to limit ourselves to a smaller set. For this, in *proteus.conf*, the sequence/conformational space of selected protein and/or ligand residues can be restricted to particular types and/or rotamers. For example, in the redesign of the compstatin peptide, in the energy matrix calculation, we set all 15 ligand

positions to be active and all protein sidechains to be inactive; subsequently, in proteus, we optimized the sequence of just a two-residue extension; the other peptide positions were not allowed to mutate. The default option corresponds to a full scale exploration of all possible amino acid types and rotamers for each active and inactive position (*see* **Note 5**).

3. Choose among the mean field, heuristic, and Monte Carlo sequence/structure exploration methods, and assign the relevant parameters. For example, if the MC method is employed, we might use a high initial temperature (given in $k_B T$ units) to overcome local energy barriers, and run several long simulations [millions of steps; (*see* **Notes 5–7**)]. By default, the simulation starts from a random sequence/structure combination and uses the Metropolis criterion to evaluate the successive moves in sequence and rotamer space. The exploration is performed using single and/or double moves, improving the sampling of coupled sidechains. The frequency of each type of move during the simulation is also controlled by the occurrence probability of each mutation type; a small sequence/structure move ratio (1:10 or 2:10) allows the system to relax its structure slightly in the presence of the new amino acid type (*see* **Note 6**).
4. All exploration parameters mentioned in **steps 2** and **3** are set up via a simple, user-editable configuration file (*proteus.conf*), which is read as the standard input by the proteus executable.
5. After the exploration step, proteus is run again in post-processing mode, to convert the resulting solutions into a more readable (fasta-like) format. The output file *proteus.rich* reports each solution by the sequence of: (a) amino acid types, (b) residue numbers, and (c) rotamer numbers. The Perl script *analyze_proteus_sequences.pl* sorts the solutions (combinations of sequences and rotamers) by their frequency of occurrence and calculates the minimum, maximum, and average folding free energies.

3.4.2 Structure Optimization

After large-scale sequence exploration, it can be desirable to do more extensive rotamer exploration for selected sequences.

1. Repeat the above steps for a chosen subset of designed sequences. Keep each protein and ligand sequence invariant, and explore its conformational space through rotamer optimization. Compute the statistical average of the folding free energy over all sampled conformations, to improve the energy estimate for the chosen sequences.
2. Use the Perl script *rot_distrib_proteus.pl* to compute the rotamer distribution of all residues from the pseudo-trajectory obtained during optimization, to characterize the flexibility of each sidechain.

3. Cluster the protein and ligand conformations based on selected sidechains, and reconstruct the minimum energy conformation of each cluster to get a set of "good" conformations.

3.5 pK_α Calculations

In some applications, we wish to determine sidechain protonation states through pK_α calculations. For each titratable sidechain, the energy will include a pH-dependent term, $E_{corr}{}^{X}$, where X is the sidechain type.

1. First, compute the correction energy term $E_{corr}{}^{X}$ at pH = 7 (*see* Eq. 7), by evaluating the energy GX^{model} of the model compound in solution with Eq. 3, and replace the values representing the unfolded state energy from the diagonal matrix elements with $-G_X^{model}$.
2. Modify the proteus configuration file to restrain the mutation space of each active-titratable residue to its two or three ionization states (ASP/ASH, GLU/GLH, CYS/CYM, HID/HIE/HIP, TYR/TYD, LYS/LYN); restrict the other positions to their native type (or make them inactive during the energy matrix calculation).
3. Run a proteus MC simulation, to identify optimum combinations of sequences (protonation states) and structures at the specified pH. Start with one million equilibration steps at high temperature $(k_B T = 1\,\text{kcal}/\text{mol})$, extract the final state and continue with ten million production steps at room temperature; use a relatively small sequence-to-structure move ratio (1:10), to allow the system to relax after protonation moves.
4. At the end of the MC simulation, compute the probabilities of each protonated state at each active, titratable position (*see* **Note 8**).
5. Run a full pH scan by increasing progressively the pH from 0 to 15 and repeating **steps 1–4**.
6. Fit the fractional occupancy of the protonated state to the modified Hill equation (*see* Eq. 8) for each titratable sidechain using the Perl script *evalpka.pl*; extract the pK_α value with the corresponding Hill coefficient at the mid-point of the sigmoidal curve.

Table 1 (*adapted from* Ref. 42) shows pK_α calculations for nine proteins and 130 titratable groups with sufficient sidechain type diversity (35 Asp, 34 Glu, 13 Tyr, 28 Lys, and 20 His). Overall, the agreement with experiment is good, with an rms deviation of just 1.1 pH units, for reasonable protein dielectric constants of four and eight. For sidechains with large pK_α shifts, ≥ 2, the rms error with our method is 1.8, compared to 2.6 with the Null model (and 1.1 with the specialized PROPKA program).

Table 1
Comparing large and small pK_α shifts

			[a]MC		
Experimental range	**Number of sidechains**	**[a]Null model**	$\varepsilon_p = 4$	$\varepsilon_p = 8$	**[a]PROPKA3**
$\lvert \Delta pK_\alpha \rvert < 1$	85	0.5	0.9	1.0	0.6
$1 \le \lvert \Delta pK_\alpha \rvert < 2$	34	1.7	1.3	1.2	1.0
$2 \le \lvert \Delta pK_\alpha \rvert$	11	2.6	1.8	1.8	1.1
All	130	1.1	1.1	1.1	0.8

[a]Rms deviations between computed and experimental *pK_α* shifts

An application example involves the chemical and structural optimization of a complex between the MHC class II protein HLA-DQ8 and the vinculin epitope [43]. Since the structure of the specific complex was not known, we started from the X-ray structure of the HLA-DQ8 complex with an insulin peptide. MHC class II proteins bind various peptides in the endosome, where the pH ranges from 4.5 to 6.0; therefore, in the initial setup we determined the ionization state of titrating groups by pK_α calculations with Proteus. The binding site (residues within 8 Å of the peptide) contains 23 titrating sidechains (3 Lys, 3 His, 2 Asp, 6 Glu and 9 Tyr residues, out of 98 residues). Arginines were excluded, since they titrate well outside the pH range of interest ($4.0 \le \text{pH} \le 7.0$). We focused on a group of residues near the first anchor position (P1) of the binding groove, where αGlu31, βGlu86, αHis24, and αArg52 form a strong interaction network. Between αGlu31, αHis24, and P1 there is also an important crystallographic water. The two gluatamic acids are 4.1 Å apart $\left(C_{.} - C_{.}\right)$ and their titrating behavior is coupled. The net charge of this group of residues could not be verified by X-ray crystallography [44], and was a matter of discussion in subsequent studies of HLA-DQ8 and MHC class II proteins [45, 46]. We performed pK_α calculations with two dielectric constants, $\varepsilon_p = 4$ and 8, both in the absence and the presence of the vinculin peptide; and compared our results with the empirical Propka model. For extracellular pH values around 7, Proteus calculations with $\varepsilon_p = 4$ and Propka predict a neutral histidine and a protonated αGlu31. The pK_α of the other glutamic acid, βGlu86, is overestimated by Proteus, but becomes better at $\varepsilon_p = 8$. Similar pK_α values are obtained for the complex and the free protein. Figure 3 shows a superposition of the reconstructed optimum conformation (vinculin) and the template X-ray structure (insulin). Setting the appropriate ionization state for αGlu31 promotes a successful sidechain placement of all key residues that take part in binding (*see* Fig. 3). Structure preparation as performed by preliminary pK_α calculations and sidechain placement is an important byproduct of Proteus.

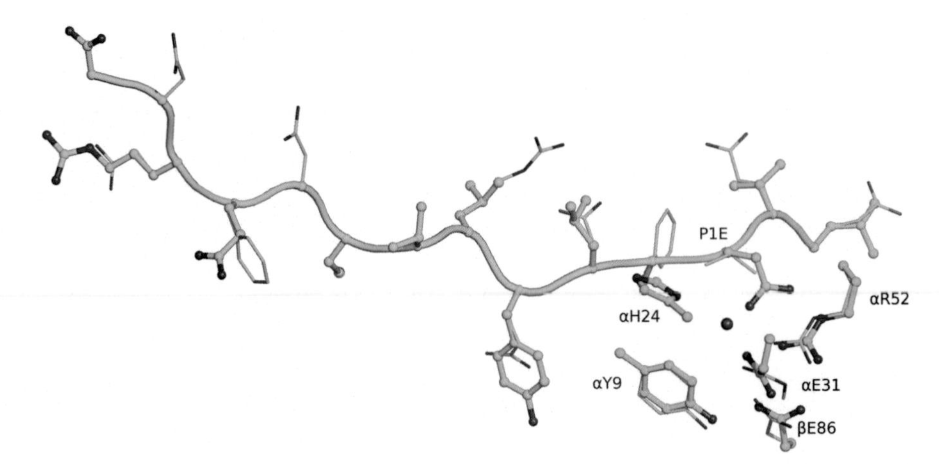

Fig. 3 Superposition of the starting X-ray structure of the insulin complex (*ball*-and-*stick view*) and the optimized conformation of the vinculin complex (*thick lines*)

3.6 Specificity Calculations by Ligand Titration

In many applications, we want to discover sequences that favor one ligand over another, and design for specificity. One approach is to make two or more ligands compete for a single binding site. By gradually increasing the concentration of one ligand, we gradually displace the other(s), and can extract the relative binding free energy from the titration curve. This can be done with the protein sequence fixed or variable. Here, for simplicity, we describe an application where the protein sequence is fixed, and we focus on the relative binding strength of two ligands.

1. Set all or part of the ligand to be active, with two or more types; say, X_{nat} (natural ligand) and X_{mut} (alternative, or "mutant" ligand). The protein and any remaining ligand positions are inactive. To speed up the calculation, constrain the rotamer space of distant residues (further than 8 Å, say, from the active position) to their native conformation (*see* **Note 5**).
2. Assign a correction term to the mutant ligand (*see* Eq. 4), to reflect a low initial, relative concentration. This term has two parts. The first part is $k_B T \ln\left(L_{X_{mut}} / L_{X_{nat}}\right)$. The second part is the energy difference between the two unbound ligands, computed with Eq. 3. The first contribution can be set to $-5\,\text{kcal}/\text{mol}$; this corresponds to the case where the native ligand is represented in the mixture at a much higher concentration than the mutant type, favoring the native ligand binding.
3. Run a short equilibration stage (500,000 steps) at high temperature, followed by a long production stage (ten million steps) at room temperature starting from the final state of equilibration.

4. Count the number of steps with the mutant ligand present and deduce the population fraction with a bound mutant ligand.
5. Repeat **steps 1–4** while gradually increasing the relative concentration term of the mutant ligand from −5 to +5 kcal / mol . As we increase the concentration, $L_{X_{mut}}$ gradually replaces $L_{X_{nat}}$ in the binding site.
6. Fit the data to the appropriate titration curve (*adapted from* Eq. 5) and obtain the binding free energy difference from the mid-point, where the populations of the bound mutant and native ligands are equal.

A ligand titration example: This example involves the redesign of the cyclic 13-residue peptide compstatin, which regulates the function of protein C3 of the complement system. We and our collaborators have studied extensively the binding of compstatin and its analogs to C3 by computational and experimental methods [36, 37, 47, 48]. In recent work [38, 39], we explored the addition of a two-residue extension [XY] to the N-terminal end of the compstatin double mutant Ac-Val4Trp/His9Ala ([XY]W4A9). MD simulations had suggested that this extension may increase the number of contact residues with the protein. Using a snapshot from MD simulations of the C3 complex with [RS]W4A9, we searched for extension sequences that optimized ligand binding. To determine the amino acid type preference of the two-residue extension of compstatin, we computed the binding free energy difference (*see* Eq. 2) of each amino acid type X with respect to Ala at each position of the extension. Binding affinities (relative to Ala) for various amino acid substitutions at positions −2 and −1 are summarized in Table 2. Columns 2 and 6 contain the results from design calculations at extension positions −2 and −1, respectively, in which all amino acid types are allowed to compete simultaneously; the resulting affinities are computed from the individual amino acid frequencies in the resulting solutions. Columns 3 and 7 contain the results of calculations in which only one amino acid at a time competes with Ala; the corresponding relative affinities are computed from Eq. 5. The results of the two methods agree closely. Experimentally, positions −2 and −1 can tolerate various amino acid types, without large differences in the corresponding binding free energies [38]. The design favors a positively charged Arg residue at position −2. MD simulations of the [RS]W4A9 complex with C3 suggest that an Arg residue at position −2 forms a strong electrostatic interaction with proximal residue Glu372 (*see* Fig. 4a); this interaction is captured by the Proteus design. Position −1 is predicted to not have a strong propensity for one particular sidechain type; it somewhat disfavors 14 out of

18 types, especially bulky hydrophobic sidechains. This can be explained by the fact that sidechains at position −1 are oriented toward the solvent.

7. It can be useful to reassess the designed sequences by additional calculations. In the compstatin redesign study, we performed rotamer optimization on the designed sequences and clustered the resulting conformations (based on the rotamer states of all sidechains within 8 Å of the extension). For each sequence, we reconstructed representative conformations from the ten most populated clusters, and subjected them to 100 steps of energy minimization with the Powell conjugate gradient method. During minimization, we kept the backbone fixed, to facilitate comparison with the raw design results. We then computed the binding free energy of each conformation at the end of minimization with the GBSA approximation, as the difference between the free energy of the complex and the isolated ligand and protein. The results, averaged over the ten conformations, are included in columns 4 and 8 of Table 2; the values are expressed relative to alanine. Some bulky amino acid types (Trp, Lys, Met, His, Tyr, Leu, Val, Ile) become slightly preferred at position −2 after minimization, due to enhanced van der Waals interactions with Val375 (*see* Fig. 4b). At position −1, Arg still represents the optimum sidechain after reconstruction and minimization. These predictions may still change after MD simulations of the same complexes.

4 Notes

1. The ligand can be a polypeptide segment (chain C), like the insulinB 14-mer bound to HLA-DQ8, which we treat in the same way as the protein, or a nonpeptidic molecule like the heme in hemoglobin. In that case, we need to define the topology of the new molecule and specify the necessary parameters and possibly rotamers. The new segment must be named "LIGA".
2. The file *build.inp* must be modified to match the segment names defined by the user. The file reads the amino acid sequence of each chain according to its segment name and adds disulfide bonds and terminal group patches, to generate the corresponding molecular structure. The coordinates of any missing hydrogens are assigned, and the structures are saved in the *system.psf* and *system.pdb* files.
3. The energy minimization steps done in **steps 1** and **3** from Subheading 3.3 balance to some extent the suboptimal orientations available to the sidechains due to the discrete rotamer space. The number of minimization steps can be adjusted for

Table 2
Sequence optimization, affinity, and specificity calculations in the compstatin:C3 complex, targeting the N-terminal extension of compstatin

Extension residues							
Position −2				Position −1			
	$\Delta\Delta G^a$	$\Delta\Delta G^b$	$\Delta\Delta G^c$		$\Delta\Delta G^a$	$\Delta\Delta G^b$	$\Delta\Delta G^c$
aa type	(kcal/mol)			aa type	(kcal/mol)		
R	−0.9	−2.0	−1.4	R	−0.4	0.0	−1.4
Y	−0.1	0.0	−1.7	S	0.0	0.0	−0.4
A	–	–	–	A	–	–	–
M	0.0	0.0	−1.9	N	0.0	0.0	−0.4
C	0.0	0.0	−0.6	C	0.1	0.0	−0.1
K	0.1	0.0	−1.1	T	0.3	0.5	0.2
N	0.1	0.0	−0.8	Q	0.4	0.8	−0.1
V	0.1	0.0	−0.8	M	0.5	0.9	−0.5
Q	0.1	0.0	−1.2	V	0.5	1.9	−0.3
S	0.2	0.0	0.0	K	0.5	1.3	0.0
I	0.2	0.3	−1.4	Y	0.6	1.0	−0.6
F	0.2	0.4	−0.3	W	0.7	1.5	−0.3
W	0.4	0.5	−3.4	H(N_ε)	0.8	1.5	0.0
T	0.4	0.5	0.0	H(N_δ)	0.8	1.5	−0.2
H(N_δ)	0.4	0.5	−1.8	E	0.8	1.3	−0.1
H(N_ε)	0.4	0.5	−0.7	D	0.8	1.3	−0.2
L	0.5	1.0	−1.3	F	2.0	0.9	−0.8
E	0.6	1.1	−0.8	I	1.1	2.0	−0.5
D	0.9	1.5	0.0	L	1.1	2.0	−0.3

All binding affinities computed relative to Alanine (A)
[a]Estimated from the frequency of the solutions with the corresponding amino acid in target position −2 or −1
[b]Estimated from the titration curves
[c]Estimated after reconstruction and minimization of the resulting solutions for a 100 steps with a fixed backbone. The results are averaged over the ten most populated rotamer conformations, taking into account all sidechains within 8 Å from the extension

specific cases. For several systems, extending the minimization to more than 50 steps was shown to increase computational cost without a significant improvement in the results.

4. The protein dielectric constant is an empirical parameter. Its value depends on the type of calculation and the solvation

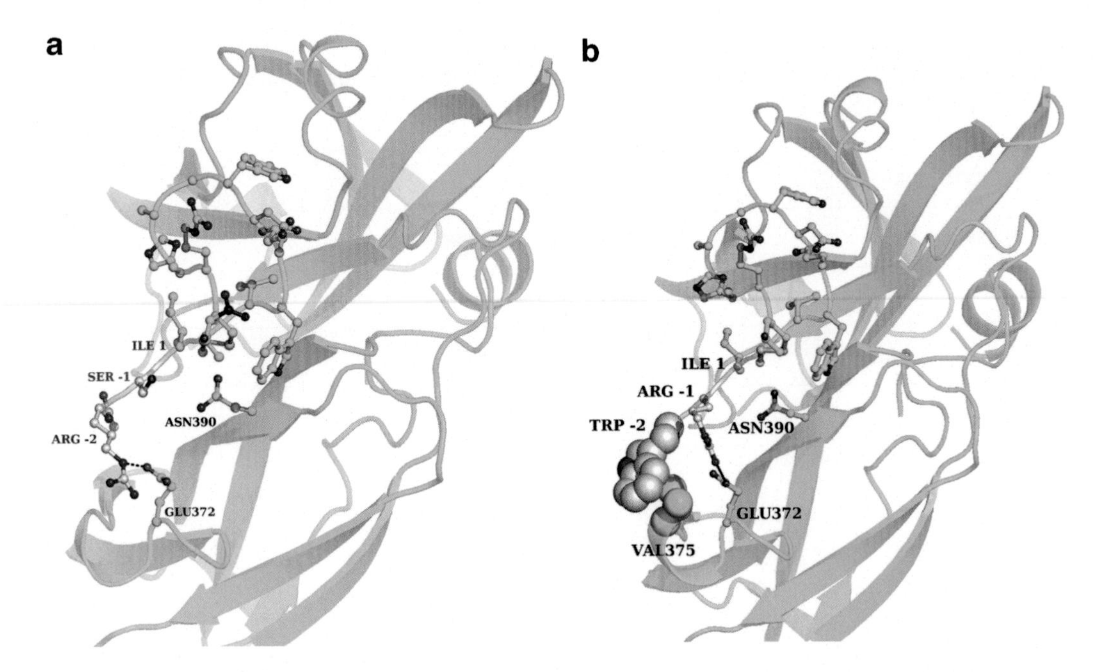

Fig. 4 3D structure of the cyclic 13-residue peptide compstatin analog W4A9 (*cyan*) and a two-residue extension to the N-terminal end (*white*) in complex with the protein C3 (*green*). (a) Starting structure used by Proteus, (**b**) minimized structure of a predicted mutant

model used. For CPD applications with a GBSA implicit solvent model, we found that low dielectric values of 4–8 give reasonable results. pK_α calculations on a large data set of titrating sites showed good accuracy for $\varepsilon_{\mathrm{p}} = 8$ [42]. For whole protein designs, a higher value such as $\varepsilon_p = 16$ may give better results [49, 50].

5. To obtain adequate sampling, we restrict the sequence/conformation space depending on the application. For the compstatin redesign, we focused on the area surrounding the peptide extension. The two extension residues are allowed to sample all amino acid types and rotamers without any restrictions, while every other sidechain within 8 Å from any atom of the extension changes only its conformation. The remaining residues are held fixed, together with the backbone, in the X-ray conformation. With these "local" space restrictions, the exploration converged within ten million steps. The quality of the sampling can be assessed by repeating the calculation with different random number seed values, or by performing both backward and forward pH or ligand concentration scans (*see* Eqs. 4 and 7). The convergence of the method can also be tested with additional simulations of increasing length.
6. With MC exploration, the relative frequency of mutation and rotamer moves (both single and double) can be adjusted by the user in the *proteus.conf* configuration file to match the

needs of a given calculation [51]. Conformational changes are usually less drastic than amino acid type changes (i.e., Ala → Arg); therefore, it is generally preferred to allow more rotamer than type moves, to allow the system to relax after a mutation.

7. With MC exploration, it is possible to run multiple simulations in parallel, with different temperatures, such that the simulations periodically exchange their temperatures. This method is known as Replica Exchange, or REMC. It is activated in the *proteus.conf* file by indicating the number of simulations (or "walkers"), their temperatures, and the interval between temperature swaps. Each walker then generates its own output files. On a multi-core machine, the simulations will run in parallel if the OpenMP library is present.
8. To calculate correctly the fractional occupancies from the Monte Carlo simulation, both accepted and rejected moves should be accounted for, since a move rejection signifies a preference for the previously occupied state.

Acknowledgements

GA, SP, and EM acknowledge financial support through a grant offered by the University of Cyprus.

References

1. Kortemme T, Baker D (2004) Computational design of protein–protein interactions. Curr Opin Chem Biol 8(1):91–97
2. Floudas C, Fung H, McAllister SR, Monnigmann M, Rajgaria R (2006) Advances in protein structure prediction and de novo protein design: a review. Chem Eng Sci 61:966–988
3. Boas EF, Harbury PB (2007) Potential energy functions for protein design. Curr Opin Struct Biol 17(2):199–204
4. Lippow SM, Tidor B (2007) Progress in computational protein design. Curr Opin Biotechnol 18:305–311
5. Das R, Baker D (2008) Macromolecular modeling with Rosetta. Biochemistry 77(1): 363–382
6. Karanicolas J, Kuhlman B (2009) Computational design of affinity and specificity at protein-protein interfaces. Curr Opin Struct Biol 13:26–34
7. Damborsky J, Brezovsky J (2009) Computational tools for designing and engineering biocatalysts. Curr Opin Struct Biol 19: 458–463
8. Mandell DJ, Kortemme T (2009) Backbone flexibility in computational protein design. Curr Opin Biotechnol 20:420–428
9. Suarez M, Jaramillo A (2009) Challenges in the computational design of proteins. J R Soc Interface 6:477–491
10. Saven JG (2010) Computational protein design: advances in the design and redesign of biomolecular nanostructures. Curr Opin Colloid Interface Sci 15:13–17
11. Pantazes RJ, Greenwood MJ, Maranas CD (2011) Recent advances in computational protein design. Curr Opin Struct Biol 21: 467–472
12. Der BS, Kuhlman B (2013) Strategies to control the binding mode of de novo designed protein interactions. Curr Opin Struct Biol 23(4):639–646
13. Moal IH, Moretti R, Baker D, Fernandez-Recio J (2013) Scoring functions for protein-protein interactions. Curr Opin Struct Biol 23(6)
14. Zanghellini A (2014) de novo computational enzyme design. Curr Opin Biotechnol 29: 132–138

15. Khoury GA, Smadbeck J, Kieslich CA, Floudas CA (2014) Protein folding and de novo protein design for biotechnological applications. Trends Biotechnol 32(2):9099–9109
16. Schmidt am Busch M, Lopes A, Mignon D, Simonson T (2008) Computational protein design: software implementation, parameter optimization, and performance of a simple model. J Comput Chem 29:1092–1102
17. Polydorides S, Amara N, Simonson T, Archontis G (2011) Computational protein design with a generalized Born solvent model: application to asparaginyl-tRNA synthetase. Proteins 79:3448–3468
18. Simonson T, Gaillard T, Mignon D, Schmidt am Busch M, Lopes A, Amara N, Polydorides S, Sedano A, Druart K, Archontis G (2013) Computational protein design: the Proteus software and selected applications. J Comput Chem 34:2472–2484
19. Brünger AT (1992) X-plor version 3.1, A System for X-ray crystallography and NMR. Yale University Press, New Haven
20. Srinivasan J, Cheatham T, Cieplak P, Kollman P, Case DA (1998) Continuum solvent studies of the stability of DNA, RNA, and phosphoramidate-DNA helices. J Am Chem Soc 120:9401–9409
21. Simonson T (2013) Protein-ligand recognition: simple models for electrostatic effects. Curr Pharm Des 19:4241–4256
22. Brooks B, Bruccoleri R, Olafson B, States D, Swaminathan S, Karplus M (1983) Charmm: a program for macromolecular energy, minimization, and molecular dynamics calculations. J Comput Chem 4:187–217
23. Cornell W, Cieplak P, Bayly C, Gould I, Merz K, Ferguson D, Spellmeyer D, Fox T, Caldwell J, Kollman P (1995) A second generation force field for the simulation of proteins, nucleic acids, and organic molecules. J Am Chem Soc 117:5179–5197
24. Pokala N, Handel TM (2005) Energy functions for protein design: adjustment with protein–protein complex affinities, models for the unfolded state, and negative design of solubility and specificity. J Mol Biol 347:203–227
25. Dahiyat BI, Mayo SL (1997) De novo protein design: fully automated sequence selection. Science 278:82–87
26. Wernisch L, Hery S, Wodak S (2000) Automatic protein design with all atom force fields by exact and heuristic optimization. J Mol Biol 301:713–736
27. Pace CN, Grimsley GR, Scholtz JM (2009) Protein ionizable groups: pKa values and their contribution to protein stability and solubility. J Biol Chem 284:13285–13289
28. Aleksandrov A, Thompson D, Simonson T (2010) Alchemical free energy simulations for biological complexes: powerful but temperamental. J Mol Recognit 23:117–127
29. Tuffery P, Etchebest C, Hazout S, Lavery R (1991) A new approach to the rapid determination of protein side chain conformations. J Biomol Struct Dyn 8(6)
30. Gaillard T, Simonson T (2014) Pairwise decomposition of an mmgbsa energy function for computational protein design. J Comput Chem 35:1371–1387
31. Koehl P, Delarue M (1994) Application of a self-consistent mean field theory to predict protein sidechain conformations and estimate their conformational entropy. J Mol Biol 239:249–275
32. Zou BJ, Saven JG (2005) Statistical theory for protein ensembles with designed energy landscapes. J Chem Phys 123:154908
33. Metropolis N, Rosenbluth AW, Rosenbluth MN, Teller AH, Teller E (1953) Equation of state calculations by fast computing machines. J Chem Phys 21:1087–1092
34. Frenkel D, Smit B (1996) Understanding molecular simulation. Academic, New York
35. Qu H, Ricklin D, Lambris JD (2009) Recent developments in low molecular weight complement inhibitors. Mol Immunol 47(2): 185–195
36. Tamamis P, Pierou P, Mytidou C, Floudas CA, Morikis D, Archontis G (2011) Design of a modified mouse protein with ligand binding properties of its human analog by molecular dynamics simulations: the case of c3 inhibition by compstatin. Proteins 79(11):3166–3179
37. Tamamis P, Lopez de Victoria A, Gorham RD, Bellows ML, Pierou P, Floudas CA, Morikis D, Archontis G (2012) Molecular dynamics in drug design: new generations of compstatin analogs. Chem Biol Drug Des 79(5):703–718
38. Gorham RD, Forest DL, Tamamis P, Lopez de Victoria A, Kraszni M, Kieslich CA, Banna CD, Bellows ML, Larive CK, Floudas CA, Archontis G, Johnson LV, Morikis D (2013) Novel compstatin family peptides inhibit complement activation by drusen-like deposits in human retinal pigmented epithelial cell cultures. Exp Eye Res 116:9096–9108
39. Gorham RD, Forest DL, Khoury GA, Smadbeck J, Beecher CN, Healy ED, Tamamis P, Archontis G, Larive CK, Floudas CA, Radeke MJ, Johnson LV, Morikis D (2015) New compstatin peptides containing n-terminal extensions and non-natural amino acids exhibit potent complement inhibition and improved solubility characteristics. J Med Chem 58(2): 814–826

40. Hawkins GD, Cramer C, Truhlar D (1997) Parameterized model for aqueous free energies of solvation using geometry-dependent atomic surface tensions with implicit electrostatics. J Phys Chem B 101:7147–7157
41. Schaefer M, Karplus M (1996) A comprehensive analytical treatment of continuum electrostatics. J Phys Chem 100:1578–1599
42. Polydorides S, Simonson T (2013) Monte Carlo simulations of proteins at constant pH with generalized born solvent. J Phys Chem B 34:2742–2756
43. van Heemst J, Jansen DTSL, Polydorides S, Moustakas AK, Bax M, Feitsma AL, Bontrop-Elferink DG, Baarse M, van der Woude D, Wolbink G-J, Rispens T, Koning F, de Vries RRP, Papadopoulos GK, Archontis G, Huizinga TW, Toes RE (2015) Crossreactivity to vinculin and microbes provides a molecular basis for HLA-based protection against rheumatoid arthritis. Nat Commun 6.1–11
44. Lee K, Wucherpfennig K, Wiley D (2001) Structure of a human insulin peptide-HLA-DQ8 complex and susceptibility to type 1 diabetes. Nat Immunol 2(6):501–507
45. Yaneva R, Springer S, Zacharias M (2009) Flexibility of the MHC class II peptide binding cleft in the bound, partially filled, and empty states: a molecular dynamics simulation study. Biopolymers 91(1):14–27
46. Henderson KN, Tye-Din JA, Reid HH, Chen Z, Borg NA, Beissbarth T, Tatham A, Mannering SI, Purcell AW, Dudek NL, van Heel DA, McCluskey J, Rossjohn J, Anderson RP (2007) A structural and immunological basis for the role of human leukocyte antigen DQ8 in celiac disease. Immunity 27(1)
47. Bellows M, Fung H, Taylor M, Floudas C, Lopez de Victoria A, Morikis D (2010) New compstatin variants through two de novo protein design frameworks. Biophys J 98(10): 2337–2346
48. Tamamis P, Morikis D, Floudas CA, Archontis G (2010) Species specificity of the complement inhibitor compstatin investigated by all-atom molecular dynamics simulations. Proteins 78(12):2655–2667
49. Schmidt am Busch M, Mignon D, Simonson T (2009) Computational protein design as a tool for fold recognition. Proteins 77: 139–158
50. Schmidt am Busch M, Sedano A, Simonson T (2010) Computational protein design: validation and possible relevance as a tool for homology searching and fold recognition. PLoS One 5(5):10410
51. Mignon D, Simonson T (2015) Sequence exploration in computational protein design with stochastic, heuristic and exact methods (in press)

Chapter 7

A Structure-Based Design Protocol for Optimizing Combinatorial Protein Libraries

Mark W. Lunt and Christopher D. Snow

Abstract

Protein variant libraries created via site-directed mutagenesis are a powerful approach to engineer improved proteins for numerous applications such as altering enzyme substrate specificity. Conventional libraries commonly use a brute force approach: saturation mutagenesis via degenerate codons that encode all 20 natural amino acids. In contrast, this chapter describes a protocol for designing "smarter" degenerate codon libraries via direct combinatorial optimization in "library space."

Several case studies illustrate how it is possible to design degenerate codon libraries that are highly enriched for favorable, low-energy sequences as assessed using a standard all-atom scoring function. There is much to gain for experimental protein engineering laboratories willing to think beyond site saturation mutagenesis. In the common case that the exact experimental screening budget is not fixed, it is particularly helpful to perform a Pareto analysis to inspect favorable libraries at a range of possible library sizes.

Key words Protein library design, Degenerate codon optimization, Rational mutagenesis, Saturation mutagenesis, Regression, Cluster expansion

1 Introduction

1.1 Expanding Computational Protein Design Horizons Using Regression

Algorithms for searching large conformational spaces tend to be iterative, evaluating one conformation at a time. Molecular dynamics simulations and conventional Monte Carlo protein structure prediction fall into this category, as do simulations that support the refinement of models to fit nuclear magnetic resonance spectroscopy or x-ray diffraction data. Even when each evaluation calculation is rapid, iterative methods are often unequal to the required conformational sampling tasks. Several grand-challenge problems in computational structural biology are intractable in part because of the inability of current methods to efficiently search through the space of protein conformations. For example, consider the problem of predicting the detailed structure of a protein, starting from the structure of a homologous protein that happens to be 2 Å root

Barry L. Stoddard (ed.), *Computational Design of Ligand Binding Proteins*, Methods in Molecular Biology, vol. 1414,
DOI 10.1007/978-1-4939-3569-7_7, © Springer Science+Business Media New York 2016

mean square deviation (rmsd) from the target structure. Even though the initial structure and the target are "close," it is difficult to find the target structure in part due to the vast number of similar protein conformations.

For some problems, including fixed-backbone protein design, it is feasible to limit the search to the combinatorial placement of discrete favored sidechain positions called rotamers [1]. In this case, finding the optimal combination is still a challenging (i.e. NP-hard) computational problem [2]. However, numerous powerful combinatorial optimization algorithms have been developed to optimize sidechain placement and protein design [3–8]. Predicting the energy for any given sequence requires a combinatorial rotamer optimization calculation.

Regression-based approximate models provide a powerful approach to circumvent this limitation. The basic strategy is to prepare an approximate model that can be used to rapidly guide more expensive search calculations to productive combinations. Much of the recent research that adopts this strategy has been elucidated and described as cluster expansion [9–14]. However, the use of regression approaches to model experimental protein library data belongs in the same category.

For example, Hahn et al. used regression to model experimental data for SH3 domains [13]. The ProteinGPS methodology of DNA2.0 also quantifies protein properties in terms of sequence variables [15]. Finally, the Arnold lab was repeatedly able to rationalize the thermostability of protein "chimeras" using only crude regression models that account for 1-body contributions from each sequence block [16–21]. A chimera is a protein composed of fragments of parent proteins joined at sequence junctions called crossover sites. Crossover sites are chosen with a variety of techniques that are designed to minimize the disruption of coherent and stable fragments within the protein [22–25]. The Arnold results suggest that protein fragments can make surprisingly modular contributions to the overall protein structure and stability. Johnson et al. recently provided an interesting exception to this trend, by characterizing a library of enzyme chimeras in which stability affects were decidedly cooperative rather than modular [25].

1.2 Protein Library Design

Whereas traditional computational protein design (CPD) calculations yield a single sequence (and structure) with minimal energy, the ultimate design target for this chapter is a library. Protein library design is a highly practical calculation with diverse protein engineering applications in industrial biotechnology, materials development, and the development of therapeutic biomolecules. To efficiently identify functional and/or optimized protein sequences, protein engineers commonly work at the level of libraries. Often, if structural information is absent and a suitable assay is available, these libraries consist of randomly mutated variants.

However, when a structure is available, there is a wide range of design options.

At one end of the continuum, a structure may be used simply to identify which residues are most likely to play an important role. Saturation mutagenesis refers to the practice of mutating such target residues to all possible amino acids. For example, a protein engineer working in the area of industrial biotechnology may seek to alter the specificity of a substrate-binding pocket. Alternately, a protein engineer working to optimize a therapeutic binding protein might wish to screen a library that diversifies the amino acids at the protein–protein interface.

At the other end of the continuum, conventional CPD methods combine explicit modeling of the structure with combinatorial optimization to predict a new low-energy sequence and structure thereof. There is interest in methods that combine the practical benefits of synthesizing a library of protein variants with benefits of structure-guided design. For example, Voigt et al. described a self-consistent mean field approach to identify low-energy amino acids for subtilisin E and T4 lysozyme [26]. There are numerous routes to merge these approaches. For example, the Arnold lab has often used structure-guided design to optimize libraries of synthetic enzymes derived via site-specific recombination [18].

Much of the effort has been to develop algorithms for the specific practical task of optimizing degenerate codons (*see* below). A variety of algorithms that have been developed use as an input a list of target sequences or a $20 \times n$ matrix that indicates the target frequency of each amino acid for each of the n design positions [27]. Enumeration, dynamic programming, and integer linear programming methods have all been described for the selection of degenerate codons to cover the desired sequence space [28–32].

Here, three such algorithms are described briefly. The LibDesign algorithm [28] begins with a set of aligned amino acid sequences and then identifies favorable degenerate codons independently for each position. A favorable degenerate codon encodes the specified amino acids with minimal degeneracy, avoiding stop codons if possible. Permutations of candidate codons are assessed via the resulting library size and the number of recovered sequences from the input alignment. Allen developed an algorithm called "Combinatorial Libraries Emphasizing and Reflecting Scored Sequences" (CLEARSS) that extends the conventional CPD approach [29]. CLEARSS begins with a list of fixed-backbone sequence designs. Possible degenerate libraries are sampled, given a list of allowed amino acids and a range of allowed library sizes, and are assessed using the ranked list of specific sequences. The overall score of a candidate library is the sum of scores for each design site, and the score for each design site is the sum of the Boltzmann weights of the sequences in the ranked list that contain a library-encoded amino acid. Finally, SwiftLib from the Kuhlman group uses dynamic programming to optimize the placement of multiple degenerate codons, obtaining very efficient libraries [32]. Notably, SwiftLib is presented as a highly accessible web server.

One limitation of such algorithms is the neglect of 2-body interactions. At the cost of significantly more difficult calculations (NP-hard optimization), this was addressed by Bailey-Kellogg and coworkers in the Optimization of Combinatorial Mutagenesis (OCoM) algorithm [30]. Another limitation is the use of a pre-calculated list of designs rather than a direct optimization in library space. It is not clear that pre-calculated lists of designs offer a balanced or thorough exploration of favorable sequence space; they may instead reflect a shallow exploration of sequence space, may feature diversity only at permissive sites, and could reflect systematic inaccuracies in the design potential. Treynor et al. performed combinatorial optimization in library space, but the 2-body potential between degenerate codons had significant drawbacks (amino acid:amino acid scores were obtained without rotamer optimization) [33]. The final relevant example is the Structure-based Optimization of Combinatorial Mutagenesis (SOCoM) algorithm reported in 2015 [14]. This last report is highly suggested reading as the SOCoM algorithm closely matches our independently developed approach.

There are several relevant figures of merit for candidate libraries. First, since these tools are intended to assist with actual experimental library design, the number of theoretical variants present in the encoded library is a key parameter. Theoretical library size is the starting point for selecting the number of clones that should be experimentally screened to obtain a target library coverage [34]. Another key parameter for a candidate library is the mean energy score ($\langle E \rangle$) according to a design scoring function. Throughout this chapter the energy function is an all-atom Rosetta energy function [35]. Scores for protein structures are reported in Rosetta energy units (REU). One possible limitation of $\langle E \rangle$ is that the folding and functionality of protein sequences is not a graded response. Therefore, it may be more relevant to estimate the number of library members with $E < E_{cutoff}$, a threshold meant to flag library members at an elevated risk of not folding.

1.3 Degenerate Codon Libraries

A conventional approach to encode focused site diversity is to use a degenerate codon, in which the synthesized DNA primer consists of a mixture of nucleotides at particular positions. A single character analogous to the pure bases (A, T, C, G) represents each mixture of bases, with W→AT, S→CG, M→AC, K→GT, R→AG, Y→CT, B→CGT, D→AGT, H→ACT, V→ACG, and N→ACGT. A codon that includes at least one degenerate nucleotide is a degenerate codon. Common degenerate codons for site saturation mutagenesis are NNK and NNS, both of which encode all 20 amino acids (with varying codon and amino acid frequency). It is also important to note that 1/32nd of the codons that are physically realized from the NNK and NNS degenerate codons (assuming equimolar nucleotide mixtures) encode stop codons.

A key limitation of saturation mutagenesis is poor scaling to multiple residue targets, due to combinatorial explosion of the size of the resulting library. Fortunately, there are opportunities to improve; NNK is only one of many possible degenerate codons, the vast majority of which are underutilized. Since there are 15 possible nucleotide mixtures (*see* above) at each of the three positions making up a codon, there are 3375 legal degenerate codons. Ignoring codon usage considerations (organism codon preferences that are the usual target of codon optimization), there are 1482 degenerate codons that encode different ratios of amino acid (and stop codon) outcomes. To further simplify, degenerate codons that specify the same sets of amino acids (with varying amino acid probability) can be eliminated. Of these 840 degenerate codons, 115 can also be discarded since they encode sets of outcomes that are redundant with another degenerate codon except for the inclusion of stop codon outcomes. Thus, there are 725 degenerate codons that encode unique sense mixtures of amino acids. The specific computational challenge addressed by this chapter is to select which of these 725 options to pick for each site within a design problem (*see* **Note 1**).

Several groups have developed methods for site-specific libraries that rely on mixing primers rather than ordering standard degenerate oligonucleotides [32, 36–41]. By taking these alternate approaches, the precise set of desired amino acids can be encoded at each site. Can the computational design framework described here be useful in such scenarios? In theory, the method should apply equally well when selecting between arbitrary amino acid sets. The critical challenge is that the unconstrained library search space is much larger. Rather than the 725 mixtures of amino acids above, any combination of the 20 amino acids might be used. The number of possible amino acid sets is large enough ($2^{20} = 1,048,576$) that the current methods would likely be impractical due to memory limitations or combinatorial optimization performance limitations.

1.4 Regression and Energy Functions

Regression is a powerful tool to uncover the relationship between a dependent variable and one or more independent variables. In the current case, the dependent variable is the output value from a calculation (particularly a computationally expensive calculation) applied to a protein structure. Meanwhile, the independent variables correspond to the binary presence (1) or absence (0) of various mutually exclusive options. For example, in a protein design calculation, the task is to select exactly one amino acid at each design site. For a protein repacking problem, the task is to select exactly one sidechain rotamer position. Necessarily, the regression model only approximates the results from the more expensive calculation. The benefit is the dramatic increase in speed, since the predicted score for any discrete combination covered by the regression model can be computed nearly instantaneously [10]. Thus, if the regression model has sufficient accuracy, it can be used to effectively search enormous solution spaces.

Energy functions are used to evaluate structures and test them for plausibility. The Rosetta energy function is a well-known example, as are the energy functions employed by molecular dynamics simulations. Both of these are scoring functions, although there are important differences. Rosetta includes "knowledge-based" terms derived from protein structure statistics that are usually eschewed by the "force fields" that contain only physics-based, molecular mechanics terms. In either case, energy functions typically take the form of a sum of terms that approximate various interactions between atoms in a protein. The complexity of energy functions used for protein design is often immense. Many mathematical terms (e.g. electrostatic interactions, bond angles, solvation energy, etc.) may be combined in an effort to improve the accuracy of an energy function. Regression models can be used to approximate results obtained with these more expensive calculations.

1.5 Cluster Expansion

The use of regression to accelerate otherwise intractable protein calculations has been popularized in recent years by Grigoryan, Keating, and coworkers as cluster expansion [9, 10]. Cluster expansion is a regression-dependent method that was initially made to study alloys [42]. Cluster expansion techniques have now been used to generate useful approximations for a variety of protein-related problems. At heart, cluster expansion relies on regression to fit an expensive calculation (e.g. the stability of a protein evaluated via repacking calculations). The terms may be 1-body (e.g. is-residue-10-an-arginine), 2-body (e.g. are-both-arginine-10-and-glutamate-18-present), 3-body, or higher-order. Regression is used to determine the value of the terms. A key benefit is that the dependent variable, the expensive calculation, can be arbitrarily sophisticated.

Commonly, the expensive calculation includes combinatorial optimization. In the case of protein design, cluster expansion serves to "integrate out" the sidechain placement problem, providing a model that predicts the post-repack energy for any sequence. Given a model with only sequence variables, new design possibilities become feasible. For example, Grigoryan et al. used integer linear programming in conjunction with the cluster expansion model to directly incorporate negative design into the design of a family of coiled coils with orthogonal specificity [11]. In the current case (protein degenerate codon library design), a sequence-level model that predicts the energy of any sequence is converted into a library-level model that predicts the mean energy of any degenerate codon library (Fig. 1). A recent report from Verma et al. demonstrates an equivalent approach, direct optimization of degenerate codons via cluster expansion [14] (*see* **Note 2**).

One drawback of cluster expansion in particular, and regression in general, is the necessity of training the model with a large set of initial calculations; typical training sets for protein design

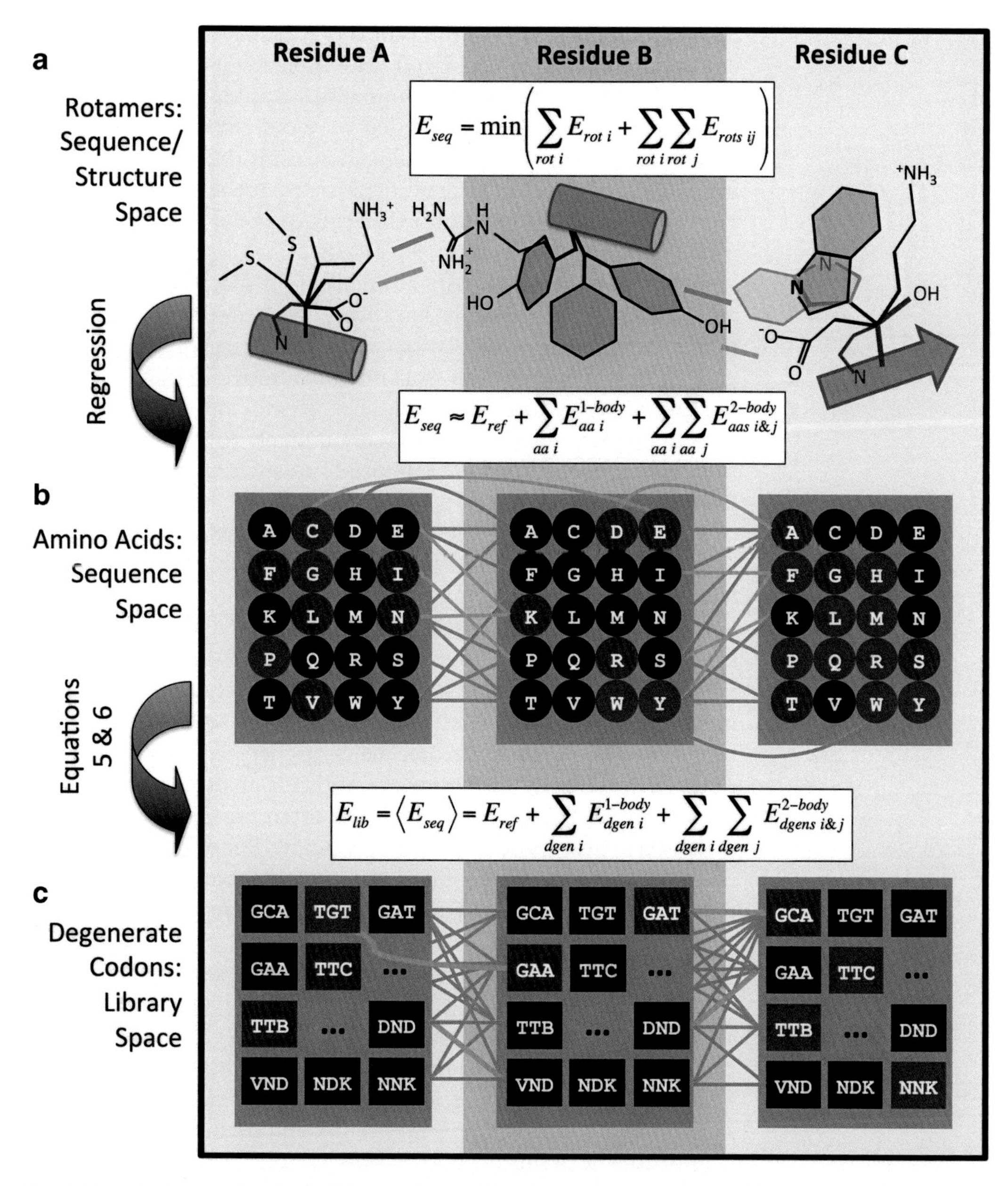

Fig. 1 At each design site, the first layer (**a**) consists of a combined sequence/structure search space with discrete alternative positions for the sidechains (rotamers). Each rotamer gets a 1-body energy due to interactions with immobile groups and 2-body energy terms due to interactions with neighboring mobile groups. These might be particularly favorable (*green edges*) or unfavorable (*orange edges*). (**b**) Integrating out the structure degrees of freedom, we arrive at a regression model that only contains sequence variables. Favorable and unfavorable 1-body terms are represented with *green* or *red tint*, while 2-body terms are again represented as edges. Finally, by applying Eqs. 5 and 6, we can construct another energy graph (**c**) in which the numerous vertices correspond to degenerate codons. Depending on the constituent amino acids, the degenerate codons may be favorable (*green tint*) or unfavorable (*orange tint*). Edges may likewise carry favorable (*green*) or unfavorable (*orange*) effects

problems contain tens of thousands of calculations. The aggregate computational expense is significant but necessary. To save computational time, a training set of minimum feasible size is preferable, but large training sets are needed to avoid over-fitting large free parameter collections. Eventually, increasing the size of the training set will not lead to improved accuracy; at this point it has been saturated, and adding additional terms to the regression model is more likely to lead to an improvement.

Perhaps surprisingly, Apgar et al. found that a more expensive calculation (with a flexible backbone) was easier to approximate than a less expensive calculation (rigid backbone) [12]. It was suggested that allowing the backbone to move minimized steric clashes between individual residues. Removing steric clashes is helpful because such interactions are unlikely to be physically realistic, and because the large amplitude of such interactions can be difficult to fit.

Ng and Snow found that lower-order terms were sufficient for the prediction of multi-body energy function scores [43]. Specifically, the AMOEBA polarizable energy function [44], which is not pairwise decomposable, was approximated for combinatorial sidechain optimization. Lower-order (1-body, 2-body, and 3-body) terms were shown to be sufficient to accurately approximate the multi-body polarization effects. In addition, sets of lower-order terms could be used to predict which higher-order terms are relevant. If the 2-body terms for amino acids at three positions had significant magnitude, it was worth attempting to add a third-order term for those three amino acids. Snow and Ng's work revealed that one could filter out (i.e. ignore) almost 80 % of 3-body terms and thereby reduce the complexity of the regression with this simple check.

The generality of the regression/cluster expansion approach is a key feature. Many hard problems in CPD benefit from an accurate model that predicts energy directly from the sequence. The current chapter describes open-source, permissively licensed software for expanding the combinatorial optimization approach to problems that may not be pairwise decomposable. To take advantage of this flexibility, Python scripts are presented for computing regression-based approximate models with the robust combinatorial optimization capacity of the open source SHARPEN software platform [45].

2 Methods

2.1 Overview

In broad strokes, the steps to take to apply the regression tools are the same regardless of the exact goal. First, an initial set of discrete combinations is "instantiated," a process that varies depending on the problem but always includes an assessment or scoring of the combination in question. For the current case studies, the discrete combination is the protein sequence, and the instantiation consists of a combinatorial optimization of sidechain rotamer positions. The set of instantiated combinations is divided into two subsets: one to train the regression model and another to test the resulting approximation. The resulting trained regression model provides a rapid approximation to the more expensive instantiation operation.

The SHARPEN package [45] provides convenient data structures to store and apply regression approximations. Specifically, EnergyGraphs efficiently store 1-body and 2-body terms in a conventional graph structure consisting of nodes and edges (a thin wrapper around the underlying Boost Graph type). More unusually, SHARPEN also provides EnergyHyperGraph data structures that can also accommodate higher-order terms such as 3-body, 4-body, or N-body effects. Either EnergyGraphs or EnergyHyperGraphs can be used with a variety of independent combinatorial optimization routines for identifying favorable combinations. Therefore, it is easy to efficiently identify discrete combinations, "targets," that are predicted to minimize the instantiated score according to the current approximation.

The value of the entire scheme is predicated on the utility of the approximation to allow the combinatorial search process to more rapidly explore enormous swaths of a combinatorial search space, and to do so with enough accuracy to discover favorable combinations. Given the astronomical search size of typical combinatorial problems, and the rapidity of search methods using the regression approximation, accelerating the sampling is likely assured. The more challenging aspect is ensuring that the regression-based approximation is sufficiently accurate. Fortunately, this chapter illustrates that 2-body regression models appear to be largely sufficient to approximate the favorable portions of the combinatorial search space, and that such approximations can be used to facilitate the optimization of degenerate codons directly in library space.

The routines described below are implemented in a set of python scripts that use methods provided by a python module, **dgen_design**. These tools use the open source SHARPEN software, and are therefore provided via the www.sharp-n.org website wiki. Other useful scripts for practical protein design calculation, described by Johnson et al. [25], are also hosted on this site.

2.2 Instantiation: Combinatorial Optimization of Sidechain Positions

1. The only requirement for an instantiation method is that it accepts a combination and produces a score. Any algorithm that can be applied to a candidate protein and produces a number could be an instantiation method. Instantiation for this work involves combinatorial optimization of the sidechains ("repacking") for a particular sequence variant to minimize the model score according to an all-atom Rosetta energy function [35]. The outcome is the score E in Rosetta energy units

(REU). Structures with lower Rosetta energy scores are more plausible protein conformations.

From the standards of CPD, the case studies presented herein are small problems (Table 1). The FasterPacker combinatorial optimization object mimics the “singles” routine from the Desmet and Lasters FASTER algorithm [4]. The FasterPacker works somewhat like a traditional Monte Carlo trajectory, except that the moves that are accepted or rejected are “batch” moves. Candidate batch moves are generated by temporarily fixing a perturbing rotamer change and then sequentially relaxing interacting sidechains to their low energy rotamer.

2. FasterPacker typically yields optimal or near optimal solutions for problems of this size. To demonstrate, 600 sequences for case B.1 (*see* below) were solved to optimality using the mixed integer linear programming program CPLEX [46] via a CplexPacker wrapper provided by SHARPEN. Because these problems are reasonably small, the CplexPacker is able to identify the global minimum energy combination (GMEC) relatively quickly (an average time of 3.7 s). In 552 of 600 cases FasterPacker found a solution within 1E-6 REU of the GMEC, but did so in an average of only 0.14 s. Notably, CplexPacker was used to optimize the sidechain rotamer positions of the initial protein model prior to any other calculations.

Table 1
The reported best low-energy testset rmsd values ($rmsd_{LET}$) correspond to the lowest value encountered for varying training set sizes. Where applicable, the best-case exponential weight (τ) and regularization parameter (k) are also noted

	Site	Active design sites	Other mobile sites	Seq space size	Lib space size	Seq/str search size	Figures	Best LET rmsd	τ	k
A.1	Core	5, 30, 43	52, 54	8000	3.8E8	1.5E11	4,6,7,8	1.8	–	1e–7
A.2						4.9E7	4,6,13	1.7	–	0.01
A.3							5,6	1.5	125	1e–6
B.1	Core	5, 30, 43, 52, 54	3, 7, 16, 45	3.2E6	2.0E14	3.9E20	9	5.0	–	0.1
B.2								4.1	50	1e–7
B.3						2.5E13	10,13	14.4	–	1
B.4							10	8.4	75	1
C.1	Surf	2, 4, 6, 8, 13, 15, 17, 19, 42, 44, 46, 48, 49, 51, 53, 55	None	6.6E20	5.8E45	1.8E35	11,13, 14	9.9	–	10
C.2						6.5E45	12	4.8	–	1

3. A pool of random combinations is instantiated via FasterPacker at the outset of the campaign. This pool serves as the source of training and test combinations. A training batch is used to kick off the regression, while the test batch (all other members of the initial pool) is held in reserve to quantify regression model quality.

For the case study problems here, combinatorial optimization is quite rapid since there are a limited number of mobile residues (Table 1). For example, the 5-site library (case B), with the default (non-minimal) rotamer generation scheme has a structure-sequence search space size of 3.9×10^{20}. Only 20 min are required to instantiate 50,000 random sequences using a 2.8 GHz Intel Core i7 CPU.

2.3 Term Selection

1. The ability of regression to produce accurate approximations is predicated on the ability of terms to stand in for more complicated processes. It is desirable to be selective when adding terms, since adding an excessive number of terms will result in overfitting. To recapitulate most physical problems, it is necessary to include at least 1-body and 2-body terms. A free constant (i.e. a 0-order term) can also be helpful, allowing the remaining parameters to adopt smaller values without degrading the overall fit. Alternately, to shrink the absolute value of the free constant, a reference energy (*Eref*) can be subtracted from each element within the instantiated score vector (Υ).
2. In the particular case of approximating the energy of protein sequences, our general recommendation is to consider the wild-type (WT) or initial protein sequence as the reference state (with $E = Eref$). Then, each individual mutation at an active design site gets a 1-body parameter. WT amino acids at the design positions do not get parameters, as their contribution is subsumed within the reference state. Similarly, only interactions between two mutations serve as 2-body parameters, since a WT:mutant pair is already accounted for in the 1-body parameter for that mutant. Ideally, regression will drive the free constant parameter toward the score of the reference state. All of the problems described below include a free constant, all 1-body terms, and all possible 2-body terms (Fig. 1). For larger problems, it could become useful to skip 2-body terms that are not likely to correspond to physical effects. For example, one could require physical proximity or more direct evidence of energetic coupling between the particular sites before adding terms.
3. Similarly, it could also be useful to identify higher-order terms (i.e. 3-body terms) to improve the accuracy of the model in recapitulating low-energy combinations. A tricky aspect to this is that sizable 3-body effects for protein design can be "frustration" effects in which three pairs of amino acids can each coexist

nicely, but the combination of all three induces an unavoidable steric clash. Modeling this effect requires a large 3-body term, which breaks the typical approximation paradigm that higher-order effects will have lower magnitude than lower-order effects.

4. A more sophisticated (and lengthy) approach to term selection was described by Hahn et al. who developed an iterative feature selection scheme with rapid cross-validation [13]. Candidate terms are individually considered and included if they make a statistically significant improvement. For the degenerate codon design problem described here, one can avoid using 3-body terms and lengthy term selection procedures due to the sufficient accuracy of the 2-body models and the technical feasibility of modeling all possible 2-body terms.

2.4 Solving Large Regularized Regression Problems

1. After the training batch is instantiated, and fitting terms are selected, regression can proceed. The regression model will ascribe values to the fitting terms so that summing the appropriate terms can approximate any combination. For the case study problems here, there are thousands of one-body and two-body terms, and thousands of training set members.
2. Each of the training set members will have a relatively small number of applicable terms, depending on which amino acids (potential 1-body terms) and pairs of amino acids (potential 2-body terms) are present at the variable sites. To tackle the resulting large sparse regression problems, our approach relies on two solvers that work with sparse matrices and are convenient for use from Python. Specifically, the CVXopt software package [47] provides a sparse matrix structure, an interface to the Cholesky factorization routines of the CHOLMOD package [48], and functions for solving sparse sets of linear equations. Alternately, one can use the LSMR package [49], which is integrated into scipy [50]. For the following code snippets, *Y* is the vector of instantiation scores, *k* is a regularization parameter (*see* below, Eq. 1), and *spX* is a sparse matrix that encodes which terms apply to which training set combinations.

```
import scipy.sparse.linalg
results=scipy.sparse.linalg.lsmr(spX, Y, damp=k)
or
import cvxopt
from cvxopt import spmatrix, spdiag, cholmod
B=spX.T * cvxopt.matrix(Y)
XT_X=spX.T * spX
ridge=k * cvxopt.spdiag([1] * len(B))
cvxopt.cholmod.linsolve(XT_X+ridge, B)
```

3. Given the large number of fitting parameters that arise when 2-body or 3-body terms are included, overfitting is a serious concern. To combat the tendency for overfitting, use regularized regression. In both code snippets above the core calcula-

tion consists of regularized regression, also known as ridge regression (Eq. 1) or Tikhonov regression [51]. This technique penalizes terms that deviate from zero. The regularization parameter, k, serves to restrain the magnitude of the fitting parameters, β. The matrix X specifies which fitting terms contribute to each combination, with the vector Y holding instantiated scores and I as the diagonal identity matrix.

$$\left(X^{\mathrm{T}}X + kI\right)\dagger = X^{\mathrm{T}}Y \tag{1}$$

Ridge regression can be useful to suppress overfitting. It is important, however, to setup the problem so that the value of the terms should indeed be small numbers.

2.5 Weighted Regression

1. Weighting is a useful optional strategy to increase the accuracy of the regression model for some of the combinations. Typically, the performance of the approximation is much more important for favorable combinations than unfavorable combinations. It is recommended to sacrifice the *overall* fit in favor of higher accuracy for the *favorable* combinations. A matrix W has weights along the diagonal, w_i, that are selected using the following scheme intended to resemble Boltzmann weighting. The adjustable parameter τ sets the energy scale that defines the favorable sequences of interest.

$$X^{\mathrm{T}}WX\dagger = X^{\mathrm{T}}WY \tag{2}$$

$$w_i = \min\left[\mathrm{e}^{-\frac{(y_i - \min(y))}{\tau}}, 0.02\right] \tag{3}$$

2. If the exponential weighting parameter $\tau = 100$, then training set combinations that are 10, 25, 50, 100, and 200 REU less favorable than the minimum REU combination in the training set will have weights of 0.90, 0.78, 0.61, 0.37, and 0.14. Equation 3 assumes that the more important combinations have the lower scores. If necessary, the sign of the scores can be flipped. The use of a minimum weight (0.02 above) ensures that the regression model cannot entirely neglect high-energy combinations.

2.6 Quantifying Regression Model Performance

1. To quantify the performance of a regression model, one can compute the root mean square deviation (rmsd) between the predicted E scores for the testset with the actual repacked E scores. However, also consider the possibility that the regression model predictions may have a systematic bias (e.g. a slope of 1.5 or a non-zero intercept). If such a bias is consistent, it could be corrected by fitting a line. Therefore, before computing rmsd one should correct systematic deviations using the scipy.stats.linregress function to compute the slope and intercept.

2. It is not recommended to equally favor all combinations. The explicit goal is to maximize accuracy for the more favorable, low-energy combinations. Rough prediction of high-energy combinations is sufficient; clashes need not be precisely quantified if they can be avoided. Accordingly, one might quantify accuracy for the favorable members of the test set, with $E_i < \min(E) + 100$ REU. Hereafter, this figure of merit will be termed $rmsd_{LET}$, the rmsd for the low-energy testset. One may also use rmsd to quantify the extent to which library <E> predictions match directly sampled <E> values.

2.7 Using the Approximation to Select Targets

1. Once a regression model has been trained, it can be used to efficiently identify combinations that are predicted to be favorable upon instantiation. Such combinations are termed "targets." If only 1-body terms are present in the model, optimal targets are trivially easy to identify; one need only select the best score for each mutually exclusive choice. More generally, a low-scoring target combination is found using combinatorial optimization routines, typically the FasterPacker described above, or the SimulatedAnnealingPacker. A SimulatedAnnealingPacker implements a Monte Carlo trajectory over combinations with a gradually reducing temperature value.
2. These combinatorial optimization methods are generally intended to find individual favorable combinations, possibly the GMEC. When target diversity is critical, the output combination from the SimulatedAnnealingPacker or FasterPacker can serve as the initial combination for subsequent MonteCarloPacker sampling. To ensure a diverse pool of target combinations, one can generate multiple Monte Carlo trajectories at escalating temperature. Temperature is increased in repeated Monte Carlo runs until a minimum number of distinct combinations are found. This method of target selection is still effective if higher-order terms are present in the regression model. Also, the use of an escalating temperature should render this protocol somewhat robust when applied to different problems with varying intrinsic energy scales.
3. A related strategy can come into play at the outset of a learning process. In the case of building a training set to use regression to approximate the AMOEBA energy function (Ng, 2011), the model was trained using rotamer combinations that were highly diverse, but not the purely random combinations that usually result in van der Waal clashes. To do so, an initial approximate energy model that included only strong van der Waal clashes was built. Then, to generate a maximally diverse pool of combinations, excluding unrealistic high-energy combinations, Ng ran a series of Monte Carlo trajectories with the temperature set to zero. These trajectories began from random

combinations and executed a downhill walk, thereby preserving maximal diversity while attempting to avoid unphysical clashes. This allowed the regression process to "learn" about more interesting effects than the steric clashes.

2.8 Calculating the Properties of Degenerate Codon Libraries

1. For small degenerate codon libraries (e.g. case A with three design sites), it is feasible to enumerate the predicted E values for each of the encoded sequences. Then, the expectation value of the Rosetta energy for any library can be readily calculated by computing the mean energy, <E>, of the constituent sequences. However, for library design problems with more sites, precise calculation of <E> can be overly time-consuming. Therefore, our code instead estimates <E> by sampling the value of n random combinations drawn from the library. One can compute the standard error of the mean ($\sigma\langle_{\mathrm{E}}\rangle$) given the standard deviation of the E values in the sample (σ), using the finite sample correction for libraries with N members:

$$\sigma_{\mathrm{E}} = \frac{\sigma}{\sqrt{n}}\sqrt{\frac{N-n}{N-1}} \tag{4}$$

2. Relatively modest samples ($n = 400$) are sufficient to assess the correlation between the predicted library <E> from the regression model and the sampled <E>. To estimate the number of library combinations with E below a threshold, compute the fraction below the threshold for the sample and multiply by the library size. The library size, N_{lib}, refers to the theoretical number of distinct sequences and is calculated as the geometric product of the number of amino acids Naa^{site} encoded at each design site: $N_{\mathrm{lib}} = \prod_{\text{design sites}} N_{aa}^{\mathrm{site}}$.

2.9 Combinatorial Design in Degenerate Codon Library Space

1. The technical details above cover the preparation, tuning, and validation of regression models that map specific protein sequences to predicted post-instantiation Rosetta energy scores. However, a larger goal for this chapter is to demonstrate how such models can be used to efficiently select degenerate codon libraries. Specifically, the goal is to execute a search for favorable degenerate codon libraries directly in "library space." To enable library design via the various combinatorial optimization algorithms provided by SHARPEN (e.g. the Packers mentioned above), one need only prepare an EnergyGraph or EnergyHyperGraph in which the nodes no longer correspond to mutually exclusive amino acid choices, but instead correspond to mutually exclusive degenerate codon choices (Fig. 1).
2. The regression models described above make this possible. In the notation below, $E_i^{1\text{-body}}$ is the 1-body regression term for amino acid outcome i, and $E_{ij}^{2\text{-body}}$ is the 2-body regression term for the simultaneous selection of amino acids i and j. First, com-

pute 1-body terms for each degenerate codon by computing the expectation 1-body term for the constituent sense amino acids according to the regression model (*see* **Note 3**). The relative frequency of the constituent amino acids, *pi*, can serve as weights to compute the expectation value. Alternately, *pi* values can be set to model equally probable amino acid outcomes. The latter approach is adopted here under the assumption that all variants within the library might be isolated and characterized (ignoring the different frequencies of encountering these variants).

$$E_{\text{dgen A}}^{\text{1-body}} = \sum_{\text{A } aa\ i} p_i \cdot E_i^{\text{1-body}} \tag{5}$$

3. Similarly, compute the expectation value for the 2-body interaction between two degenerate codons (A and B):

$$E_{\text{dgens A and B}}^{\text{2-body}} = \sum_{\text{A } aa\ i} \sum_{\text{B } aa\ j} p_i \cdot p_j \cdot E_{ij}^{\text{2-body}} \tag{6}$$

4. For a closer look at the calculation of an EnergyGraph that embodies the library design landscape, *see* the *M_score_dgen_codon_sets.py* script.

2.10 Sampling Diverse Libraries via Combinatorial Optimization

1. Library size is a key consideration when it is time to select a library for experimental testing. However, the feasible experimental library size is rarely in practice a strict cutoff. Instead, it is valuable to illustrate what candidate libraries look like as a function of library size before making final decisions.

All other factors being equal, use libraries with a lower <E>, since those libraries are the most likely to be highly folded, stable, and functional. However, the global minimum energy library (GMEL) will have exactly one sequence and that sequence will correspond to the GMEC. This is true since each individual amino acid is an option within the 725 amino acid sets that can be encoded. If libraries are sampled thoroughly, the libraries with minimal <E> are going to contain relatively few sequences. The more useful task is to identify libraries with minimal <E> for every library size. This will allow the protein engineers involved to select the library size that offers the best <E> yet fits within the assay screening budget. Therefore, when performing combinatorial optimization directly in library space, an explicit bias favoring larger libraries may help to sample diverse library options.

2. To sample larger libraries, implement a bias favoring degenerate codons that encode more amino acids. First, compute log(Naa^{site}) for each degenerate codon. Then, when assessing candidate degenerate codon combinations, the total library size is $e^{\left[\sum_{\text{design sites}} \log\left(N_{aa}^{\text{site}}\right)\right]}$. Each candidate degenerate codon choice contributes additively to the predicted library E via its 1-body

and higher-order regression terms, and contributes additively to the library size (N_{lib}) via the log(Naa^{site}). Iteratively perform combinatorial optimization according to the energy model, but in each round r_i increment a cumulative 1-body bias favoring degenerate codons that encode more amino acids: $r_i \cdot \varepsilon \cdot \log\left(N_{aa}^{site}\right)$, where ϵ is a small weight factor. For a closer look at this iterative library sampling scheme, *see* the *P_iterlib_sample.py* script.

3. Manually adjust ϵ so that libraries of the largest interesting size are sampled by the end of 100 rounds of combinatorial optimization. The largest interesting library size is problem-dependent. For three design sites, full NNK saturation may be worth considering. Given more numerous design sites, the maximum interesting library size will likely be limited by the screening capacity. Even in vitro methods such as mRNA display or ribosome display have limits (e.g. 10^{14} variants) [52].

2.11 Selecting Advantageous Degenerate Codon Libraries via Pareto Analysis

1. Seek to identify libraries that are Pareto optimal [53] for minimal <E> and large library size. In other words, if candidate library 1 has a higher <E> and a smaller size than candidate library 2, then candidate library 1 can be discarded from consideration. To accelerate this process, our code uses a divide and conquer approach. *See* the *Q_calc_pareto_stats.py* script for more details.

Due to threshold protein stability effects, a library with 90 % favorable sequences and 10 % very unfavorable sequences may be preferable to a library consisting entirely of mediocre sequences. Unfortunately, the <E> could be lower for the latter library. Therefore, a preferable Pareto analysis scheme identifies libraries that have the greatest (predicted) number of sequences with scores below a threshold, while otherwise having the smallest total library size. Generally, this threshold should be set to a value such that combinations exceeding the threshold would be at risk of not being functional.

2. After either Pareto analysis is complete, the remaining set of libraries (the Pareto front) includes only the libraries that are most worthy of consideration. Inspection of the resulting plots should help when weighing the tradeoffs between selecting small libraries with favorable energy statistics and larger libraries with less favorable statistics.

3 Example Tests and Results

3.1 Model Design Problems

This section describes results for several illustrative degenerate codon design problems (Fig. 2) using protein G (pdb entry 1pgb). Table 1 defines which amino acids are design positions and which other amino acids that are allowed to move.

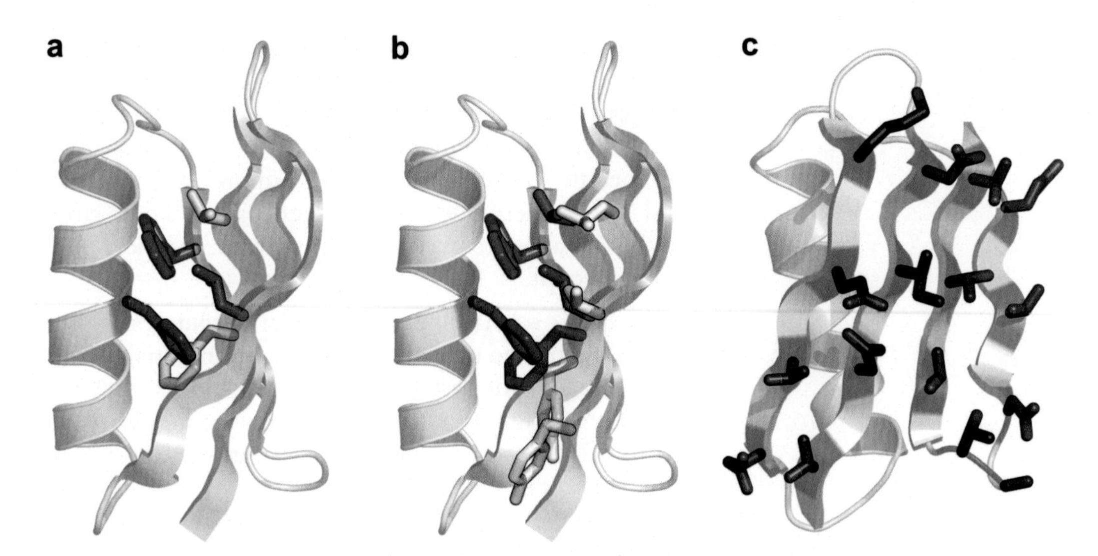

Fig. 2 Case study design problems. Design position (*gray*) and mobile sidechains (*white*) are shown in sticks. (**a**) Case A has three design sites in the hydrophobic core. (**b**) Case B has five design sites in the hydrophobic core. (**c**) Case C has 16 design sites on the exposed surface of the beta sheet

1. **Case A** is intended to provide the smallest possible interesting problem (Fig. 2a). In this case, by limiting the number of design positions to 3, it is possible to optimize and score each of the 8000 possible sequences via sidechain optimization. Note, however, that even very rapid calculations become time-consuming when applied to 381 million candidate degenerate codon libraries (725^3).
2. **Case B** is intended to provide an example of a realistic use scenario for these tools (Fig. 2b). For many experimental assays, it would be impractical to experimentally screen a site saturation library at 5 design positions. However, by using tailored degenerate codons it may be possible to obtain a library that is small enough to screen, and will consist of a higher fraction of favorable sequences. Thus, by degenerate codon design, one could make the most of the available screening capacity (e.g. ten 96-well plates). For example, consider a size constrained hydrophobic site. Rather than using all 20 amino acids, the degenerate codon "VTM" would provide just Ile, Leu, and Val, and would help reduce combinatorial explosion of the library size.
3. **Case C** is intended to demonstrate performance when applying the approach to a larger design problem (*see* **Note 4**). Case C constitutes the redesign of an entire surface face of a beta sheet (Fig. 2c). Saturation mutagenesis of such a 16-site library is out of reach for experimental screening, but tailored codons might be used to identify favorable libraries small enough to be screened for binding properties via a high-throughput approach (e.g. fluorescence-activated cell sorting).

4. All three of these cases are suitable for conventional CPD. The case A GMEC (wild-type Leu5, Phe30, and Trp43) was found using SimulatedAnnealingPacker, FasterPacker, or CplexPacker in 0.6, 13, or 87 s respectively. For case B.1, design results in a double mutation W43T, V54I. The GMEC was found using SimulatedAnnealingPacker, FasterPacker, or CplexPacker in 1.8, 104, or 795 s respectively. Finally, FasterPacker and CplexPacker found the GMEC for case C.1, which had 15 surface mutations (T2R, K4E, I6E, N8R, K13E, E15R, T17Y, E19W, E42L, T44R, A48N, T49R, T51R, T53I, and T55I). In this last case, the FasterPacker and CplexPacker required 5 and 803 s, respectively.
5. To illustrate the performance determinants for the presented methods, variant calculations were performed (Table 1) to assess the effects of rotamer density and the use of weighting. Specifically, for cases A.2, A.3, B.3, B.4, and C.1 rotamers were reduced to base Dunbrack rotamer options [54]. For these cases, the sequence/structure search space size was reduced (Table 1) and combinatorial sidechain optimization was more rapid. Cases A.3, B.2, and B.4 use weighted regression. Table 1 indicates which subsequent figures apply to each case and highlights the best-case prediction accuracy ($rmsd_{LET}$).

As described above, the first step for each of these model design problems is to perform thousands of combinatorial sidechain optimization calculations (instantiation) for random sequences. For case A, all 8000 variant sequences were instantiated. For cases B and C, 50,000 and 80,000 combinations were instantiated, respectively. For all three cases, a large fraction of the sequence space achieves a low score when optimized via FasterPacker. As expected, the cases with more generous rotamer provisioning (Fig. 3abc) reach lower energy values.

The score for the wild-type sequence with fully optimized sidechain rotamers is −112.5 REU. For case A.1 the mean (median) E is −70.9 (−99.6) REU. For case B.1 the mean (median) E is −75.8 (−104.6) REU. For case C.2 the mean (median) E is -58.7 (−93.0) REU. The median values are lower than the mean values due to the outsized influence of high-energy sequences on the mean. Given these values, sequences with a predicted E above −105 REU were flagged as having an elevated risk of being unfolded.

3.2 Case A: High Accuracy Approximation of a 3-Site Library

1. Despite the close physical interaction of the design site residues for case A, it was possible to very accurately fit the post-repacking energy of the 8000 possible sequences via regression (Fig. 4a–c). There are 1141 fitting parameters in this case, consisting of 1 free constant, 57 one-body terms (the 19 possible mutations for each of the three sites), and 1083 two-body terms (double mutations). If all 8000 sequences are fit, the rmsd is only 1.9 REU. A better test, however, is to train the regression

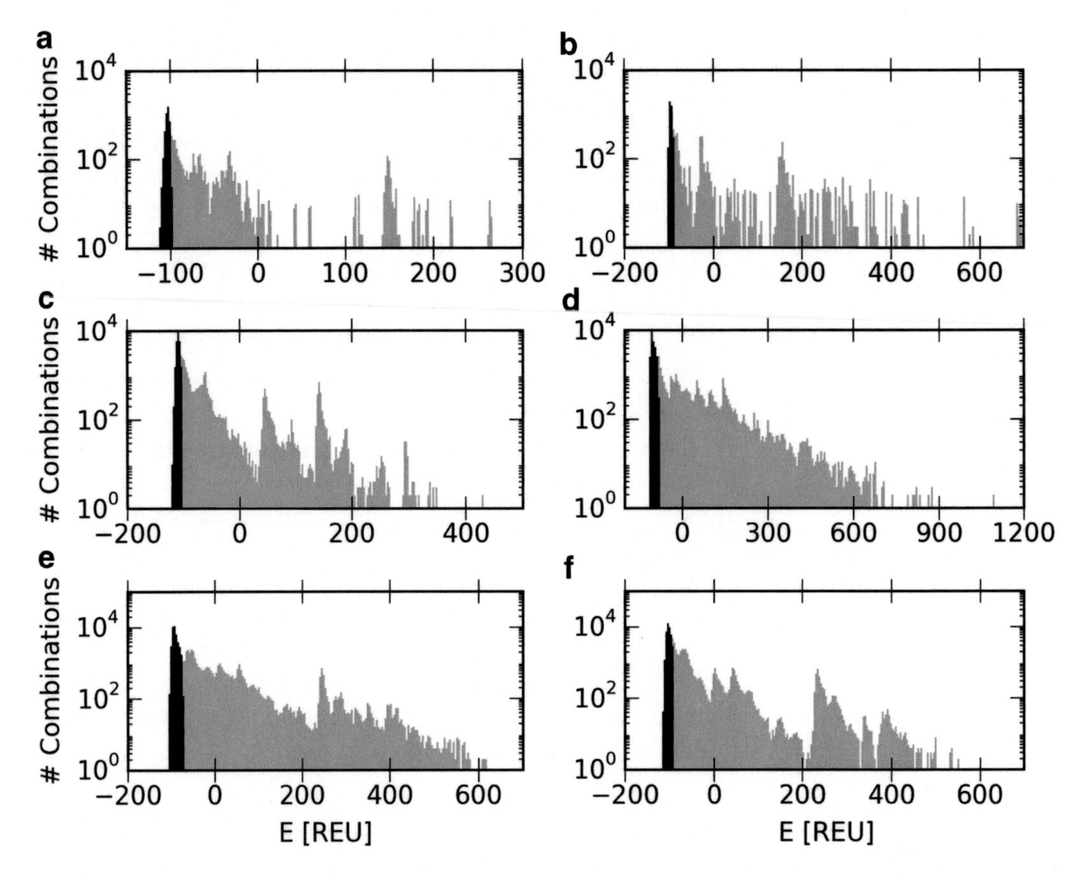

Fig. 3 Instantiated scores for random combinations. *Black* (*orange*) *bars* are combinations with E lower (higher) than the median. (**a**) All 8000 sequences for case A.1. (**b**) All 8000 sequences for case A.2 (minimal rotamers). (**c**) 50,000 random sequences for case B.1. (**d**) 50,000 random sequences for case B.3 (minimal rotamers). (**e**) 80,000 random sequences for case C.2. (**f**) 80,000 random sequences for case C.1 (minimal rotamers)

model using portions of the 8000-sequence pool and to assess the quality of the resulting approximate energy model using the remaining sequences as a test set. To illustrate the effect of training set size and regularization parameter, Fig. 4a shows how approximation accuracy depends on these parameters.

2. For case A.1, regularization was not critical. Scanning the training set size and the regularization parameter (Fig. 4a), the best $rmsd_{LET}$ was an impressively low 1.8 REU. The best performance came when using a 7500-member training set with the smallest test regularization parameter (k = 1E−7). Running regression without regularization produced the same results.

Surprisingly, reducing the number of rotamers (case A.2) does not reduce the performance of the regression model. Instead, $rmsd_{LET}$ actually decreased from 1.8 to 1.7 REU. One difference between case A.1 and case A.2 comes for small training sets (approximately 1000 combinations) and low regularization penalty (k < 1E−3). The slight shoulder in the case A.1 parameter scan

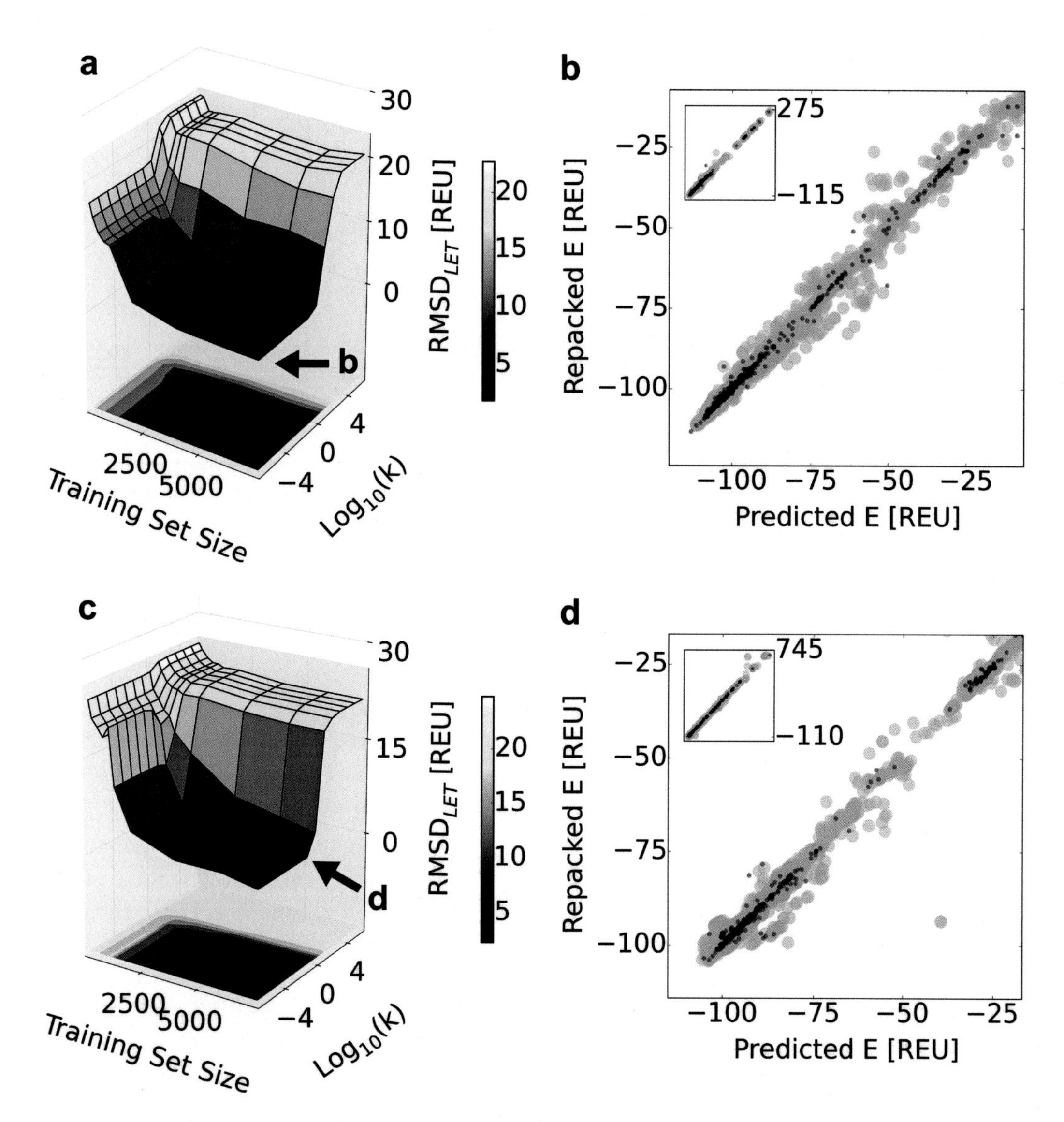

Fig. 4 Case A.1 and A.2 approximation performance. Training set combinations are partial transparent orange points while test set combinations are black points. (**a**) $rmsd_{LET}$ versus training set size and the regularization parameter for **case A.1**. The best performance from this scan is shown in (**b**), where a random training set (7500 combinations) was used to fit 1141 parameters with regularization ($k = 1e-07$) resulting in training set recapitulation (rmsd = 1.8). Performance for low-energy combinations was excellent ($rmsd_{LET} = 1.8$ REU) for the 468 test set combinations within 100 REU of the minimum test set combination (−113.1 REU). The entire test set (500 combinations) was predicted with rmsd = 3.0 REU (inset). (**c**) $rmsd_{LET}$ versus training set size and the regularization parameter for **case A.2** (minimal rotamers). (**d**) A random training set (7500 combinations) was used to fit 1141 parameters with regularization ($k = 0.01$) resulting in training set recapitulation (rmsd = 4.2 REU). Performance for low-energy combinations was excellent ($rmsd_{LET} = 1.7$ REU) for the 390 test set combinations within 100 REU of the minimum test set combination (−103.9 REU). The entire test set (500 combinations) was predicted with rmsd = 2.9 REU (inset)

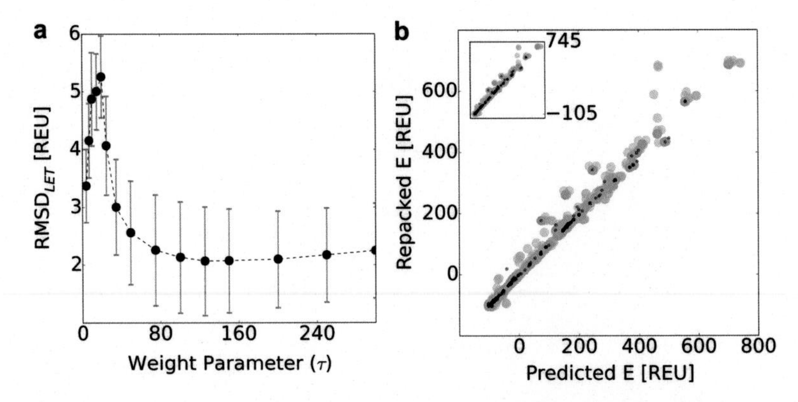

Fig. 5 Case A.3. (a) Scanning the expweight parameter (τ in Eq. 3). *Error bars* reflect the standard deviation from 20 trials with random 7500-member training sets. (**b**) With $\tau = 125$, a random training set (7500 combinations) was used to fit 1141 parameters with regularization ($k = 1\text{e}-06$) resulting in training set recapitulation (rmsd = 11.0). Performance for favorable test set combinations was good (**rmsd$_{LET}$ = 1.5 REU**) for the 410 combinations within 100 REU of the minimum test set combination (−103.7 REU). The entire test set (500 combinations) was predicted with rmsd = 8.7 REU (inset)

surface (Fig. 4a) becomes a distinct peak for case A.2 (Fig. 4c). This peak represents a counterintuitive result; *decreased* prediction performance for a *larger* training set. This result will be discussed below in the Overfitting Trends section.

3. Case A.3 attempts to improve the case A.2 performance with weighting. Keeping the training set and regularization parameter fixed (7500 training set members and $k = 1\text{E}{-}6$), the exponential weighting parameter τ was varied to determine which value gave the lowest rmsd$_{LET}$ (Fig. 5a). $\tau = 125$ was most effective (Fig. 5a). Compared to the non-weighted case A.2 (Fig. 4d), Fig. 5b demonstrates slightly improved rmsd$_{LET}$ ($1.7 \rightarrow 1.5$ REU), with a significant concomitant sacrifice of global fit rmsd ($2.9 \rightarrow 8.7$ REU).

3.3 Predicting <E> for Case A Libraries

1. The regression models described above predict the instantiated Rosetta energy for any sequence within the 8000-sequence search space. For each of the cases above, 1000 random degenerate codon libraries were selected. For each degenerate codon library, the <E> was computed using the pre-calculated energies for constituent sequences.
2. The library <E> predictions are quite accurate (Fig. 6), with <E> prediction rmsd values lower than the rmsd for the prediction of E for individual sequences. The rotamer-rich case A.1 accuracy was good globally (rmsd = 0.38 REU) and for the 559 libraries with predicted <E> within 30 REU of the −105.5 REU minimum (rmsd = 0.37 REU). The case A.2 accuracy was comparable (0.5 REU globally, 0.35 REU for the 256 libraries with predicted <E> within 30 REU of the −98.7 REU mini-

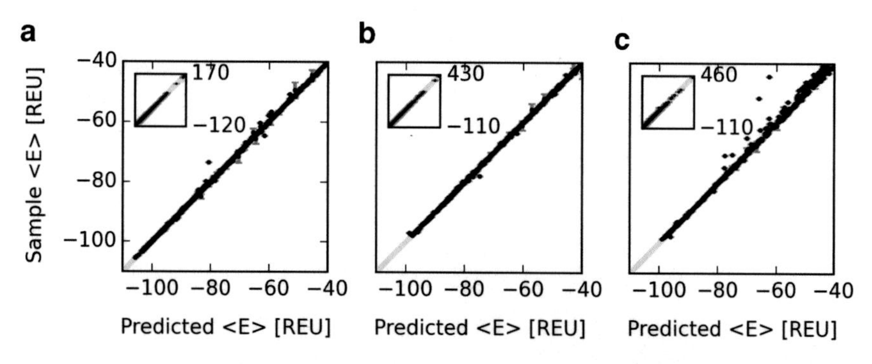

Fig. 6 Predicted library <E> versus instantiated sample <E>. *Vertical error bars* reflect $\sigma\langle E\rangle$ (Eq. 4). An identity line is orange. Global (or best 30 REU) rmsd values for (**a**) Case A.1, (**b**) Case A.2, and (**c**) Case A.3 were 0.38 (0.37), 0.5 (0.35), and 2.6 (0.73) REU respectively

mum). Finally, weighting (case A.3) degraded the <E> prediction, with 2.6 REU rmsd for the global library <E> prediction, and 0.73 REU for the 240 libraries within 30 REU of the −99.0 REU minimum. Despite the counterproductive effect of weighting, these levels of precision for library <E> prediction performance are encouraging. It was particularly gratifying that the minimal rotamer case A.2 performed so well, since all 8000 sequences can be instantiated in less than 4 s in this case.

3.4 Sampling Case A Libraries

1. Random sampling is neither a systematic nor a satisfying solution for efficiently identifying libraries that are maximally appealing for experimental testing. A systematic approach is preferable. Since 1-body and 2-body scoring terms for the degenerate codons are stored in a SHARPEN EnergyGraph, a variety of combinatorial optimization algorithms are readily available to assist with sampling.

For case A, the total library search space of 381 million is small enough for enumeration, albeit via a relatively expensive calculation (approximately 26 min and 8 Gb memory). Therefore, for a 3-site library the BruteForcePacker object from SHARPEN is feasible. The BruteForcePacker can be configured to retain a ranked queue of the best combinations encountered. For case A, a large priority queue was needed to retain the 4.6E5 libraries with a predicted <E> < −105 REU (Fig. 7). For a closer look at this enumeration-based sampling scheme, *see* the *sample_libs*_via_*enum.py* script.

2. For larger libraries, enumeration is not going to be a practical option. Instead, it would be better to identify the potentially numerous low-<E> libraries with an inexpensive calculation. Another example script launches and pools parallel Monte Carlo trajectories to rapidly collect a set of unique libraries. Almost 4E5 libraries predicted to be low energy were collected in less than 5 min (Fig. 7b). For a closer look at this Monte Carlo sampling scheme, *see* the *sample_libs*_via_*MC.py* script.

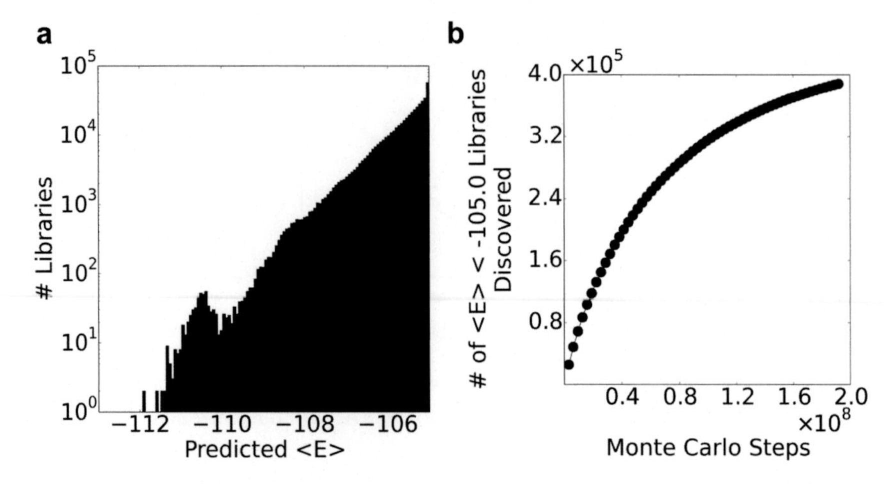

Fig. 7 High density of low <E>-prediction libraries for Case A.1. (**a**) Distribution of predicted <E> for the 3-site libraries with predicted <E> < –105 REU. (**b**) Discovery of libraries with predicted <E> -105 REU using parallel MonteCarloPacker trajectories

3. With so many candidate libraries there is a clear need for effective methods for identifying the most favorable options at a range of library sizes. Iterative library design with an increasing bias favoring larger libraries was used to compile a thorough list of favorable case A.1 libraries. For each library, tabulated energy values for all 8000 possible case A sequences were used to compute the library <E> and the number of library members with E < –105 REU.

3.5 Selection of Case A Libraries via Pareto Analysis

1. Pareto analysis helps identify interesting library candidates that are worth consideration given two or more competing quality metrics. For a given library size, libraries with lower <E> are preferable. For a given <E>, libraries with larger size are preferable. Several illustrative example libraries are described in Table 2.

2. In the absence of a high-throughput assay, a library that is highly enriched for stable sequences may be a superior option. In this scenario, a strong case A candidate library consists of the degenerate codons CTG:DVC:NNK (Table 2). These encode a Leu for residue 5, Ala/Cys/Asp/Gly/Asn/Ser/Thr/Tyr for residue 30, and all 20 amino acids (and a stop codon) for residue 43. This library is a nice example of how the design approach can end up providing suggestions that are quite different from traditional saturation mutagenesis; only residue 43 gets full amino acid diversity while residue 30 gets a tailored amino acid palette and residue 5 is left as the wild-type. This library has $1 \times 8 \times 20 = 160$ sense outcomes and a total library size of 168. The library <E> is –106 REU, and 136 of the sequences have E < -105 REU. The 24 less favorable (E > –105) variants include Cys 30 paired with (Lys, Arg, Ile, Gly, His, Cys, Asn, Asp, or Leu 43), Gly 30 paired with (His,

Table 2
Library size refers to all distinct outcomes including genes that include stop codons. Stop codons are encoded by DND, NNK, THW, and NDK

Degenerate codons	<E>, REU	#E<−105 REU	# Sense	Library size	Amino acid outcomes
CTG:DVC:NNK	−106	136	160	168	Just L: ACDGNSTY : All 20
VND:DND:NNK	−90.2	665	5440	6048	All but CFWY: All but QHP: All 20
NNK:NNK:NNK	−70.9	761	8000	9261	All 20: All 20: All 20
TTA:THW:NDK	−108	68	68	90	Just L: FLSY: All but APT
VND:TTC:VND	−94.7	143	256	256	All but CFWY: F : All but CFWY

Asn, Cys, Gly, Asp, or Leu 43), Thr 30 with Leu 43, and all 8 variants with Pro 43.

3. Rather than asking what fraction of a library is predicted to be low-energy, it may be helpful to turn the question around and ask what fraction of the low-energy sequence space the library captures. For case A, low-energy sequences could be identified exactly via enumeration. The CTG:DVC:NNK library captures only 18 % of the 761 case A sequences with instantiated E<−105 REU.
4. To capture a larger share of the low-energy sequence space, be willing to test libraries with a larger risky sequence fraction. In this latter scenario, the Pareto analysis can still provide guidance. For example, the library encoded by VND:DND:NNK (Fig. 8a) is likely a better choice than the next library on the Pareto front (NNK:DHN:NNK), which only excludes Arg, Cys, Gln, Gly, His, Pro, and Trp30 from full saturation mutagenesis.
5. A second illustrative example of a tailored library is TTA:THW:NDK (Fig. 8c). This library has a low <E> of −108, and all 68 of its sense outcomes are low-energy variants (Table 2). As with the other low-energy library described above, this one uses most of the "diversity budget" for residue 43, fixes residue 5 as Leu, and uses a more tailored degenerate codon for residue 30.

The second Pareto analysis (Fig. 8b) identifies libraries with the largest predicted number of sub-threshold sequences (E<−105 REU for case A.1) for a given library size. Full saturation mutagenesis is required to capture all 761 of the low-energy sequences (Fig. 8b). A close inspection of the smaller libraries (Fig. 8d) shows that the fraction of the library that consists of low-energy sequences drops dramatically as the library size exceeds 256. The VND:TTC:VND library is an appealing option since 56 % of its members consist of low-energy sequences and there are no stop codons (Table 2).

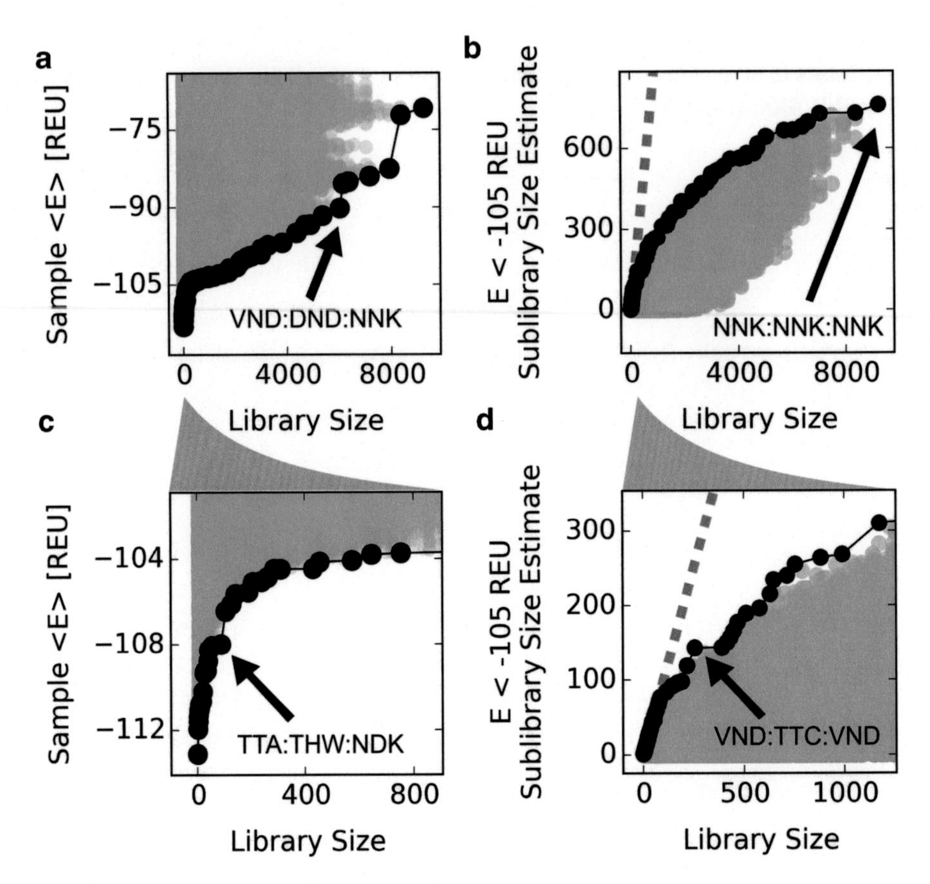

Fig. 8 Case A.1 library selection Pareto analysis. All 450838 libraries sampled by the iterative bias scan are shown as *orange dots*. Pareto optimal libraries are *black dots*. (**a**) The Pareto front for those libraries that have a low <E> and a high library size. (**b**) The Pareto front for libraries with a high estimated no. of sequences with E<−105 REU and a low total library size (including nonsense members). (**c**) Small library close inspection of panel a. (**d**) Small library close inspection of panel b

6. Note that the calculated <E> values for these libraries weight all constituent sequences equally, rather than reflecting the statistical likelihood of observing each sequence (*see* **Note 5**).

3.6 Case B: The 5-site Core Redesign Library

1. The five-site library is intended to serve as a realistic model for the kind of library where degenerate codon optimization is valuable. Thorough sampling of five-site full saturation mutagenesis libraries is beyond the reach of most experimental assays. Thus, codon tailoring is advisable prior to undertaking experimental construction and characterization of a library.

Several trends are consistent with the 3-site library (Table 1). The best $rmsd_{LET}$ (4.1 REU) was for case B.2 with plentiful rotamers and optimized exponential weighting ($\tau = 50$, Eq. 3). The choice of regularization penalty (k) was somewhat important when using smaller training data sets (Fig. 9c). The best case B.2 $rmsd_{LET}$ was obtained (Fig. 9d) when using a large training set (32,000

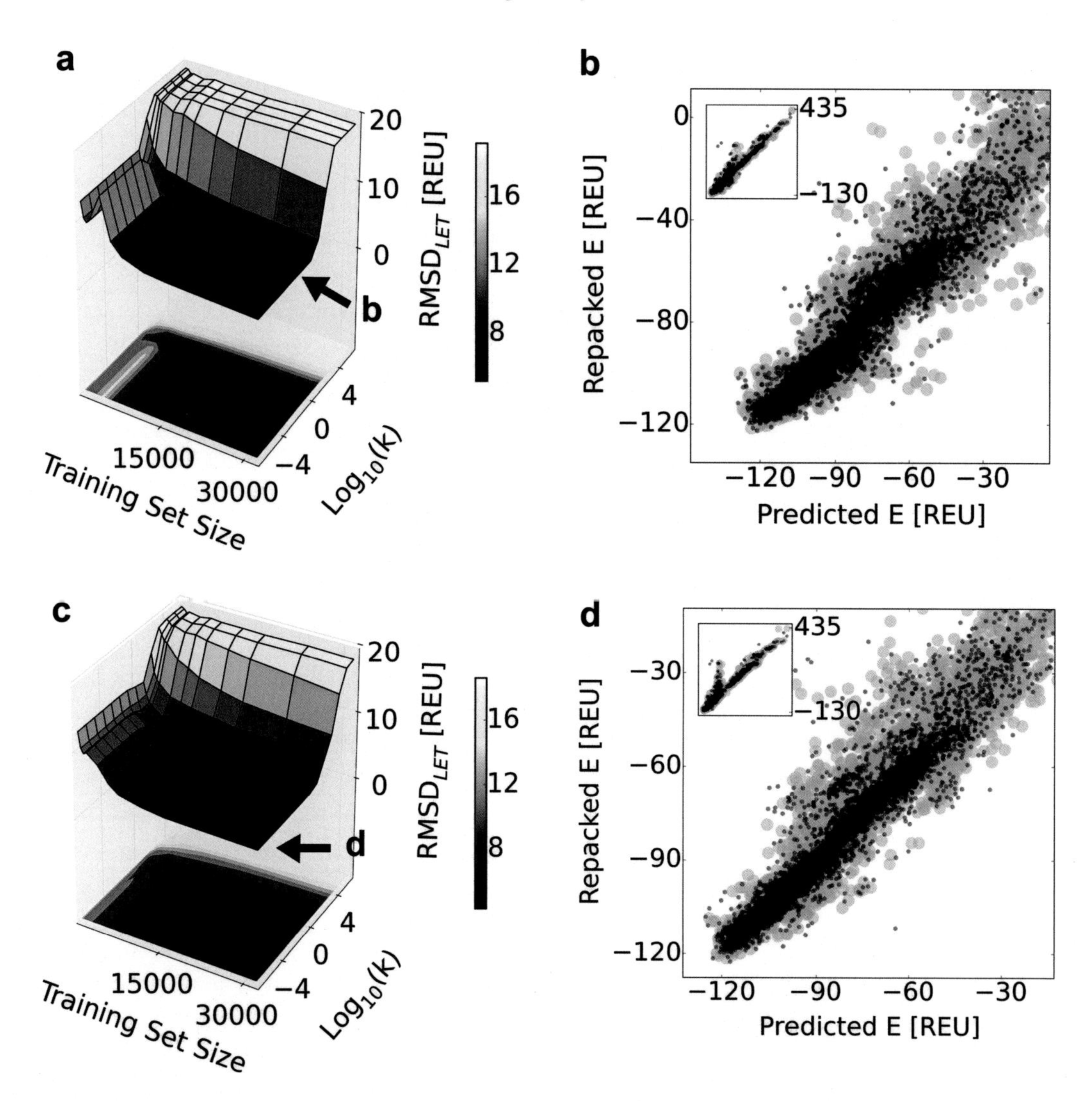

Fig. 9 Case B.1 and B.2 approximation performance. Training set combinations are partial transparent orange points while test set combinations are black points. (**a**) **Case B.1** $rmsd_{LET}$ versus training set size and the regularization parameter *k*. The best prediction was (**b**), when a random training set (32,000 combinations) was used to fit 3706 parameters with regularization ($k = 0.1$) resulting in training set recapitulation (rmsd = 5.95 REU). Performance for favorable test set combinations was reasonable (**$rmsd_{LET}$ = 5.0 REU**) for the 15,984 combinations within 100 REU of the minimum test set combination (−122.5 REU). The entire test set (18,000 combinations) was predicted with rmsd = 7.5 REU (inset). (**c**) $rmsd_{LET}$ versus training set size and the regularization parameter *k* for **Case B.2** (weighted regression with a tuned $\tau = 50$ REU). The best prediction was (**d**), when a random training set (32,000 combinations) was used to fit 3706 parameters with regularization ($k = 1e-07$) resulting in training set recapitulation (rmsd = 7.36 REU). Performance for favorable test set combinations was reasonable (**$rmsd_{LET}$ = 4.1 REU**) for the 15,984 combinations within 100 REU of the minimum test set combination (−122.5 REU). The entire test set (18,000 combinations) was predicted with rmsd = 8.3 REU (inset)

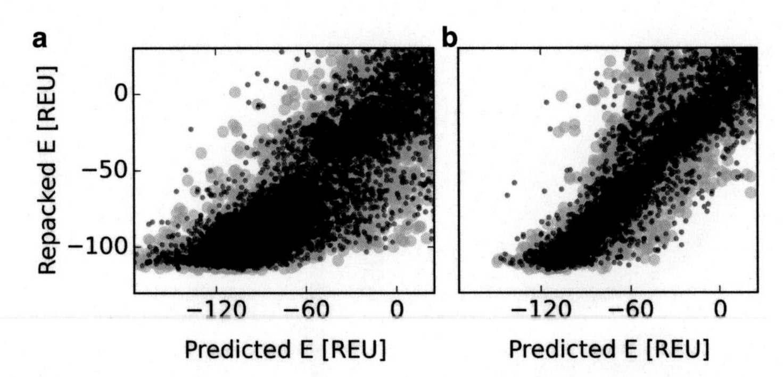

Fig. 10 Benefits of weighted regression. (**a**) **Case B.3** versus, (**b**) **Case B.4**. Using weighted regression (with an optimized parameter $\tau = 75$) cut rmsd_{LET} almost in half, from 14.4 REU (case B.3) to 8.4 REU (case B.4)

random combinations) with minimal regularization parameter ($k = 1\text{E}{-7}$).

2. Without the exponential weighting (case B.1), performance is still good but the regularization parameter (k) is more important still (Fig. 9a). The lowest rmsd_{LET} (5.0 REU, Fig. 9b) was obtained when using a large training set (32,000 random combinations) with a significant regularization parameter ($k = 0.1$).
3. Pruning rotamers (cases B.3 and B.4) results in significant performance degradation. Given minimal rotamers, the effect of exponential weighting was more dramatic. Without exponential weighting (case B.3), rmsd_{LET} was fairly high (14.4 REU, Fig. 10a). However, with minimal rotamers and exponential weighting (case B.4), rmsd_{LET} was greatly improved (8.4 REU, Fig. 10b). This result highlights the potential utility of weighting.

3.7 Case C.1: A 16-Site Surface Library

1. Case C has 16 design positions. Unlike cases A and B, these amino acids are on the protein surface which may significantly change the roughness of the design energy landscape (*see* Fig. 3f versus d). Two specific scenarios are illustrated. In case C.1, the 16 surface sites are provided only with base Dunbrack rotamers for all 20 possible amino acids resulting in a combinatorial search problem of 1.8E35 sequence/structures. In case C.2, provision of standard rotamers increases the search size to 6.5E45.
2. Broadly speaking, the regression performance trends are similar to the trends from the smaller libraries. Case C.1 prediction performance suffers for overly large regularization parameter (Fig. 11a, $k \gg 10$), but a modest penalty of ten yields the best prediction performance ($\text{rmsd}_{\text{LET}} = 9.9$ REU). An eyecatching feature of the regression model training performance plot (Fig. 11a) is the large peak (poor prediction performance)

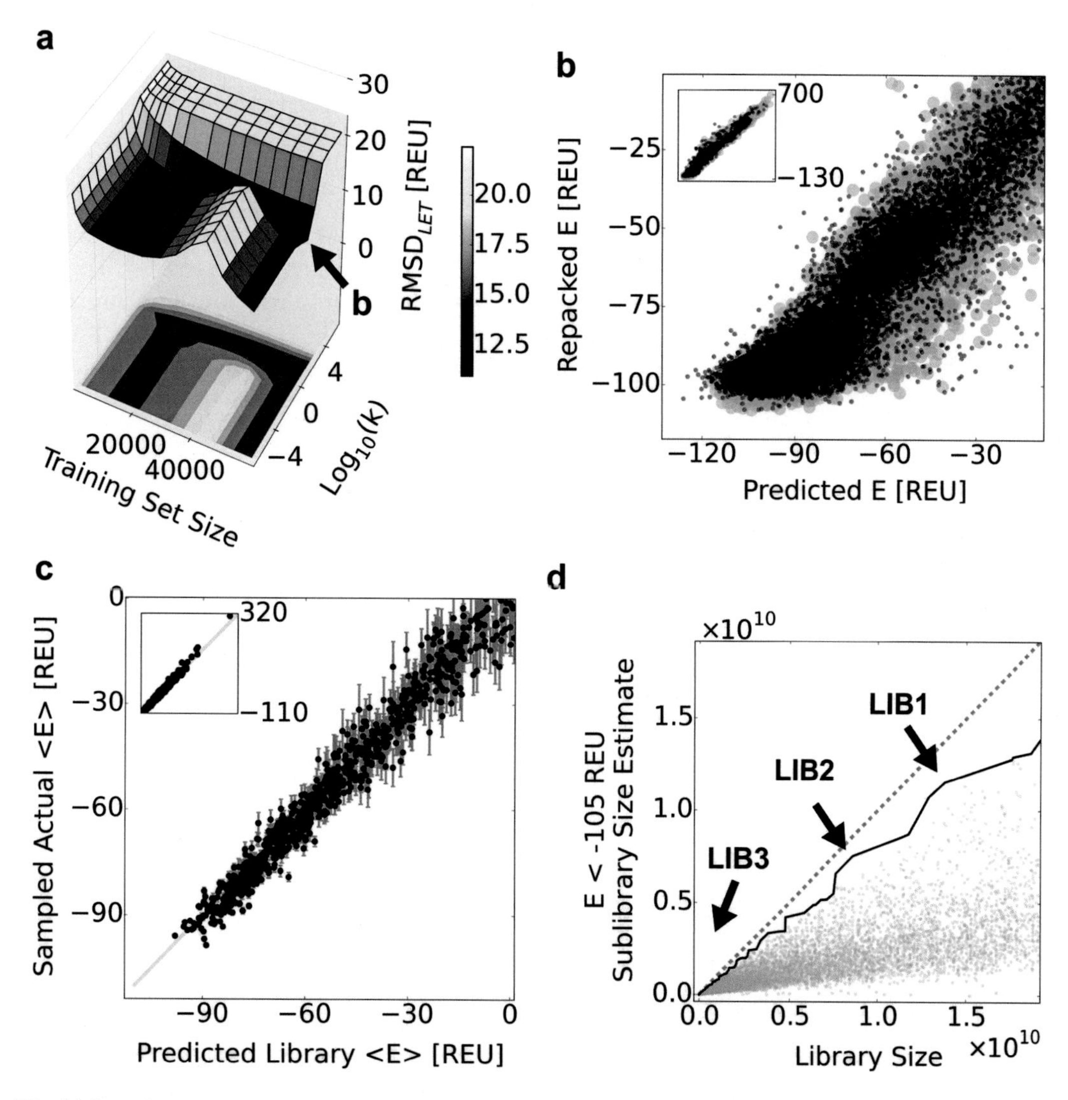

Fig. 11 Case C sequence and library approximation performance. (**a**) Test set rmsd versus training set size and the regularization parameter *k*. (**b**) A random training set (60,000 combinations) was used to fit 43,625 parameters with regularization ($k=10$) resulting in training set recapitulation (rmsd = 10.46 REU). Performance for favorable test set combinations was reasonable (**rmsd$_{LET}$ = 9.9 REU**) for the 15,265 combinations within 100 REU of the minimum test set combination (−107.9 REU). The entire test set (20,000 combinations) was predicted with rmsd = 16.1 REU (inset). (**c**) 1000 random degenerate codon libraries were selected. From each, either the full library or a sample of 400 sequences were optimized via sidechain repacking and scored. The estimated <E> values of the sample sequences were fairly well predicted. *Error bars* reflect $\sigma\langle E\rangle$ (Eq. 4). The full range (inset) was predicted with rmsd = 5.5 REU, while the 257 libraries with predicted <E> within 30 REU of the minimum (−98.1) had **rmsd = 3.6 REU.** (**d**) Pareto analysis of libraries sampled via the iterative combinatorial optimization with an escalating large-library bias. *See* below for library descriptions

when using a low regularization parameter ($k<0.01$) and fairly large training sets (30,000–50,000 combinations). This apparent overfitting pathology is present to varying degree in the previous cases (Figs. 4ac and 9ac). This phenomenon is investigated below in the Overfitting Trends section.

3. Pareto analysis (Fig. 10d) suggests that it is easy to find large case C libraries that are largely composed of low-energy members ($E < -105$ REU). For example, the libraries that are marked LIB1, LIB2, and LIB3 all are predicted to have an 83 % or greater low-energy fraction (Table 3). Notably, the predicted <E> for LIB1 is on par with the original wild-type sequence (-112 REU). LIB2 is slightly smaller (8.6E9 total variants) and has a higher low-energy fraction (88 %). To get a library that is predicted to fall entirely below the threshold, much smaller libraries are necessary (i.e. LIB3 with 8.96E6 sequences). The predicted <E> is well below the original wild-type sequence (-127 REU) thanks to accrued mutations that Rosetta assesses as stabilizing. Six sites are fixed while other sites have up to eight amino acids. The encoded amino acid sets for LIB3 are:

```
L:NT:G:R:GILRSV:DGHNRS:FSY:IKLMQR:FV:EGKMRV:F:CDGHNRSY
:H:DEGHKNQRS:S:EGIKLQRV
```

As the library size decreases further, it becomes easy to find libraries that are predicted to fall entirely below the −105 REU threshold. Almost all of the 288 sampled libraries with size below two million sequences fall into this category. In principle, to differentiate between these candidates it could be helpful to introduce another evaluation criterion such as <E>, or to assess the fraction of constituent sequences that meets a more stringent stability threshold (e.g. $E < -120$ REU).

3.8 Case C.2: With Additional Rotamers

1. The largest combinatorial problem for this chapter is case C.2 (Table 1). As above, the additional rotamers make a significant improvement in performance (Fig. 12). It may be surprising to note that the $rmsd_{LET}$ is lower (4.8 REU) than the comparable calculation for case B ($rmsd_{LET} = 5.0$). This can be rationalized by noting that the design positions for case C are all surface-exposed sites, where it is easier for amino acid combinations to avoid clashes (given sufficient rotamer flexibility). Thus, there are fewer legitimate higher-order frustration effects encountered in this scoring landscape (Fig. 3), and it is possible to obtain a high accuracy fit.

Presumably, this fit could be further improved using weighting or perhaps the introduction of 3-body terms. However, the current level of accuracy seems quite sufficient to assist with the design of degenerate codon libraries (Fig. 12c). The rmsd between regression model <E> predictions and directly sampled <E> estimates was only 4.1 (or 1.7) REU for all (or $\langle E \rangle < -76$ REU) libraries. When chasing high precision, it is important to recall that the underlying scoring function is itself a fairly crude approximation of the biophysical effects in play.

Table 3
Example libraries of interest for Case C. Library size refers to all distinct outcomes including genes that include stop codons

Case	Name	Codons	Library size	<E> [REU]	% with E < −105 REU
C.1	LIB1	CNT:KYC:GGT:CDD:KYA:ASC:TBS:ABG:RYG:VDA:MDK:VDS:YDY:SAA:VBB:MWA	1.4E10	−112	84
C.1	LIB2	WTG:DSG:GGT:VDG:WYW:HDT:SDR:WSG:RYG:KNT:TTC:VDW:GVA:GVA:VDG:CRS	4.7E9	−113	88
C.1	LIB3	TTA:AMC:GGT:AGG:VKY:VRT:THY:MDR:KTC:RDG:TTC:NRC:CAT:VRS:AGC:VDA	9.0E6	−127	100
C.2	LIB4	BBG:TAC:WYS:NKB:ABG:HKY:SAA:MWA:VKY:RNR:VRS:CWR:VDS:CDG:RHR:ADB	1.1E11	−110	88
C.2	LIB5	BBG:TAC:WYS:NKB:ABG:SRA:SAA:MWA:VKY:RNR:VRS:CWR:VDS:CDG:RHR:ADB	7.3E10	−111	89
C.2	LIB6	BTT:AGG:GTA:AGG:VAK:DYR:RHS:ARS:AHH:AAD:RHS:AMC:AGG:RBH:RDG:VKY	1.6E8	−119	100
C.2	SAT	NNK:NNK:NNK:NNK:NNK:NNK:NNK:NNK:NNK:NNK:NNK:NNK:NNK:NNK:NNK:NNK	6.6E20	−59	15

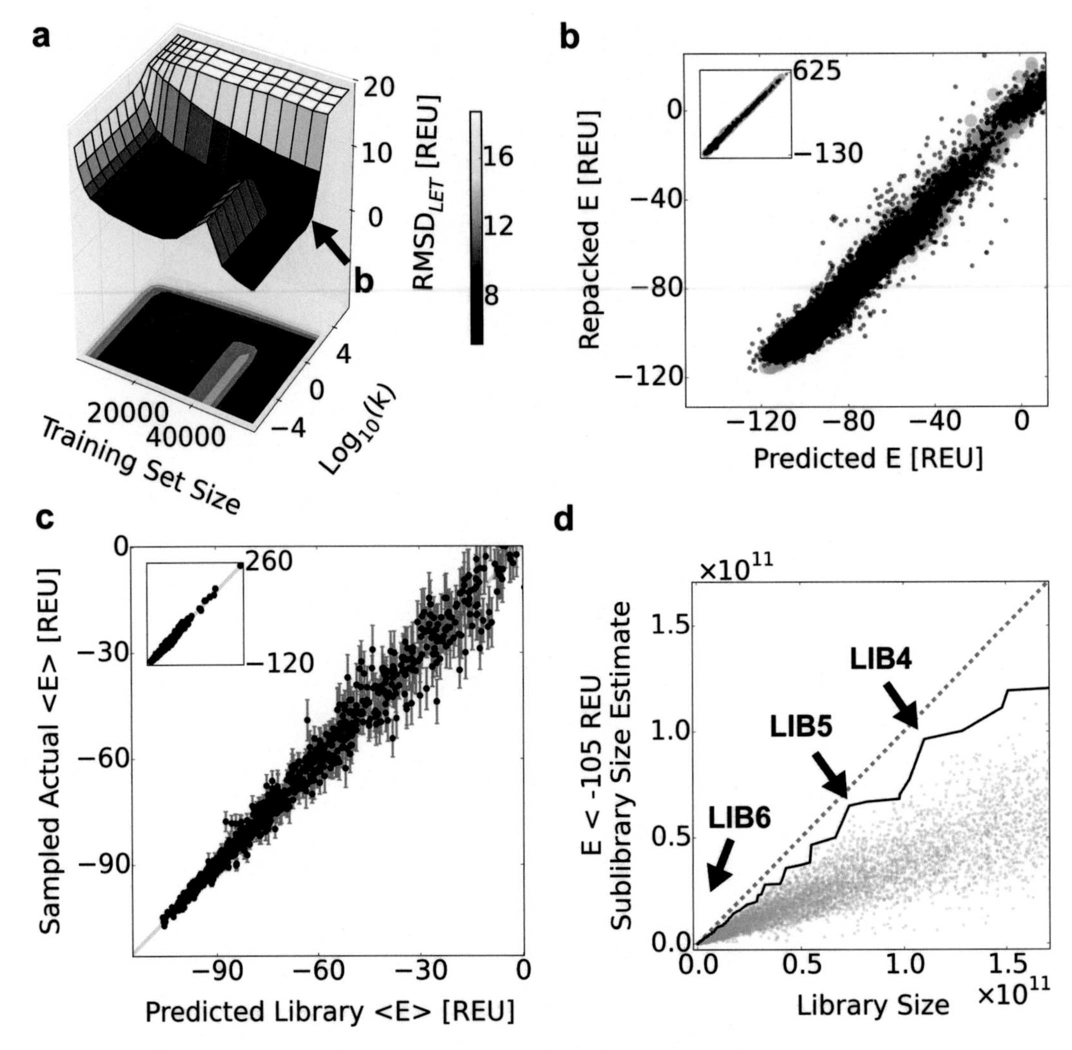

Fig. 12 Case C.2 sequence and library approximation performance. (**a**) Test set rmsd versus training set size and the regularization parameter *k*. (**b**) A random training set (60,000 combinations) was used to fit 43,625 parameters with regularization ($k = 1$) resulting in training set recapitulation (rmsd = 1.7 REU). The test set (20,000 combinations) was predicted with rmsd = 6.6 REU. Performance for favorable combinations was reasonable (**$\mathbf{rmsd_{LET}}$ = 4.8 REU**) for the 16,996 test set combinations within 100 REU of the minimum test set combination (−117 REU). (**c**) 1000 random degenerate codon libraries were selected. From each, either the full library or a sample of 400 sequences were optimized via sidechain repacking and scored. The estimated <E> values of the sample sequences were fairly well predicted (rmsd = 4.1 REU over the full range, inset). *Error bars* reflect $\sigma\langle E\rangle$ (Eq. 4). Performance for the 457 libraries with predicted <E> within 30 REU of the minimum (−106.5 REU) was better still (**rmsd = 1.7 REU**). (**d**) Pareto analysis of libraries sampled via the iterative combinatorial optimization with an escalating large-library bias. *See* below for LIB descriptions

2. As above, use iterative bias sampling to collect optimized libraries of varying size and proceed with Pareto analysis. Illustrative Pareto analysis of 755,912 candidate case C.2 libraries (Fig. 12d) suggests that it is even easier to find large case C.2 libraries that are highly enriched for low-energy members (Table 3).

This is not surprising since the additional rotamers result in more highly optimized structures with lower Rosetta scores. LIB6 is the largest library that is predicted to have 100 % low-energy constituents. The encoded amino acid sets for LIB6 are:

FLV:R:V:R:DEHKNQ:AILMSTV:ADEIKMNTV:KNRS:IKNT:KN:ADEIKMNTV:NT:R:AGIRSTV:EGKMRV:GILRSV

These libraries compare favorably to a brute force saturation mutagenesis approach. For one, library size can be matched to the available transformation and screening capacity whereas the size of the NNK library exceeds feasible screening size. Also, the NNK library has a high <E> (−59 REU) and a large fraction of the constituent sequences are unfavorable (85 % have E > −105 REU).

3.9 Overfitting Trends

1. The purpose of this section is to investigate the counterintuitive overfitting behavior noted above. Inferior prediction performance for models derived from larger training sets occurred repeatably for varying input training sets and using either the lsmr or cvxopt regression tools. This effect was most prominent for three regression model variants (cases A.2, B.3, and C.1) with negligible regularization (k = 1E−7) (Fig. 13a). The effect was largest for case C.1. Therefore, to illustrate the fitting pathology two regression models can be compared: the fit for case C.1 with training sets of 24,000 (Fig. 13b) versus 45,000 (Fig. 13c).

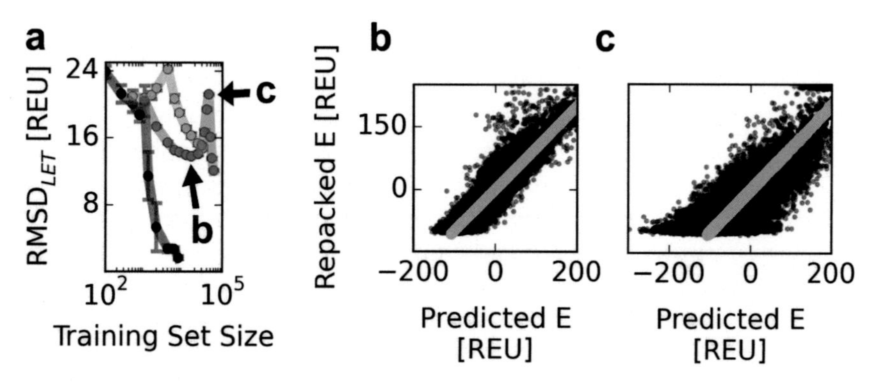

Fig. 13 Overtraining. (**a**) Apparent overfitting with "free" regression (regularization k = 1E−7) was prominent for training case A.2 with ~1000 combinations (black), case B.2 with ~4000 combinations (*orange*), and case C.1 with ~45,000 combinations (*green*). *Error bars* reflect the standard deviation among 5 random training set replicates. We use case C.1 to further illustrate the degradation of prediction performance: (**b**) Prediction performance (*black*) is reasonable ($rmsd_{LET}$ = 13.9 REU) when trained with 24,000 random combinations. The training set (*orange*) is recapitulated exactly (rmsd = 0.0). (**c**) Prediction performance (*black*) is degraded ($rmsd_{LET}$ = 21.2 REU) when trained with 45,000 random combinations. The training set (*orange*) is still recapitulated nearly exactly (rmsd = 1.4 REU)

2. First, a random training set (24,000 combinations) is used to fit 43,625 parameters with regularization (k = 1e–07). Given the over-abundance of fitting parameters, it was not surprising that the training set was recapitulated *exactly* (rmsd = 0.00, orange dots). In contrast, the entire test set was predicted with rmsd = 20.7 REU (Fig. 13b). The 42,745 test set combinations within 100 REU of the minimum test set combination (–107.9 REU) could be predicted with $rmsd_{LET}$ = 13.9 REU (Fig. 13b).
3. For comparison, a larger random training set (45,000 combinations) was used to fit 43,625 parameters with regularization (k = 1e–07) resulting in *near exact* training set recapitulation (rmsd = 1.4 REU, orange dots). In contrast, the entire test set was predicted with rmsd = 42.4 REU (Fig. 13c). The 26,756 test set combinations within 100 REU of the minimum test set combination (-107.8 REU) were only predicted with $rmsd_{LET}$ = 21.2 REU.

3.10 Recap the Pertinent Observations

1. Scenarios with fewer rotamers (e.g. cases A.2, B.3, and C.1) have a greater tendency to experience overfitting.
2. Overfitting can be suppressed by regularization (e.g. Fig. 11a)
3. The best regularization parameter seems to grow with the problem size (k = 0.01, 1, and 10 for cases A.2, B.3, and C.1).
4. Prediction performance is most degraded when the fit has a certain number of training examples: ~1000 for A.2, ~4000 for B.3, and ~45,000 for C.1 (Fig. 13a). These numbers are similar to the number of fitting parameters: respectively 1141, 3706, and 43,625.
5. Despite training the case C.1 model with 45,000 combinations, the training set combinations (orange points) are still clearly being overfit (compare to Fig. 11)
6. Given these observations, it may be that having a small number of training instances (relative to the number of fit parameters) serves to restrain the magnitude of the fit parameters, much as the regularization process penalizes large fit parameter magnitude. When the number of training instances is comparable to the number of parameters, fit parameters are more likely to adopt large magnitude values to fit the training set data. Meanwhile, minimal rotamer cases make the energy landscape rougher (compare Fig. 3bdf to Fig. 3acd). The rougher training set energy landscape will also result in more extreme fit parameters that degrade the testset prediction performance. The effectiveness of regularization, which attempts to keep parameters near 0, supports the idea that the fitting pathology is tied to the magnitude of the fitting terms.

To investigate, examine the fit parameters for the two case C1 regression models (Fig. 13bc). The superior model trained with

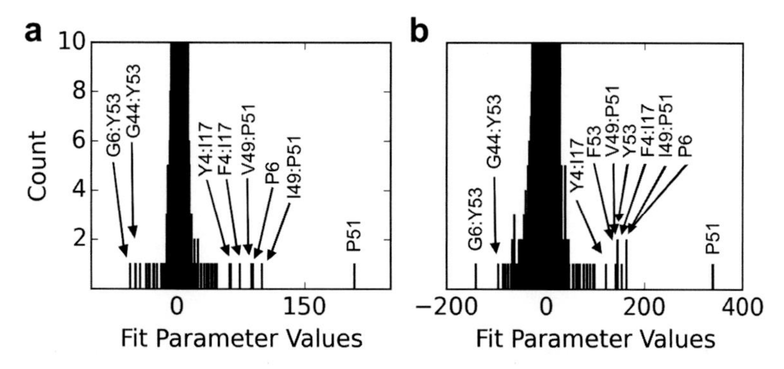

Fig. 14 Comparison of Case C.1 fit coefficients. (**a**) The superior model was trained with 24,000 sequences (*see* also Fig. 13b) and had parameters of lesser magnitude, while (**b**) the inferior model was trained with 45,000 combinations (*see* also Fig. 13c) and had fit parameters of greater magnitude

24,000 sequences has a sum of absolute parameter values of 60,831, while the inferior model trained with 45,000 combinations has a sum of absolute parameter values of 301,104.

7. The dramatic fivefold shift in the magnitude of the fit parameters also appears in histograms of the fitting parameter values (Fig. 14). There is some qualitative consistency between the two models. For example, the most unfavorable term for both models is the 1-body effect of T51P (unsurprising since T51 is within a beta strand), and the most favorable term is the 2-body effect of I6G:T53Y which can be rationalized as a bump/hole interaction between these adjacent residues on the beta sheet surface. However, it is important to note that the actual values of the coefficients are not stable; attempts to glean additional insight from inspection of the fit coefficients may be problematic.

One take home lesson is that regression model performance can be difficult to anticipate (and may be strongly dependent on regularization) unless the training set size significantly exceeds the number of fitting parameters. Any scientist preparing a regression model of this type should carefully scan the training set size and regularization parameter to ensure optimal model quality (*see* **Note 6**).

3.11 Computational Time

The three design problems discussed here can be framed as conventional CPD calculations. Searching the sequence-structure space directly, optimal solutions can be found using either CPLEX or FasterPacker (*see* above). However, the 725 possible mixtures of amino acids possible at each site dwarfs the 20 possible amino acids. Sampling in library space, therefore, is significantly more challenging. With the exception of case A, it is impractical to tabulate the instantiated energy of all the sequences encoded by each library. A brute force search instantiating all sequence combinations (assuming generous rotamers) requires only 38 seconds for case A.1 to a projected 8E11 years for case C.2.

Table 4
This Table summarizes the Python scripts used for the calculations. Time represents the elapsed wall clock time in seconds necessary to complete the calculations on a single CPU

Time (s)	Script name	Script purpose
3	A_prep.py	Process the input PDB model and save as a CHOMP System
219	B_fill_energy_graph.py	Setup the design problem, fill, and save the EnergyGraph
10	C_pick_initial_set.py	Select a set of combinations to serve as an initial pool
3077	D_score_initial_set_multi.py	Instantiate: run repacking calculations on the initial pool
5	E_split_to_training_samples.py	Randomly divide the initial pool into training and test sets
nd	*F_try_regression.py*	*Do a quick regression test to ensure things are working so far*
2190	G_scan_training_params.py	Repeatedly run regression varying the training set size and regularization parameter
nd	*H_gen_figure_trainsize_vs_ridgeparam_vs_testrmsd.py*	*Illustrate test set performance* versus *training parameters*
nd	*I_gen_fig_example_train_test.py*	*Illustrate the training set and test set fit for the best case parameters*
1978	M_score_dgen_codon_sets.py	Convert the amino acid level regression model to a degenerate codon level model
13,563	N_sample_random_libraries.py	Randomly select a set of random degenerate codon libraries and perform the requisite instantiation (repacking)
nd	*O_lib_predictE_vs_actualE.py*	*Illustrate the correlation between predicted and actual (sampled) <E> for each library*
1490	P_iterlib_sample.py	Collect libraries by 100 rounds of combinatorial optimization for low <E> with an escalating bias favoring larger libraries.
22,918	Q_calc_pareto_stats.py	First lookup or predict the energy for a sequence sample from each candidate library. Then calculate <E> and the expected number of variants with E < a threshold. Finally, use a divide and conquer approach to compute the two Pareto fronts of interest.
nd	R_plot_pareto_ok.py	Plot the Pareto front (black) and other libraries (orange)

In comparison, the aggregate calculation time for the steps described above is attractive. All reported calculations could be performed on a single 2.8 GHz Intel Core i7 machine over several days (*see* Table 4 for case C.2 calculation time table). Most of the time-consuming calculations were parallelized across 8 threads using the Python multiprocessing module. Distributing bottleneck calculations beyond the cores of a single CPU could easily further reduce wall time.

One easy way to limit the CPU time while retaining the power of the library-space optimization approach would be to prune the set of degenerate codons considered at each design site. One pragmatic approach might be to design a limited number of amino acid sets, guided by biophysical intuition (e.g. hydrophobic, large hydrophobic, small, large, charged, aromatic, etc.). Selecting several hundreds of these useful amino acid sets would make the library design code more efficient than the current search over 725 possibilities. Similarly, with repeated use of the current 725-member design palette, it may be possible to identify which degenerate codons are rarely useful and eliminate them from consideration.

4 Notes

1. The presented methods are flexible, and amenable for modification. One such modification that might be particularly desirable would be to enable optimization of amino acid bias. At the outset, the degenerate codon search space was defined to be the 725 degenerate codons that produce unique sense mixtures of amino acids. It is worthwhile to note, however, that the formalism presented here would also work if the design palette consists of the 1439 degenerate codons that produce unique sense *ratios* of amino acids. If outcome amino acid probabilities are included when creating the degenerate codon energy model (Eqs. 5 and 6), the resulting optimization target $\langle E \rangle$ will reflect the expectation REU score for clones pulled at random from the experimental library. This additional level of design could prove useful. Optimizing amino acid frequencies could further increase library fitness by decreasing $\langle E \rangle$. For example, given a particular site that favors leucine over phenylalanine, combinatorial library optimization might select a degenerate codon like YTD that encodes a 5:1 ratio of Leu to Phe rather than TTB that encodes a 1:2 ratio of Leu to Phe.
2. The illustrative examples presented in this chapter provide another example of regression-based approximations successfully capturing more expensive calculations with sufficient accuracy to guide an otherwise infeasible search problem. By "integrating out" the structure variables, and providing an essentially instantaneous lookup of the predicted energy for

any given sequence, it becomes feasible to execute a combinatorial search directly in "library space." This approach was recently reported in the context of cluster expansion [14]. Readers of this chapter who are preparing to design a codon library are therefore encouraged to review Verma et al.

3. In principle, a penalty could also be levied for stop codons by giving stop codon outcomes a large 1-body energy term. The goal would be to ensure that non-sense outcomes have large unfavorable scores commensurate with other likely unfolded sequences.
4. It is worth noting that there are certain technical challenges to performing library-space optimization for case C. With 725 possible degenerate codons and 16 design sites, the library search space has 725^{16} combinations, or 5.8E45. Building a graph of the codon:codon scores (Fig. 1) required 30 minutes. Storing the graph in binary form on disk requires nearly a gigabyte.
5. In practice, some sequences will be more frequent than others. For example, the MKD degenerate codon yields an arginine 5 times more frequently than a serine. If desired, it would be easy to instead calculate the expectation value <E> for sequences drawn from the library with the actual amino acid frequency weights (Eqs. 5 and 6) rather than assuming equal representation. The former approach may be preferable if the planned approach is to build a large experimental library and characterize only a random subset thereof.
6. Additional caution and careful regularization parameter tuning is recommended if pursuing high-accuracy regression models including 3-body terms, since the high number of possible 3-body terms may make it difficult to prepare models with a large excess of training data.

References

1. Ponder JW, Richards FM (1987) Tertiary templates for proteins. Use of packing criteria in the enumeration of allowed sequences for different structural classes. J Mol Biol 193: 775–791
2. Pierce NA, Winfree E (2002) Protein design is NP-hard. Protein Eng 15:779–782
3. Desmet J, De Maeyer M, Hazes B, Lasters I (1992) The dead-end elimination theorem and its use in protein side-chain positioning. Nature 356:539–542
4. Desmet J, Spriet J, Lasters I (2002) Fast and accurate side-chain topology and energy refinement (FASTER) as a new method for protein structure optimization. Proteins 48:31–43
5. Canutescu AA, Shelenkov AA, Dunbrack RL (2003) A graph-theory algorithm for rapid protein side-chain prediction. Protein Sci 12:2001–2014
6. Kingsford CL, Chazelle B, Singh M (2005) Solving and analyzing side-chain positioning problems using linear and integer programming. Bioinformatics 21:1028–1039
7. Allen BD, Mayo SL (2006) Dramatic performance enhancements for the FASTER optimization algorithm. J Comput Chem 27: 1071–1075
8. Hallen MA, Keedy DA, Donald BR (2012) Dead-end elimination with perturbations (DEEPer): A provable protein design algorithm

with continuous sidechain and backbone flexibility., Proteins

9. Zhou F, Grigoryan G, Lustig SR et al (2005) Coarse-graining protein energetics in sequence variables. Phys Rev Lett 95:148103
10. Grigoryan G, Zhou F, Lustig SR et al (2006) Ultra-fast evaluation of protein energies directly from sequence. PLoS Comput Biol 2, e63
11. Grigoryan G, Reinke AW, Keating AE (2009) Design of protein-interaction specificity gives selective bZIP-binding peptides. Nature 458: 859–864
12. Apgar JR, Hahn S, Grigoryan G, Keating AE (2009) Cluster expansion models for flexible-backbone protein energetics. J Comput Chem 30:2402–2413
13. Hahn S, Ashenberg O, Grigoryan G, Keating AE (2010) Identifying and reducing error in cluster-expansion approximations of protein energies. J Comput Chem 31(6):2900–2914
14. Verma D, Grigoryan G, Bailey-Kellogg C (2015) Structure-based design of combinatorial mutagenesis libraries. Protein Sci 24: 895–908
15. Liao J, Warmuth MK, Govindarajan S et al (2007) Engineering proteinase K using machine learning and synthetic genes. BMC Biotechnol 7:16
16. Otey CR, Landwehr M, Endelman JB et al (2006) Structure-guided recombination creates an artificial family of cytochromes P450. PLoS Biol 4, e112
17. Li Y, Drummond DA, Sawayama AM et al (2007) A diverse family of thermostable cytochrome P450s created by recombination of stabilizing fragments. Nat Biotechnol 25: 1051–1056
18. Heinzelman P, Snow CD, Wu I et al (2009) A family of thermostable fungal cellulases created by structure-guided recombination. Proc Natl Acad Sci U S A 106:5610–5615
19. Heinzelman P, Snow CD, Smith MA et al (2009) SCHEMA recombination of a fungal cellulase uncovers a single mutation that contributes markedly to stability. J Biol Chem 284:26229–26233
20. Heinzelman P, Komor R, Kanaan A et al (2010) Efficient screening of fungal cellobiohydrolase class I enzymes for thermostabilizing sequence blocks by SCHEMA structure-guided recombination. Protein Eng Des Sel 23:871–880
21. Smith MA, Rentmeister A, Snow CD et al (2012) A diverse set of family 48 bacterial glycoside hydrolase cellulases created by structure-guided recombination. FEBS J 279:4453–4465
22. Silberg JJ, Endelman JB, Arnold FH (2004) SCHEMA-guided protein recombination. Methods Enzymol 388:35–42
23. Endelman JB, Silberg JJ, Wang ZG, Arnold FH (2004) Site-directed protein recombination as a shortest-path problem. Protein Eng Des Sel 17:589–594
24. Pantazes RJ, Saraf MC, Maranas CD (2007) Optimal protein library design using recombination or point mutations based on sequence-based scoring functions. Protein Eng Des Sel 20:361–373
25. Johnson LB, Huber TR, Snow CD (2014) Methods for library-scale computational protein design. Methods Mol Biol 1216: 129–159
26. Voigt CA, Mayo SL, Arnold FH, Wang ZG (2001) Computational method to reduce the search space for directed protein evolution. Proc Natl Acad Sci 98:3778
27. Wang W, Saven JG (2002) Designing gene libraries from protein profiles for combinatorial protein experiments. Nucleic Acids Res 30:e120
28. Mena MA, Daugherty PS (2005) Automated design of degenerate codon libraries. Protein Eng Des Sel 18:559–561
29. Allen BD, Nisthal A, Mayo SL (2010) Experimental library screening demonstrates the successful application of computational protein design to large structural ensembles. Proc Natl Acad Sci 107:19838–19843
30. Parker AS, Griswold KE, Bailey-Kellogg C (2011) Optimization of combinatorial mutagenesis. J Comput Biol 18:1743–1756
31. Chen TS, Palacios H, Keating AE (2013) Structure based re-design of the binding specificity of anti-apoptotic Bcl-xL. J Mol Biol 425:171–185
32. Jacobs TM, Yumerefendi H, Kuhlman B, Leaver-Fay A (2015) SwiftLib: rapid degenerate-codon-library optimization through dynamic programming. Nucleic Acids Res 43:e34
33. Treynor TP, Vizcarra CL, Nedelcu D, Mayo SL (2007) Computationally designed libraries of fluorescent proteins evaluated by preservation and diversity of function. Proc Natl Acad Sci U S A 104:48–53
34. Patrick WM, Firth AE, Blackburn JM (2003) User-friendly algorithms for estimating completeness and diversity in randomized protein-encoding libraries. Protein Eng 16:451–457
35. Rohl CA, Strauss CEM, Misura KMS, Baker D (2004) Protein structure prediction using Rosetta. Methods Enzymol 383:66–93

36. Hughes MD, Nagel DA, Santos AF et al (2003) Removing the redundancy from randomised gene libraries. J Mol Biol 331:973–979
37. Tang L, Gao H, Zhu X et al (2012) Construction of "small-intelligent" focused mutagenesis libraries using well-designed combinatorial degenerate primers. Biotechniques 52:149–158
38. Kille S, Acevedo-Rocha CG, Parra LP et al (2013) Reducing codon redundancy and screening effort of combinatorial protein libraries created by saturation mutagenesis. ACS Synth Biol 2:83–92
39. Ashraf M, Frigotto L, Smith ME et al (2013) ProxiMAX randomization: a new technology for non-degenerate saturation mutagenesis of contiguous codons. Biochem Soc Trans 41:1189–1194
40. Tang L, Wang X, Ru B et al (2014) MDC-Analyzer: a novel degenerate primer design tool for the construction of intelligent mutagenesis libraries with contiguous sites. Biotechniques 56:301–302, 304, 306–308, passim
41. Nov Y, Segev D (2013) Optimal codon randomization via mathematical programming. J Theor Biol 335:147–152
42. Sanchez JM, Ducastelle F, Gratias D (1984) Generalized cluster description of multicomponent systems. Physica A 128:334–350
43. Ng AH, Snow CD (2011) Polarizable protein packing. J Comput Chem 32:1334–1344
44. Ponder JW, Wu C, Ren P et al (2010) Current status of the AMOEBA polarizable force field. J Phys Chem B 114:2549–2564
45. Loksha IV, Maiolo JR 3rd, Hong CW et al (2009) SHARPEN-systematic hierarchical algorithms for rotamers and proteins on an extended network. J Comput Chem 30: 999–1005
46. IBM ILOG CPLEX Optimization Studio 12.6.2. IBM
47. Andersen MS, Dahl J, Vandenberghe L (2013) CVXOPT: a python package for convex optimization, version 1.1.6
48. Davis TA (2009) User guide for CHOLMOD: a sparse Cholesky factorization and modification package
49. Fong D, Saunders M (2011) LSMR: an iterative algorithm for sparse least-squares problems. SIAM J Sci Comput 33:2950–2971
50. Jones E, Oliphant T, Peterson P, others (2001) SciPy: open source scientific tools for Python
51. Levine HA (1979) Review: A. N. Tikhonov and V. Y. Arsenin, solutions of ill posed problems. Bull Am Math Soc 1:521–524
52. Amstutz P, Forrer P, Zahnd C, Plückthun A (2001) In vitro display technologies: novel developments and applications. Curr Opin Biotechnol 12:400–405
53. He L, Friedman AM, Bailey-Kellogg C (2012) A divide-and-conquer approach to determine the Pareto frontier for optimization of protein engineering experiments. Proteins 80: 790–806
54. Dunbrack RL Jr, Karplus M (1993) Backbone-dependent rotamer library for proteins. J Mol Biol 230:543–574

Chapter 8

Combined and Iterative Use of Computational Design and Directed Evolution for Protein–Ligand Binding Design

Meng Wang and Huimin Zhao

Abstract

The advantages of computational design and directed evolution are complementary, and only through combined and iterative use of both approaches, a daunting task such as protein–ligand interaction design, can be achieved efficiently. Here, we describe a systematic strategy to combine structure-guided computational design, iterative site saturation mutagenesis, and yeast two-hybrid system (Y2H)-based phenotypic screening to engineer novel and orthogonal interactions between synthetic ligands and human estrogen receptor α (hERα) for the development of novel gene switches.

Key words Computational design, Directed evolution, Protein–ligand interaction, Gene switch

1 Introduction

Protein–ligand interaction is a universal and key aspect of all biological processes ranging from feedback regulation of enzyme catalysis in metabolic pathways to ligand-mediated signal transmission [1]. In addition, protein–ligand interaction is the basic mode of action of many pharmaceutical compounds, which has been heavily explored by both academia and pharmaceutical industry. Therefore, the ability to engineer protein–ligand interactions on demand is a long-sought goal.

There are two main strategies widely used to reach that goal, including (1) alteration of the ligand specificity of naturally occurring protein–ligand interactions and (2) de novo computational design of proteins to bind desired ligands. In the first strategy, directed evolution is one of the most powerful tools that enable the creation and fine-tuning of novel protein–ligand interactions that are orthogonal to the native protein–ligand pairs. However, due to the vast search space and limited throughput of a screening method, directed evolution by only using a random mutagenesis library has its limitations. In the past decade, with increased availability of crystal structures of the native protein–ligand complexes

Barry L. Stoddard (ed.), *Computational Design of Ligand Binding Proteins*, Methods in Molecular Biology, vol. 1414,
DOI 10.1007/978-1-4939-3569-7_8, © Springer Science+Business Media New York 2016

[2] and rapid development of computer programs that can analyze, predict, and simulate protein–ligand interactions [3, 4], the advantages of computational design and directed evolution were combined in an iterative fashion to significantly improve the success rate of the engineering of desired protein–ligand interactions. For example, Arnold and coworkers used a structure-guided directed evolution strategy to engineer a P450 enzyme to selectively bind dopamine and serotonin, which could be used as magnetic resonance imaging sensors [5, 6]. Ligand-dependent bacterial regulatory proteins are another category that a combined computational design and directed evolution strategy has been successfully used to alter the native protein–ligand interactions to suit the researcher's demand [7–9]. For instance, the Ara-C regulatory protein of the *Escherichia coli ara* operon was engineered to recognize d-arabinose instead of the native ligand l-arabinose [10], and XylS from the TOL pathway of *Pseudomonas putida* mt2 was evolved to increase the induction level toward benzoate ligands [11]. In addition, structure-guided directed evolution was also applied to improve the binding affinity of glucose and a glucose/galactose-binding protein for the development of a glucose biosensor [12, 13]. To take the full advantage of computational design, several brilliant strategies have been developed to create and screen a focused smart library in directed evolution. In particular, iterative saturation mutagenesis is a widely used strategy to improve the efficiency of directed evolution. Based on this strategy, Reetz and coworkers developed the combinatorial active-site saturation test (CAST) method to improve enzyme catalytic properties including substrate specificity, regioselectivity, and stereoselectivity [14] and the B-factor iterative test (B-FIT) method to improve protein thermal stability [15]. We also adopted a similar strategy to engineer novel and orthogonal interactions between synthetic ligands and human estrogen receptor α (hERα) for the development of novel gene switches [16].

On the other hand, de novo computational design of proteins that bind to desired ligands has become possible in the recent years [3, 4]. Tinberg and coworkers developed a robust computation method to create a steroid digoxigenin (DIG)-binding protein [17]. However, existing computational design tools are far from perfection. In most of such endeavors, directed evolution is still a powerful and indispensable tool to improve the performance of proteins generated by computational design [18, 19]. The desired goal can often be reached only through iterative cycles of computational design and directed evolution.

Here we use structure-guided directed evolution of a specific hERα–ligand pair as an example to illustrate the experimental procedures [16]. We provide a systematic strategy to engineer receptor proteins with significantly altered selectivity toward a target synthetic ligand 4,4′-dihydroxybenzil (DHB). Structure-based

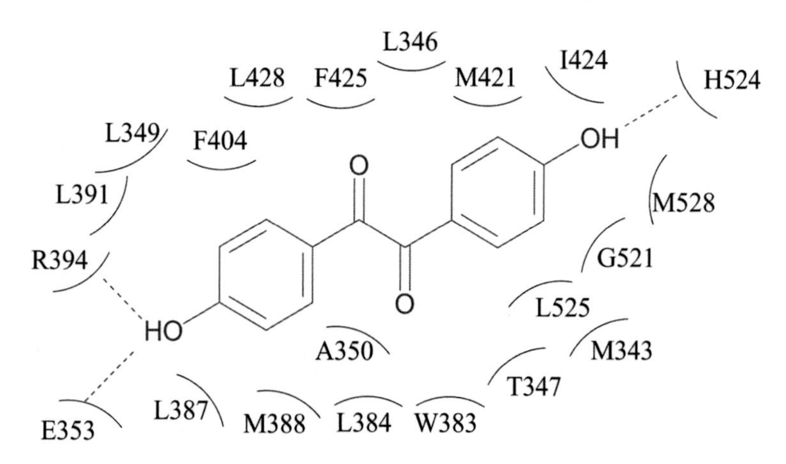

Fig. 1 Two-dimensional depiction of DHB and its surrounding residues when docked into the hERα ligand-binding pocket. Twenty-one residues were identified to be within 4.6 Å of DHB. The A ring and D ring analogues of DHB are indicated. *Dashed lines* denote hydrogen bonds

computational design was first employed for the identification of potential ligand-contacting residues (Fig. 1). Of the 21 identified residues, 14 were subjected to iterative site saturation mutagenesis, accompanied by yeast two-hybrid system (Y2H)-based phenotypic screening for variants with enhanced target ligand selectivity. Finally, a random point mutagenesis library coupled with phenotypic screening was performed to further improve the target ligand selectivity. The resulting gene switches were further evaluated in *Saccharomyces cerevisiae* and human endometrial cancer (HEC-1) cells. This same strategy was successfully used to create orthogonal receptor–ligand pairs in a single protein scaffold [20].

2 Materials

2.1 Molecular Modeling

1. Molecular Operating Environment (MOE) (Chemical Computing Group, Montreal).

2.2 Library Creation by Single-Site Saturation Mutagenesis

1. 10× PCR reaction buffer.
2. 25 mM $MgCl_2$.
3. 10 mM dNTP mix.
4. *Taq* DNA polymerase.
5. *PfuTurbo* DNA polymerase.
6. PTC-200 thermocycler.
7. Concentrated (50×) stock solution of TAE buffer: Weigh 242 g of Tris base (MW = 121.14), and dissolve it in approximately 750 mL of ddH_2O. Carefully add 57.1 mL of glacial acetic acid and 100 mL of 0.5 M EDTA, and adjust the solution to a final

volume of 1 L. This stock solution can be stored at room temperature. The pH of this buffer is not adjusted and should be about 8.5.

8. Working solution (1×) TAE buffer: Dilute the stock solution by 50-fold with ddH_2O. Final solute concentrations are 40 mM Tris–acetate and 1 mM EDTA.
9. 1 % agarose gel in 1× TAE buffer: Add 1 g of agarose into 100 mL of 1× TAE buffer, and microwave until agarose is completely melted. Cool the solution to approximately 70–80 °C. Add 5 μL of ethidium bromide into the solution, and mix well. Pour 25–30 mL of solution onto an agarose gel rack with a 2-well comb.
10. DNA gel purification kit.
11. *Dpn*I restriction enzyme.
12. 10× CutSmart buffer.

2.3 Library Creation by Random Mutagenesis

1. 10× mutagenic buffer: Add 0.0569 g of $MgCl_2$, 1.491 g of KCl, 0.0485 g of Tris–HCl and 0.040 g of gelatin into 40 mL of ddH_2O. Adjust pH to 8.3 with 8 M HCl. Store at −20 °C (*see* **Note 1**).
2. 5 mM $MnCl_2$: First make 250 mM $MnCl_2$ by dissolving 0.0495 g of $MnCl_2$ in 1 mL of ddH_2O. Then dilute 30 μL of 250 mM $MnCl_2$ solution into 1470 μl of ddH_2O to make 5 mM $MnCl_2$.
3. 100 mM dCTP, 100 mM dTTP, 100 mM dATP, 100 mM dGTP.
4. 10× EPdNTP: Add 50 μL of dCTP, 50 μl of dTTP, 10 μL of dATP, and 10 μL of dGTP into 380 μL of ddH_2O.

2.4 Yeast Transformation

1. *S. cerevisiae* YRG2 (MATα *ura3-52 his3-200 ade2-101 lys2-801 trp1-901 leu2-3 112 gal4-542 gal80-538* LYS2::UASGAL1-TATAGAL1-HIS3 URA3::UASGAL4 17 mers(×3)-TATA CYC1-lacZ).
2. YPAD medium: Dissolve 6 g of yeast extract, 12 g of peptone, 12 g of dextrose, and 60 mg of adenine hemisulfate in 600 mL of ddH_2O. Autoclave at 121 °C for 15 min.
3. Synthetic complete dropout medium lacking tryptophan (SC-Trp): Dissolve 3 g of ammonium sulfate, 1 g of yeast nitrogen source without ammonium sulfate and amino acids, 1.14 g of synthetic complete (SC) dropout media minus tryptophan, 26 mg of adenine hemisulfate, and 12 g of dextrose in 600 mL of ddH_2O, and adjust the pH to 5.6 by NaOH. Autoclave at 121 °C for 15 min.
4. SC-Trp-agar: SC-Trp medium and 20 g/L of agar.

5. 1 M lithium acetate: Dissolve 5.1 g of lithium acetate dihydrate in 50 mL of ddH_2O and filter-sterilize the solution.
6. Lithium acetate (0.1 M LiAc): Add 5 mL of 1 M LiAc into 45 mL of ddH_2O.
7. 50 % w/v PEG MW 3350: Add 25 g of PEG 3350 to about 15 mL of ddH_2O in a 100 mL beaker. Stir until it dissolves. Make up the volume to 50 mL and mix thoroughly. Filter-sterilize the solution (*see* **Note 2**).
8. 2 mg/mL single-stranded carrier DNA: Dissolve 200 mg of salmon sperm DNA in 100 mL of sterile TE (10 mM Tris–HCl, 1 mM Na_2EDTA, pH 8.0) using a magnetic stir plate at 4 °C. Aliquot 1 mL into 1.5 mL Eppendorf tubes and store at −20 °C. Denature the carrier DNA in a boiling water bath for 5 min, and chill immediately in an ice/water bath before use.
9. pGAD424-SRC1 plasmid [21].

2.5 Library Cloning and Transformation

1. pBD-Gal4-Cam plasmid.
2. *Eco*RI and *Sal*I restriction enzymes.
3. *S. cerevisiae* YRG2 strain carrying pGAD424-SRC1 plasmid (Subheading 3.4).
4. Synthetic complete dropout medium lacking leucine and tryptophan (SC-Leu-Trp): Dissolve 3 g of ammonium sulfate, 1 g of yeast nitrogen source without ammonium sulfate and amino acids, 1.06 g of synthetic complete (SC) dropout media minus leucine and tryptophan, 26 mg of adenine hemisulfate, and 12 g of dextrose in 600 mL of ddH_2O, and adjust the pH to 5.6 by NaOH. Autoclave at 121 °C for 15 min.
5. SC-Leu-Trp agar plate: SC-Leu-Trp medium and 20 g/L of agar.
6. 4,4′-Dihydroxybenzil (DHB): Synthesized as described in Ref. 22.
7. Synthetic complete dropout medium lacking histidine, leucine, and tryptophan (SC-His-Leu-Trp): Dissolve 3 g of ammonium sulfate, 1 g of yeast nitrogen source without ammonium sulfate and amino acids, 0.996 g of synthetic complete (SC) dropout media minus histidine, leucine, and tryptophan, 26 mg of adenine hemisulfate, and 12 g of dextrose in 600 mL of ddH_2O, and adjust the pH to 5.6 by NaOH. Autoclave at 121 °C for 15 min.
8. SC-His-Leu-Trp agar plate containing appropriately concentrated target ligand (DHB).

2.6 Y2H System-Based Screening

1. Round-bottom 96-well plates.
2. Sterile flat-bottom 96-well microtiter plates.

3. 17β-Estradiol (E_2).
4. 1000× 17β-estradiol (E_2): Dissolve appropriate amount of E_2 in ethanol to make 1000× stock solution.

2.7 Ligand Dose-Response Assay (Yeast Transactivation Profiles)

1. 20× ligand stock solution.
2. SpectraMax 340PC plate reader.

2.8 Subcloning of Evolved hERα LBDs

1. pCMV5-ERα plasmid [23].
2. *Hind*III, *Kpn*I, and *BamH*I restriction enzymes.
3. 10× T4 ligation buffer and T4 DNA ligase.
4. Chemically competent *E. coli* DH5α.
5. LB medium: Add 20 g of LB broth into 1 L of ddH_2O. Autoclave at 121 °C for 15 min.
6. Ampicillin stock solution: Dissolve 1 g of ampicillin powder in 10 mL of ddH_2O and filter-sterilize the solution.
7. LB-Amp^+ medium: LB medium plus 100 μg/mL ampicillin.
8. LB-Amp^+ agar plates: LB-Amp^+ medium and 20 g/L agar.
9. QIAprep Miniprep Kit.
10. 14 mL round-bottom tube.

2.9 Mammalian Transfection and Luciferase Assays (Mammalian Cell Transactivation Profiles)

1. HBSS: Hanks' balanced salt solution.
2. Pre-/post-transfection medium: Add 5 % (v/v) charcoal dextran-treated calf serum into phenol red-free improved minimum essential medium (MEM).
3. Transfection media: Serum-free improved MEM medium.
4. 24-well plates.
5. Transfection solution A (each well): Mix 5 μL of lipofectin, 16 μL of transferrin, and 54 μL of HBSS.
6. Transfection solution B (each well): 0.5 μg of pCMV β-gal, 1 μg of 2ERE-pS2-Luc, 100 ng of ER expression vector, add HBSS up to 150 μL.
7. 1000× ligand stock solution.
8. PBS.
9. 5× reporter lysis buffer.
10. Opaque 96-well plate.
11. 1 M K_2HPO_4 stock solution: Dissolve 87.09 g of K_2HPO_4 into 0.5 L of ddH_2O.
12. 1 M KH_2PO_4 stock solution: Dissolve 68.045 g of KH_2PO_4 into 0.5 L of ddH_2O.

13. 0.1 M potassium phosphate buffer, pH = 7.0: Mix 61.5 mL of 1 M K_2HPO_4 stock solution with 38.5 mL of 1 M KH_2PO_4 stock solution, and add ddH_2O up to 1 L.
14. 4 mg/mL ONPG (*o*-nitrophenyl-β-d-galactopyranoside) substrate. Dissolve 80 mg of ONPG into 20 mL of 0.1 M potassium phosphate buffer, pH = 7.0.
15. 1 M sodium carbonate: Dissolve 105.99 g of sodium carbonate in 1 L of ddH_2O.
16. Luciferase Assay Reagent.
17. LJL Biosciences Analyst HT plate reader.
18. CO_2-containing incubator.

3 Methods

3.1 Molecular Modeling

1. Load the hERα-diethylstilbestrol (DES) structure (PDB ID code: 3ERD) into Molecular Operating Environment (MOE) (Chemical Computing Group, Montreal).
2. Apply the force field MMFF94s [24], then add hydrogen atoms, and assign partial charges to all atoms. The structure is subsequently energy-minimized by using a sequential combination of steepest descent, conjugate gradient, and truncated Newton algorithms.
3. Draw a docking box with a grid consisting of 47 × 30 × 27 points around the DES ligand to specify the boundaries for the movement of the ligand to be docked. In this orientation, the box includes the entire DES ligand and a few atoms of the interacting residues. Delete the DES ligand subsequently from the structure, and dock the 4,4′-dihydroxybenzil (DHB) ligand (which has been assigned partial charges and minimized by using the MMFF94s force field) into the docking box by using a simulated annealing algorithm [25] with the following parameters: initial temperature 12,000 K, 25 runs involving six cycles per run, and 20,000 iterations per cycle.
4. Compare the five structures with the best docking score (lowest overall energy) from these docking runs, and ensure them within a root-mean-square deviation (rmsd) of 0.5 Å from each other. The lowest energy of these five is then subjected to energy minimization as described earlier, to determine the most favorable conformation and orientation of DHB in the ligand-binding pocket. Residues within 4.6 Å of the docked DHB are considered to be in contact with the ligand for purposes of receptor engineering.
5. To gauge the individual role played by the A350M and M388Q mutations, make the appropriate amino acid substitutions to

the docked DHB-hERα structure, and energy-minimize the resulting structure.

6. Superimpose the energy-minimized E_2-hERα crystal structure (PDB ID code: 1GWR) on the docked and energy-minimized DHB-hERα structure by using the align function in MOE.

3.2 Library Creation by Single-Site Saturation Mutagenesis

1. PCR-amplify the 5′ portion and 3′ portion of the hERα ligand-binding domain (LBD) gene containing the NNS substitution at the codon of interest. Four primers are involved in the library creation. The two primers flanking the hERα-LBD are CamL-ERα, 5′-CGACATCATCATCGGAAGAG-3′, and CamR-ERα, 5′-GCTTGGCTGCAGTAATACGA-3′. Two exactly complementary degenerate primers incorporate the residue to be mutated (one primer for generating the sense strand and the other for generating the antisense strand). The two degenerate primers incorporating the randomized amino acids substitute the codon corresponding to the target residue with the sequence NNS (*see* **Note 3**) and contain 9–10 additional bases on either side (5′ and 3′). Standard PCR reaction: 5 μL of 10× PCR reaction buffer, 3 μL of 25 mM $MgCl_2$, 1 μL of 10 mM dNTP mix, 25 pmol forward primer, 25 pmol reverse primer, 5 ng of template plasmid, 0.6 U of *Taq* DNA polymerase, and 0.6 U of *PfuTurbo* DNA polymerase. Adjust the volume to 50 μL with ddH_2O.
2. PCR condition: Fully denature at 94 °C for 30 s, followed by 25 cycles of 94 °C for 30 s, 55 °C for 30 s, and 72 °C for 1 min, with a final extension at 72 °C for 10 min.
3. Load the 50 μL of PCR products onto 1 % agarose gels and perform electrophoresis at 120 V for 20 min.
4. Gel-purify PCR products.
5. *Dpn*I digestion to remove any residual methylated template: 5 μL of 10× CutSmart buffer, 1 μg of PCR product, 1 μl of *Dpn*I, add ddH_2O to 20 μL, incubate at 37 °C overnight.
6. Gel-purify digestion products.
7. Use overlap extension PCR [26] to combine the 5′ portion and 3′ portion of the hERα LBD gene containing the NNS substitution at the codon of interest to generate the full-length gene. PCR reaction: 5 μL of 10× PCR reaction buffer, 3 μL of 25 mM $MgCl_2$, 1 μL of 10 mM dNTP mix, 100 ng of 5′ portion PCR product, 100 ng of 3′ portion PCR product, 0.6 U of *Taq* DNA polymerase, and 0.6 U of *PfuTurbo* DNA polymerase. Adjust the volume to 50 μL with ddH_2O.
8. Overlap extension PCR condition: Fully denature at 94 °C for 30 s, followed by ten cycles of 94 °C for 1 min, 55 °C for 1 min, and 72 °C for 3 min, with a final extension at 72 °C for 10 min.

9. PCR-amplify the mutagenized full-length hERα LBD. Primers used: CamL-ERα and CamR-ERα. Standard PCR reaction: 5 μL of 10× PCR reaction buffer, 3 μL of 25 mM $MgCl_2$, 1 μL of 10 mM dNTP mix, 25 pmol forward primer, 25 pmol reverse primer, 5 μL of overlap extension PCR products from **step 8**, 0.6 U of *Taq* DNA polymerase, and 0.6 U of *PfuTurbo* DNA polymerase. Adjust the volume to 50 μL with ddH_2O.
10. PCR condition: Fully denature at 94 °C for 30 s, followed by 25 cycles of 94 °C for 30 s, 55 °C for 30 s, and 72 °C for 1 min, with a final extension at 72 °C for 10 min.
11. Gel-purify PCR products.

3.3 Library Creation by Random Mutagenesis

1. Generate randomly mutagenized hERα LBD genes using error-prone PCR. Primers used: CamL-ERα and CamR-ERα. PCR reaction: 10 μL of 10× mutagenic buffer, 3 μL of $MnCl_2$ (*see* **Note 4**), 20 ng of template plasmid, 5 U of *Taq DNA* polymerase, 10 μL of 10× EPdNTP, 25 pmol forward primer, and 25 pmol reverse primer. Adjust the volume to 100 μL with ddH_2O.
2. PCR condition: Fully denature at 94 °C for 30 s, followed by 15 cycles of 94 °C for 30 s, 50 °C for 30 s, and 72 °C for 1 min, with a final extension at 72 °C for 10 min.
3. Gel-purify PCR products.

3.4 Yeast Transformation

1. Inoculate a single colony of *S. cerevisiae* YRG2 strain into 3 mL of YPAD medium, and grow overnight in a shaker at 30 °C and 250 rpm (*see* **Note 5**).
2. Measure the OD_{600} of the seed culture and inoculate the appropriate amount to 50 mL of fresh YPAD medium to obtain an OD_{600} of 0.2.
3. Continue growing the 50 mL of culture for approximately 4 h to obtain an OD_{600} of 0.8.
4. Spin down the yeast cells at room temperature, 4000 ×*g* for 5 min and remove the spent medium.
5. Use 50 mL of ddH_2O to wash the cells once and centrifuge again.
6. Discard water, add 1 mL of 0.1 M LiAc to suspend the cells, and move them to a sterile Eppendorf tube.
7. Spin down the cells using a benchtop centrifuge for 30 s at 4500 ×*g*.
8. Remove liquid and resuspend cells in 500 μL of 0.1 M LiAc (for ten individual transformations).
9. Transfer 50 μL resuspended cells to each sterile Eppendorf tube for each transformation.

10. Spin down the cells using a benchtop centrifuge for 30 s at 4500 × *g* and remove the liquid.
11. Add the following components in the following order: (1) 240 μL of 50 % w/v PEG, (2) 36 μL of 1 M LiAc, (3) 50 μL of 2.0 mg/mL SS-DNA, (4) X μL DNA, and (5) 34-X μL sterile ddH_2O.
12. Vortex each tube vigorously until the cell pellet has been completely mixed.
13. Heat shock in a water bath at 42 °C for 40 min.
14. Spin down the cells using a benchtop centrifuge for 30 s at 4500 × *g*. Remove liquid by pipetting.
15. Pipette 1000 μL of sterile water into each tube and resuspend the pellet by pipetting it up and down gently.
16. Plate 200 μl onto Sc-Trp agar plate.
17. Incubate the plates at 30 °C for 2–3 days until colonies appear.

3.5 Library Cloning and Transformation

1. Linearize pBD-Gal4-Cam by *Eco*RI and *Sal*I digestion. Digestion condition: 1 μg of pBD-Gal4-Cam, 2 μL of 10× buffer, 1 μL of *Eco*RI, 1 μL of *Sal*I, add ddH_2O to final volume of 20 μL. Digest at 37 °C for 3 h.
2. Gel-purify the linearized plasmid.
3. Use *S. cerevisiae* YRG2 strain carrying pGAD424-SRC1 plasmid as a parental strain for library cloning. For individual site saturation mutagenesis library, 20 ng of linearized plasmid is cotransformed with 20 ng of previously obtained mutagenized hERα LBD PCR product (Subheading 3.2) by using the previously described transformation method (Subheading 3.4).
4. Plate all saturation mutagenesis library transformants onto a SC-Leu-Trp agar plate (*see* **Note 6**).
5. For error-prone PCR libraries, cotransform 150 ng of linearized plasmid with 150 ng of previously obtained error-prone PCR product by using the previously described transformation method (Subheading 3.4) (*see* **Note** 7).
6. Plate error-prone PCR library transformants onto a SC-His-Leu-Trp agar plate containing an appropriate concentration of the target ligand (DHB) for screening.

3.6 Y2H System-Based Screening

1. Add 50 μL of SC-Leu-Trp minimal liquid media in each well of round-bottom 96-well plates.
2. Pick transformants from the individual site saturation mutagenesis library plates or the error-prone PCR library plates with sterile toothpicks to individual wells of 96-well plates, and incubate them overnight (16–20 h) at 30 °C.

3. As a control, inoculate one well in every microtiter plate with a yeast colony expressing the parental hERα LBD construct.
4. After this overnight incubation, add 250 μL of sterile ddH_2O to every well. Mix well by pipetting.
5. Add 200 μL of SC-His-Leu-Trp media with an appropriate concentration of either target ligand (DHB) or E_2 to two identical sterile flat-bottom 96-well microtiter plates.
6. Transfer 5 μL of each diluted culture to the corresponding wells of two identical 96-well plates.
7. Incubate these ligand-containing microtiter plates at 30 °C for 24 h.
8. Identify mutants with strengthened response toward the target ligand (higher cell density than parental mutant control) and weakened response toward E_2 (lower cell density than parent) using visual check.
9. Streak mutants that appear to be more selective for the target ligand relative to E_2 onto SC-Leu-Trp agar plates, and incubate at 30 °C for 2 days.
10. Pick single colonies from these streaked plates and subject them to a yeast ligand dose-response assay.

3.7 Ligand Dose-Response Assay (Yeast Transactivation Profiles)

1. Pick single colonies from the abovementioned streaked plates into individual wells of round-bottom 96-well plates containing 50 μL of SC-Leu-Trp liquid medium, and grow at 30 °C for overnight.
2. After this overnight incubation, add 250 μL of sterile ddH_2O to every well. Mix well by pipetting.
3. Transfer 5 μL of each diluted culture to sterile flat-bottom 96-well microtiter plates containing 200 μL of SC-His-Leu-Trp medium to obtain a diluted culture with final OD_{600} of ~0.002.
4. Transfer 190 μl of each diluted culture to the corresponding wells of flat-bottom 96-well plates.
5. Add 10 μl of 20× ligand stock solution.
6. Incubate these ligand-containing microtiter plates at 30 °C for 24 h.
7. Mix culture in each well by pipetting and determine their OD_{600} values by using a SpectraMax 340PC plate reader.

3.8 Subcloning of Evolved hERα LBDs

1. PCR-amplify the evolved LBD genes using primers ERα-pCMV5-5KpnI (5′CCGGTACCCCATGACCATGAC-3′) and ERα-BamHI-C (5′AGCTCTGGATCCTCAGACTGTGGCAGGGAAAC-3′) (*see* Subheading 3.2).
2. Gel-purify PCR products.

3. Linearize the pCMV5-ERα plasmid with *Hind*III using conditions described in Subheading 3.5.
4. Gel-purify 1 kb fragment of the linearized plasmid.
5. The two purified DNA fragments share an overlap region of approximately 100 bp. Perform overlap extension PCR to generate full-length mutant hERα genes. PCR reaction: 5 μL of 10× PCR reaction buffer, 3 μL of 25 mM $MgCl_2$, 1 μL of 10 mM dNTP mix, 100 ng of PCR product, 100 ng of linearized plasmid, 0.6 U of *Taq* DNA polymerase, and 0.6 U of *PfuTurbo* DNA polymerase. Adjust the volume to 50 μL with ddH_2O.
6. Overlap extension PCR condition: Fully denature at 94 °C for 30 s, followed by ten cycles of 94 °C for 1 min, and 72 °C for 4 min, with a final extension at 72 °C for 10 min.
7. PCR-amplify the full-length mutant hERα genes. Primers used: ERα-pCMV5-5KpnI and ERα-BamHI-C. Standard PCR reaction: 5 μL of 10× PCR reaction buffer, 3 μL of 25 mM $MgCl_2$, 1 μL of 10 mM dNTP mix, 25 pmol forward primer, 25 pmol reverse primer, 5 μL of overlap extension PCR products, 0.6 U of *Taq* DNA polymerase, and 0.6 U of *PfuTurbo* DNA polymerase. Adjust the volume to 50 μL with ddH_2O.
8. Overlap extension PCR condition: Fully denature at 94 °C for 30 s, followed by 25 cycles of 94 °C for 1 min, and 72 °C for 4 min, with a final extension at 72 °C for 10 min.
9. Gel-purify PCR product.
10. Digest PCR product with *Kpn*I-*Bam*HI using condition described in Subheading 3.5.
11. Digest pCMV5-ERα plasmid with *Kpn*I-*Bam*HI using condition described in Subheading 3.5.
12. Ligation reaction: 100 ng of *Kpn*I-*Bam*HI digested pCMV5-ERα, 180 ng of *Kpn*I-*Bam*HI digested PCR product, 1 μL of 10× T4 ligation buffer, 0.25 μL of T4 ligase. Incubate at 16 °C overnight.
13. Transform *E. coli* DH5α using a heat shock method: Add 4 μL of ligation reaction mixture to 50 μL of *E. coli* DH5α chemically competent cells in a sterile Eppendorf tube.
14. Heat shock *E. coli* DH5α at 42 °C for 30 s and add 1 mL of LB medium.
15. Transfer the suspended cells to a 14 mL round-bottom tube and grow at 37 °C for 1 h.
16. Spread 250 μL on a LB-Amp$^+$ plate and incubate at 37 °C overnight.
17. Inoculate single colonies to 4 mL of LB-Amp$^+$ medium and grow with shaking at 37 °C overnight.

18. Purify plasmids from each 4 mL of culture.
19. Confirm the identity of the plasmids by DNA sequencing.

3.9 Mammalian Transfection and Luciferase Assays (Mammalian Cell Transactivation Profiles)

1. Seed human endometrial cancer (HEC-1) cells in 1 mL of pre-/post-transfection medium in each well of the 24-well plates. Incubate the plate at 37 °C for 24 h.
2. Preheat transfection medium at 37 °C.
3. In each well, add 75 μL of transfection solution A and 75 μL of transfection solution B. Then add 350 μL of transfection medium. Incubate at 37 °C in a 5 % CO_2-containing incubator for 5 h.
4. Preheat pre-/post-transfection medium at 37 °C.
5. Remove the medium in each well via vacuum and wash with 1 mL of pre-/post-transfection medium.
6. Add 1 mL of pre-/post-transfection medium and 1 μL of 1000× ligand stock solution to each well. Incubate the plate at 37 °C in a 5 % CO_2-containing incubator for 24 h.
7. Wash cells with 500 μL of PBS twice.
8. Add 100 μL of reporter lysis buffer and freeze at −70 °C.
9. Thaw cells and transfer cells to a round-bottom 96-well plate. Centrifuge at 4000 × *g* for 5 min at room temperature.
10. For the β-galactosidase assay, transfer 20 μL of supernatant to a flat-bottom 96-well plate.
11. Add 200 μL of ONPG substrate mixture to each well in 96-well plates. Develop color at room temperature until a faint yellow color appears.
12. Stop the reaction by adding 150 μL of 1 M sodium carbonate. Mix by pipetting the contents of each well.
13. Read the absorbance of the samples at 405 nm in a SpectraMax 340PC plate reader.
14. For the luciferase assay, transfer 20 μL of supernatant from **step 9** to each well in an opaque 96-well plate.
15. Add 100 μL of Luciferase Assay Reagent per well.
16. Read samples with a plate reader.

4 Notes

1. Aliquot 10× mutagenic buffer into Eppendorf tubes for storage to avoid multiple cycles of freeze and thaw.
2. Make sure the container of PEG solution is tightly sealed. The transformation efficiency is highly dependent on the PEG concentration.

3. The choice of the substitution NNS allows the incorporation of all 20 amino acids while keeping the total number of codon possibilities low, at 32.
4. Mutation rate can be adjusted via changing $MnCl_2$ concentration between 0.1 and 0.2 mM.
5. *S. cerevisiae* competent cell need to be prepared freshly every time.
6. Plate a series of different amount of the transformation mixture (10, 50, 100 μL) on different plates to determine the transformation efficiency.
7. In order to obtain sufficient transformants for the random mutagenesis library, it might be necessary to perform large-scale transformation. In such case, components of transformation mixture and plasmids (Subheading 3.4, **step 11**) can be premixed and aliquoted (360 μl each) into Eppendorf tubes containing *S. cerevisiae* cells to perform the heat shock step.

References

1. Weatherman RV, Fletterick RJ, Scanlan TS (1999) Nuclear-receptor ligands and ligand-binding domains. Annu Rev Biochem 68:559–581
2. Anand P, Nagarajan D, Mukherjee S, Chandra N (2014) PLIC: protein-ligand interaction clusters. Database (Oxford) 2014:bau029
3. Damborsky J, Brezovsky J (2014) Computational tools for designing and engineering enzymes. Curr Opin Chem Biol 19:8–16
4. Feldmeier K, Höcker B (2013) Computational protein design of ligand binding and catalysis. Curr Opin Chem Biol 17:929–933
5. Brustad EM, Lelyveld VS, Snow CD, Crook N, Jung ST, Martinez FM, Scholl TJ, Jasanoff A, Arnold FH (2012) Structure-guided directed evolution of highly selective p450-based magnetic resonance imaging sensors for dopamine and serotonin. J Mol Biol 422:245–262
6. Shapiro MG, Westmeyer GG, Romero PA, Szablowski JO, Kuster B, Shah A, Otey CR, Langer R, Arnold FH, Jasanoff A (2010) Directed evolution of a magnetic resonance imaging contrast agent for noninvasive imaging of dopamine. Nat Biotechnol 28:264–270
7. de Las HA, Carreno CA, Martinez-Garcia E, de Lorenzo V (2010) Engineering input/output nodes in prokaryotic regulatory circuits. FEMS Microbiol Rev 34:842–865
8. Collins CH, Arnold FH, Leadbetter JR (2005) Directed evolution of Vibrio fischeri LuxR for increased sensitivity to a broad spectrum of acyl-homoserine lactones. Mol Microbiol 55:712–723
9. Scholz O, Kostner M, Reich M, Gastiger S, Hillen W (2003) Teaching TetR to recognize a new inducer. J Mol Biol 329:217–227
10. Tang SY, Fazelinia H, Cirino PC (2008) AraC regulatory protein mutants with altered effector specificity. J Am Chem Soc 130:5267–5271
11. Vee Aune TE, Bakke I, Drablos F, Lale R, Brautaset T, Valla S (2010) Directed evolution of the transcription factor XylS for development of improved expression systems. Microb Biotechnol 3:38–47
12. Amiss TJ, Sherman DB, Nycz CM, Andaluz SA, Pitner JB (2007) Engineering and rapid selection of a low-affinity glucose/galactose-binding protein for a glucose biosensor. Protein Sci 16:2350–2359
13. East AK, Mauchline TH, Poole PS (2008) Biosensors for ligand detection. Adv Appl Microbiol 64:137–166
14. Reetz MT, Bocola M, Carballeira JD, Zha D, Vogel A (2005) Expanding the range of substrate acceptance of enzymes: combinatorial active-site saturation test. Angew Chem Int Ed Engl 44:4192–4196
15. Reetz MT, Carballeira JD, Vogel A (2006) Iterative saturation mutagenesis on the basis of B factors as a strategy for increasing protein thermostability. Angew Chem Int Ed Engl 45:7745–7751

16. Chockalingam K, Chen Z, Katzenellenbogen JA, Zhao H (2005) Directed evolution of specific receptor-ligand pairs for use in the creation of gene switches. Proc Natl Acad Sci U S A 102:5691–5696
17. Tinberg CE, Khare SD, Dou J, Doyle L, Nelson JW, Schena A, Jankowski W, Kalodimos CG, Johnsson K, Stoddard BL, Baker D (2013) Computational design of ligand-binding proteins with high affinity and selectivity. Nature 501:212–216
18. Khare SD, Kipnis Y, Greisen P Jr, Takeuchi R, Ashani Y, Goldsmith M, Song Y, Gallaher JL, Silman I, Leader H, Sussman JL, Stoddard BL, Tawfik DS, Baker D (2012) Computational redesign of a mononuclear zinc metalloenzyme for organophosphate hydrolysis. Nat Chem Biol 8:294–300
19. Giger L, Caner S, Obexer R, Kast P, Baker D, Ban N, Hilvert D (2013) Evolution of a designed retro-aldolase leads to complete active site remodeling. Nat Chem Biol 9:494–498
20. McLachlan MJ, Chockalingam K, Lai KC, Zhao H (2009) Directed evolution of orthogonal ligand specificity in a single scaffold. Angew Chem Int Ed Engl 48:7783–7786
21. Ding XF, Anderson CM, Ma H, Hong H, Uht RM, Kushner PJ, Stallcup MR (1998) Nuclear receptor-binding sites of coactivators glucocorticoid receptor interacting protein 1 (GRIP1) and steroid receptor coactivator 1 (SRC-1): multiple motifs with different binding specificities. Mol Endocrinol 12:302–313
22. Katzenellenbogen JA, Johnson HJ, Carlson KE, Myers HN (1974) Photoreactivity of some light-sensitive estrogen derivatives. Use of an exchange assay to determine their photointeraction with the rat uterine estrogen binding protein. Biochemistry 13: 2986–2994
23. Chen Z, Katzenellenbogen BS, Katzenellenbogen JA, Zhao H (2004) Directed evolution of human estrogen receptor variants with significantly enhanced androgen specificity and affinity. J Biol Chem 279:33855–33864
24. Halgren TA (1999) MMFF VI. MMFF94s option for energy minimization studies. J Comput Chem 20:720–729
25. Hart TN, Read RJ (1992) A multiple-start Monte Carlo docking method. Proteins 13:206–222
26. Ho SN, Hunt HD, Horton RM, Pullen JK, Pease LR (1989) Site-directed mutagenesis by overlap extension using the polymerase chain reaction. Gene 77:51–59

Chapter 9

Improving Binding Affinity and Selectivity of Computationally Designed Ligand-Binding Proteins Using Experiments

Christine E. Tinberg and Sagar D. Khare

Abstract

The ability to de novo design proteins that can bind small molecules has wide implications for synthetic biology and medicine. Combining computational protein design with the high-throughput screening of mutagenic libraries of computationally designed proteins is emerging as a general approach for creating binding proteins with programmable binding modes, affinities, and selectivities. The computational step enables the creation of a binding site in a protein that otherwise does not (measurably) bind the intended ligand, and targeted mutagenic screening allows for validation and refinement of the computational model as well as provides orders-of-magnitude increases in the binding affinity. Deep sequencing of mutagenic libraries can provide insights into the mutagenic binding landscape and enable further affinity improvements. Moreover, in such a combined computational–experimental approach where the binding mode is preprogrammed and iteratively refined, selectivity can be achieved (and modulated) by the placement of specified amino acid side chain groups around the ligand in defined orientations. Here, we describe the experimental aspects of a combined computational–experimental approach for designing—using the software suite Rosetta—proteins that bind a small molecule of choice and engineering, using fluorescence-activated cell sorting and high-throughput yeast surface display, high affinity and ligand selectivity. We illustrated the utility of this approach by performing the design of a selective digoxigenin (DIG)-binding protein that, after affinity maturation, binds DIG with picomolar affinity and high selectivity over structurally related steroids.

Key words Computational design, Rosetta macromolecular modeling, Affinity optimization, Binding selectivity, Steroid binding, Protein-small molecule interactions

1 Introduction

Computational de novo design of protein function has seen remarkable success in recent years, enabling, for example, the construction of enzymes for catalyzing reactions that are not natively catalyzed by natural enzymes [1, 2], protein binders against pathogenic proteins [3], and, more recently, the design of small-molecule binding proteins with high affinity and programmable selectivity [4]. In all cases, the initial hits obtained from the computational design

Barry L. Stoddard (ed.), *Computational Design of Ligand Binding Proteins*, Methods in Molecular Biology, vol. 1414, DOI 10.1007/978-1-4939-3569-7_9,

approach were weakly active, and the use of high-throughput experimental characterization to screen and improve designed proteins was critical for success. Many of the limitations of computational design methodology, including force field inaccuracies, lack of explicit modeling of solvent and properties such as protein solubility, and, more generally, our limited understanding of protein sequence–function relationships [5], were, at least in part, overcome by screening tens of computationally designed proteins using sensitive experimental assays, identifying weakly active hits and subsequently improving their efficacies using mutagenic screening or selection techniques [6]. Conversely, the directed evolution methods used to improve activities in these efforts could be made more efficient, compared to random mutagenesis approaches, by virtue of being guided by an atomic-resolution (but partially accurate) computational model of the bound state. This iterative, combined computational–experimental strategy builds upon the strengths of these complementary methods and will continue to be a key component of various protein design applications [7].

Here, we describe the experimental strategy and protocols used in our efforts to de novo design small-molecule binding sites in proteins—these computationally designed and subsequently laboratory-evolved proteins feature affinities and selectivities that rival those of natural small-molecule binding proteins. On the computational end, we developed and used a computational design approach, in the context of the Rosetta macromolecular modeling suite, to transplant idealized binding sites for a chosen ligand—the steroid digoxigenin (DIG)—into a set of protein scaffolds. The scaffolds were remodeled to accommodate predefined interactions to DIG, and then Rosetta Design [8] was used to optimize the binding site amino acid sequences for ligand-binding affinity. A more complete description of the computational strategy and protocols used to obtain the binders can be obtained elsewhere [4]. As mentioned above, the initial hits were weak affinity binders and could be detected only with a sensitive and relatively high-throughput yeast surface display assay that conveniently allowed testing tens of computationally designed proteins (referred to as designs hereafter) and their mutagenic libraries. We focus here on the experimental assays and methods for subsequent affinity maturation as well as selectivity modulation. Results from these experimental strategies (impact of point mutations on binding) were used to both validate (or invalidate) and refine initial designs, and models of mutagenized proteins were then used to guide further optimization, for instance, by the model-guided enumeration of ligand-proximal residue positions for which mutagenic libraries were constructed and tested. The experimental data-guided design model of one of our designs was subsequently validated by the observed atomic-resolution agreement with X-ray crystallographic structures of a series of its variants [4]. Below, we describe our approach and offer some practical suggestions for the choices that are made while performing various steps.

2 Materials

Streptavidin–phycoerythrin (SAPE).

Yeast strain EBY100.

pCTCON2 or pETCON vector.

Highly avid ligand–biotin conjugate.

Monovalent ligand–biotin conjugate.

Monovalent ligand–fluorophore conjugate.

3 Methods

3.1 Overview of Approach

The overall goals of the approach are (1) to detect (initially weak) binding of the designed proteins and (2) to improve binding affinity and selectivity of the designed proteins. In the latter case, the choice of residue positions to mutate is based on the spatial proximity of these positions to the ligand in the computational model of the bound state. Typically, first-shell positions are chosen for site–saturation mutagenesis, beneficial mutations are combined (combinatorially), and these experimentally identified amino acid substitutions are used to refine or invalidate initial design model. For the optimized variant, a single-site mutagenic library at both first- and second-shell residue positions is generated, and high-throughput sequencing of screened libraries is used to guide further affinity improvements. The experimental data-guided computational model is then used to design mutations to predictively modulate the selectivity of designed proteins for the small molecule over a series of congeners.

3.2 Initial Screen of Computationally Designed Proteins

1. Designed proteins are tested for ligand binding using yeast surface display [9]. We used the vector pETCON and the *NdeI/XhoI* restriction sites in this vector to clone synthetic genes of the designs. Standard yeast surface display materials and protocols were used for growth and induction unless stated otherwise below.
2. For hydrophobic ligands (such as DIG) and designed proteins that are expected to have low affinities, it is important to guard against false-positives as exposed hydrophobic patches in proteins can nonspecifically bind the ligand with low affinity. To control for nonspecific binding, we used proteins that are both structurally and functionally unrelated to designed proteins as controls. Negative controls for binding were two tandem Z

domains of protein A (ZZ domain) [10, 11] and a mutagenic library of HIV glycoprotein (gp120) variants developed for an unrelated project.

3. The genes for the "negative control" proteins as well as designs cloned into the pETCON vector are transformed into cells of the yeast strain EBY100 using lithium acetate and polyethylene glycol [12]. Transformants are plated on selective media (C – *ura –trp*) that select for both the strain and the vector.
4. Freshly transformed cells are inoculated into 1 mL of SDCAA media [9] and grown at 30 °C, 200 rpm. After ~12 h, 1e7 cells are collected by centrifugation at 1700 × *g* for 3 min and resuspended in 1 mL of SGCAA media to induce protein expression.
5. Following induction for 24–48 h at 18 °C, 4e6 cells are collected by centrifugation and washed twice by incubation with PBSF (PBS supplemented with 1 g/L of BSA) for 10 min at room temperature. Induction times and temperatures required to obtain the highest expression levels of displayed proteins can vary and need to be empirically determined. For our system, 24–48 h at 18 °C was optimal.
6. For proteins expressed from their gene in the pETCON vector, yeast surface protein expression can be monitored by the binding of anti-cmyc-FITC antibody to the C-terminal myc-epitope tag of the displayed protein (Fig. 1a).
7. Small-molecule (in our case, DIG) binding is assessed by quantifying the phycoerythrin (PE) fluorescence of the displaying yeast population following incubation with small-molecule-biotinylated protein conjugates: DIG-BSA-biotin, DIG-RNase-biotin (Fig. 1b, c), or DIG-PEG$_3$-biotin (Fig. 1d) in our case, and streptavidin–phycoerythrin (SAPE). *See* **Note 1.**
8. Following a 2–4-h incubation at 4 °C in the dark on a rotator, cells are collected by centrifugation at 1700 × *g* for 3 min and washed with 200 μL of PBSF at 4 °C.
9. Cell pellets are resuspended in 200 μL of ice-cold PBSF immediately before use. For detecting weak affinity binders, it is important to keep the samples on ice until resuspension and resuspend immediately before use.
10. Cellular fluorescence is monitored on an Accuri C6 flow cytometer using a 488 nm laser for excitation and a 575 nm band pass filter for emission. Phycoerythrin fluorescence is compensated to minimize bleed-over contributions from the FITC fluorescence channel.
11. While negative controls are important (*see* Subheading 3.2, **step 3**), positive controls of varying affinities, if available, should be used to validate, and tune the sensitivity of, the assay.

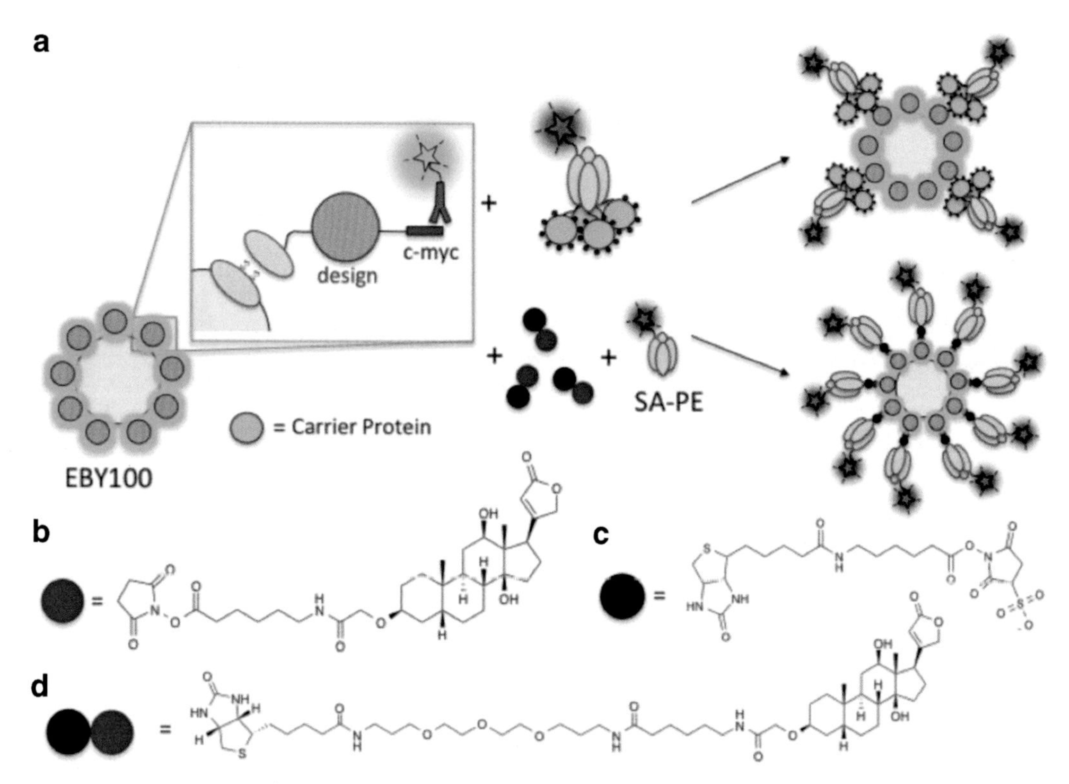

Fig. 1 Outline of assay used for detection and evolution of binding affinity of designed proteins. (**a**) Designs are expressed on the surface of yeast using the plasmid pETCON as described by Wittrup and co-workers. A c-myc tag is attached at the C-terminus of the protein to enable detection using an anti-c-myc antibody that is conjugated to a fluorophore (e.g., FITC, *green*). Binding can detected in a high-avidity format to identify initial hits (*top*) or low-avidity format to enable more sensitive detection of affinity increase during affinity maturation (*bottom*). (**b**, **c**) NHS esters of DIG and biotin that are used for conjugation to a carrier protein (e.g., BSA or RNase) in the high-avidity format. (**d**) The DIG-biotin conjugate that was used in the low-avidity format

In our case, two positive controls having different affinities for digoxigenin were used in the binding assay: a previously engineered steroid binding protein DigA16 [13] and a commercially available anti-DIG monoclonal antibody 9H27L19 (Fig. 2). Experiments using DigA16 were conducted in an identical fashion to design DIG1-17. For those employing the anti-DIG antibody, an F_c-region-binding protein, the ZZ domain (*see* Subheading 3.2, **step 3**), was displayed on the yeast cell surface, and washed cells were resuspended in 20 μL of PBSF with 2 μL of rabbit anti-DIG mAB 9H27L19. Following a 30-min incubation at 4 °C on a rotator, excess antibody was removed by washing the cells with 200 μL of PBSF. Labeling reactions were then performed as above.

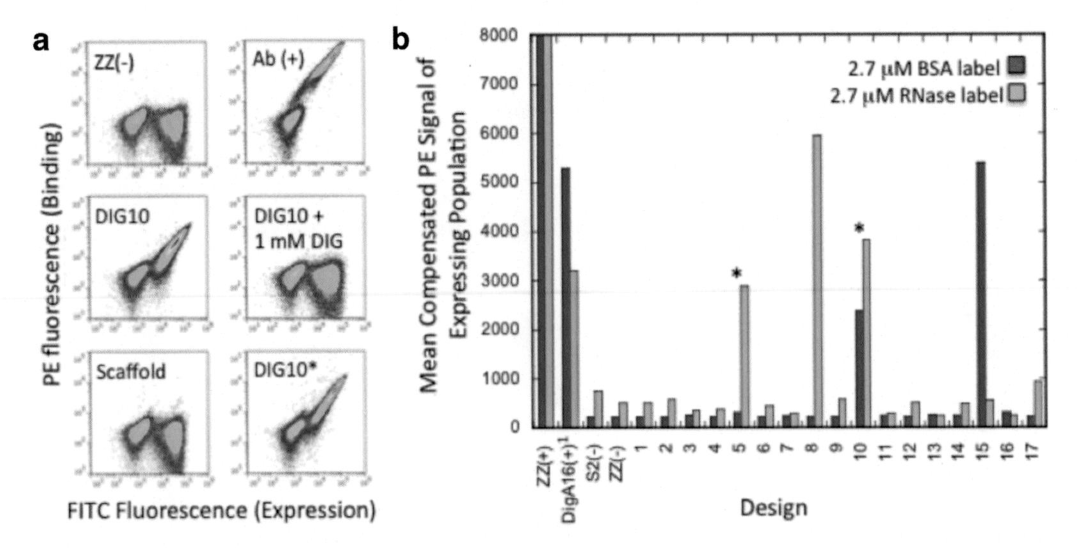

Fig. 2 Typical assay results for hits obtained in a set of computationally designed proteins. (**a**) Example results and validation experiments carried out for a hit identified from the binding assay showing no binding signal for negative control (ZZ (−)), high binding signal for positive control (Ab (+)), binding signal for the design (DIG10), no binding signal for design incubated with excess unlabeled DIG (DIG10 + 1 mM DIG), no binding signal for the wild-type scaffold protein on which the design DIG10 is based (scaffold), and similar binding signal (as DIG10) when an alternative carrier protein, RNase, is used (DIG10*). (**b**) Binding signals for controls and all 17 tested designs. Designs DIG10 showed reproducible binding signals with both carrier proteins, DIG5 and DIG8 showed high signals with RNase carrier protein but not BSA, and DIG15 showed high signals with BSA but not RNase. Tests described in (**a**) identified DIG10 and DIG5 as being specific binders to DIG. These were used for further affinity maturation

12. To test if the hits identified above are not false-positives on account of binding to other assay components (such as SAPE), it is important to perform competition experiments with the free ligand (Fig. 2a). *See* **Note 2.**
13. To further ensure specific binding to the small molecule, knockout mutagenesis of key interacting residues is performed. Residues that interact with the ligand in the computational model are mutated to amino acids that disfavor binding. This step serves to confirm that the ligand and not other assay components are binding the design as well as confirm the design model.

3.3 Affinity Improvement Using Yeast Surface Display Selections and Fluorescence-Activated Cell Sorting of Mutagenic Libraries

1. Based on the identified hits in Subheading 3.2, affinity maturation is performed using single site–saturation mutagenesis (SSM) library constructed by Kunkel mutagenesis [14] using degenerate NNK primers (Fig. 3).
2. Positions for mutagenesis are chosen based on the computational design model. Positions are chosen from the model based on the following requirements: (1) they have Cα within

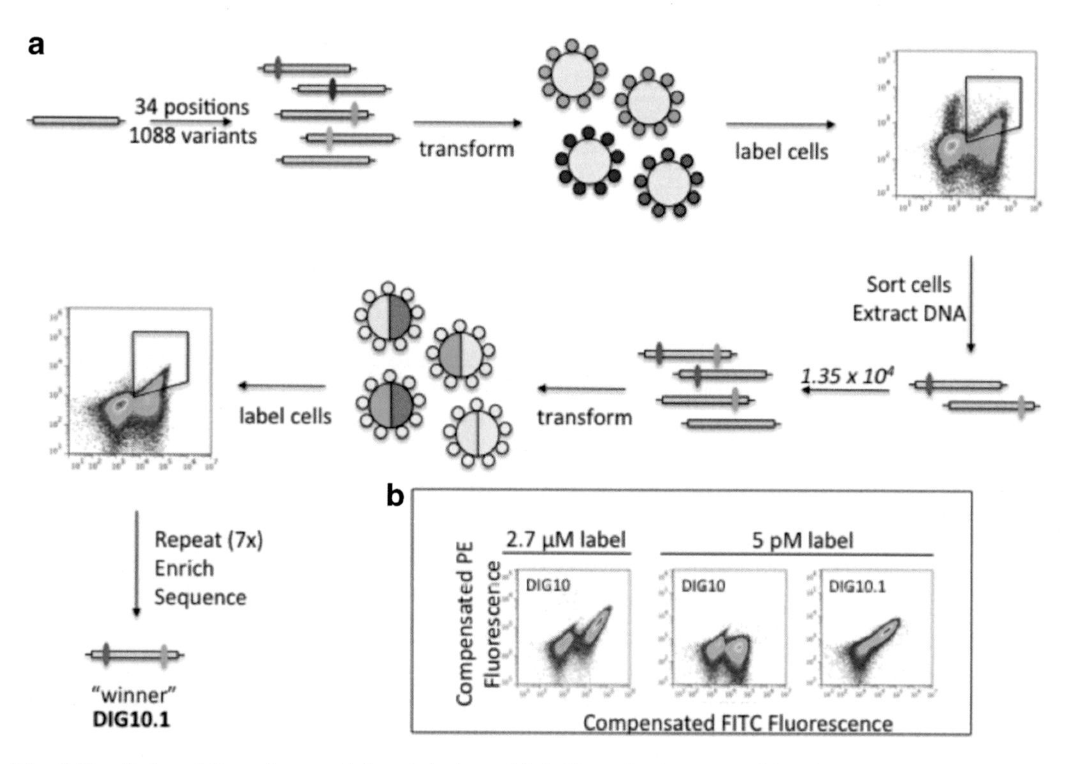

Fig. 3 Directed evolution of computational designs. (**a**) Outline of scheme used for site-directed mutagenesis of designs for affinity improvement. Several rounds of single site–saturation mutagenesis followed by combinatorial mutagenesis using identified beneficial single mutations are performed to obtain affinity improvements. (**b**) Comparison of the binding properties of the initial hit (DIG10) with the affinity matured variant (DIG10.1). High binding signals are detectable at ~6 orders-of-magnitude lower labeled ligand concentrations after affinity maturation

7 Å of any ligand heavy atom, and/or (2) they have Cα within 9 Å of any ligand heavy atom and Cβ closer to any heavy atom in the ligand than Cα. The theoretical library size can be calculated (in our case, we chose 34 positions for design DIG10 yielding a size of 1088 clones).

3. Kunkel mutagenesis of each position using mutagenic oligonucleotides is carried out independently. DNA from each reaction is dialyzed into dH_2O using a 0.025 μm membrane filter, and then the dialyzed reaction mixtures are pooled, concentrated to a volume of <10 μL using a Savant SpeedVac centrifugal vacuum concentrator, and transformed into yeast strain EBY100 using the method of Benatuil [15]. Typical yields are 1E7–1E8. *See* **Note 3.**

4. After transformation, cells are grown in 250 mL of SDCAA media for 36 h at 30 °C. Cells (5e8) are collected by centrifugation at 1700 × *g* for 4 min, resuspended in 50 mL of SGCAA media, and induced at 18 °C for 24 h.

5. Cells are subjected to multiple (we used three) rounds of permissive cell sorting to enrich for improved variants. During each round of sorting, cells are washed and then labeled with a preincubated mixture of 2.66 μM DIG-BSA-biotin, 644 nM SAPE, and anti-cmyc-FITC as noted above for single clones. During each round, the top ~10 % of cells in the PE channel are collected. It is important to sort 10–100 times the library transformation efficiency to ensure that each clone in the library is sampled during the sort. *See* **Note 4.**
6. After each round of sorting, cells are grown in SDCAA for 24 h and then induced in SGCAA for 24 h before the next sort. It is important to recover the cells in this way so that low representation clones are allowed to amplify.
7. After the final sort, an increase in the mean compensated PE fluorescence of the expressing population of the sorted cells compared to that of the original design indicates the presence of a point mutant(s) with increased binding affinity.
8. After each sort, a portion of cells are plated and grown at 30 °C. Plasmids from individual colonies are harvested and the gene is amplified by PCR. Sanger sequencing is used to sequence at least ten colonies from each population to identify mutations that increase affinity.

3.4 Combinatorial Mutagenesis Using Identified Beneficial Single-Point Mutations

1. Beneficial mutations identified in the SSM library (Subheading 3.3) are combined by Kunkel mutagenesis [14] using degenerate primers. At each mutagenized position, the original DIG10 amino acid and chemically similar amino acids to those identified in the first round of directed evolution are also allowed, resulting in a combinatorial library.
2. Four independent Kunkel reactions using different mutagenic oligonucleotide concentrations ranging from 36 to 291 nM during polymerization are performed to minimize sequence-dependent priming bias. For the same reason, oligonucleotides encoding native substitutions contain at least one codon base change.
3. Library DNA is pooled, prepared as above, and transformed into electrocompetent *E. coli* strain BL21(DE3) cells (1800 V, 200 Ω, 25 μF). Library plasmid DNA is isolated from expanded cultures. Gene insert is amplified from 10 ng of library DNA by 30 cycles of PCR (98 °C 10 s, 61 °C 30 s, 72 °C 15 s) using Phusion high-fidelity polymerase with the pCTCON2r and pCTCON2f primers. *See* **Note 5.**
4. Yeast EBY100 cells are transformed with 4.0 μg of PCR-purified DNA insert generated in the previous step and 1.0 μg of gel-purified pETCON digested with *NdeI* and *XhoI* using the method of Benatuil [15], yielding 1E7–1E8 transformants.

After transformation, cells are grown in 150 mL of low-pH SDCAA media supplemented with Pen/Strep for 48 h at 30 °C. Cells (~5e8) are collected by centrifugation at 1700 × *g* for 4 min, resuspended in 50 mL of SGCAA media, and induced at 18 °C for 24 h.

5. Cells are subjected to several rounds of cell sorting (we performed seven rounds). For the first four rounds, cells are washed and then labeled with a preincubated mixture of small-molecule BSA-biotin, SAPE, and anti-cmyc-FITC as noted above for single clones. Small-molecule-label concentrations can be decreased progressively in every round to increase the selection stringency. It is important to maintain a 4:1 (biotin/SAPE) ratio. For example, our concentrations for rounds one through four were (1) 1 μM DIG-BSA-biotin and 250 nM SAPE, (2) 750 nM DIG-BSA-biotin and 187.5 nM SAPE, (3) 50 nM DIG-BSA-biotin and 12.5 nM SAPE, and (4) 5 nM DIG-BSA-biotin and 1.25 nM SAPE. Selection stringency is increased in each round by dropping the label concentration or decreasing the avidity of the label. Note that these concentrations in this example refer to the concentration of carrier protein molecules, not DIG molecules.
6. To ensure that the identified mutations do not select for binding to the carrier protein (e.g., BSA in our case) or a specific linkage between small molecule and carrier protein or other assay components (e.g., SAPE), it is important to use an unrelated protein for labeling with small molecule (Fig. 3b). For rounds five through seven, we used DIG-RNase-biotin in a multistep labeling procedure to minimize selection for carrier protein (BSA) binding. The use of RNase also allowed a larger dynamic range in several control experiments. DIG-RNase-biotin label concentrations were 10, 5, and 5 pM (concentrations referenced to RNase) for rounds five through seven, respectively.
7. At least ten clones from each round are sequenced as noted for the SSM library. After several rounds, the library typically converges to a small number of sequences differing by a single or a few point substitutions.

3.5 Mutagenic Libraries and Deep Sequencing

1. Paired-end 151 Illumina sequencing is used to simultaneously assess the effects of mutation on binding.
2. A number of mutagenic libraries are designed, based on the distribution of mutagenized positions in and length of the gene under consideration and the optimal read length of the deep-sequencing approach being used (Fig. 4). In our case, two libraries were constructed to allow optimal probing of the mutagenic landscape using 151-bp paired-end sequencing on an Illumina MiSeq.

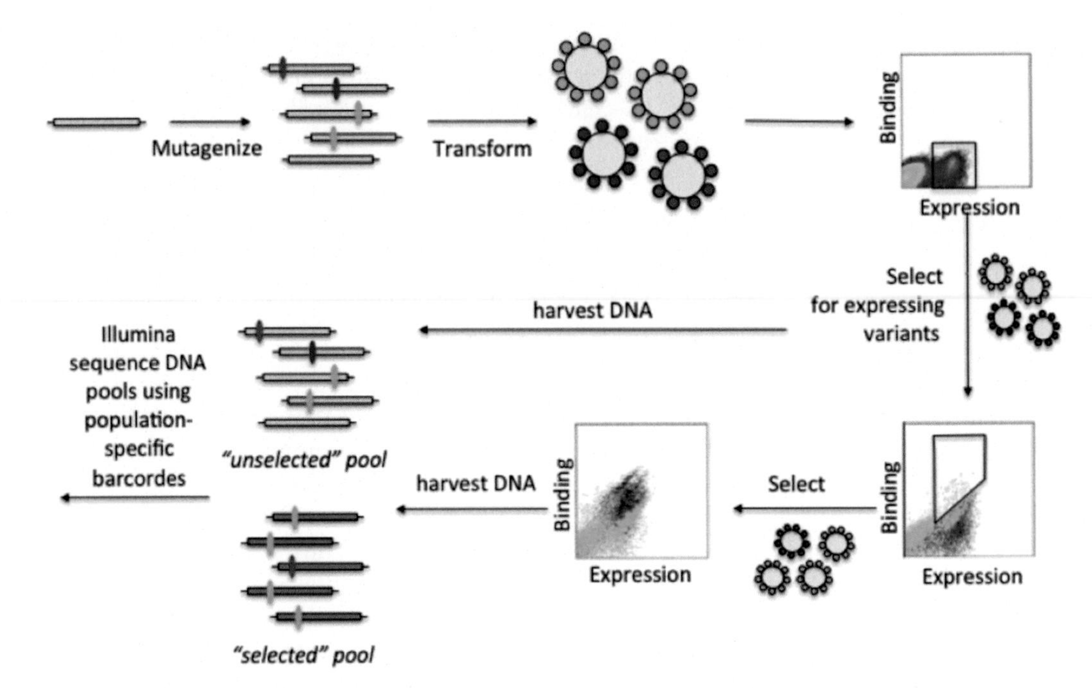

Fig. 4 Preparation for the deep sequencing-based illumination of the mutagenic landscape of binding. A mutagenic library is synthesized (see main text) and is screened first for expression and then binding. Harvested DNA at both stages is deep sequenced, and the relative frequency of individual mutations in the selected and unselected pools is used to compute the landscape

3. For each library, the full-length protein gene having additional pETCON overlap fragments at either end for yeast homologous recombination is assembled via recursive PCR. To introduce mutations, degenerate PAGE-purified oligos are used in which selected positions within the binding site are doped with a small amount of each nonnative base at a level expected to yield 1–2 mutations per gene. For this study, we ordered custom-doped oligos. *See* **Note 6.**
4. For each library assembly, overlapping oligonucleotides, including overlapping regions with the ends of the pETCON plasmid, are combined with dNTPs, 5× Phusion buffer HF, DMSO, and Phusion high-fidelity polymerase. Full-length products are assembled by PCR, and correctly assembled PCR products are amplified by a second round of PCR using oligonucleotides that overlap with the pETCON plasmid. Correct length PCR products are isolated using agarose gel electrophoresis and are purified using a Qiagen PCR cleanup kit and eluted in ddH_2O.
5. Yeast EBY100 cells are transformed with 5.4 μg of library DNA insert and 1.8 μg of gel-purified pETCON digested with *NdeI* and *XhoI* using the method of Benatuil [15], yielding ~1e6 transformants.

6. After transformation, cells are grown for 24 h in 100 mL of low-pH SDCAA media supplemented with Pen/Strep at 30 °C, passaged once, and grown for an additional 24 h under the same conditions. Cells (~5e8) are collected by centrifugation, resuspended in 50 mL of SGCAA, and induced overnight at 18 °C.
7. Induced cells (3e7) ware labeled with 4 μL of anti-cymc-FITC in 200 μL of PBSF for 20 min at 4 °C to label cells expressing full-length protein variants. Then, labeled cells are washed with PBSF and sorted. In this first round of sorting, all cells showing a positive signal for protein expression are collected.
8. Cells were recovered overnight in ~1 mL of low-pH SDCAA supplemented with Pen/Strep at 30 °C, pelleted by centrifugation at 1700×*g* for 4 min, resuspended in 5 mL of low-pH SDCAA supplemented with Pen/Strep, and grown for an additional 24 h at 30 °C.
9. Cells (~2e7) are collected by centrifugation, resuspended in 2 mL of SGCAA, and induced overnight at 18 °C.
10. Induced cells from expression-sorted libraries and two reference samples of the template protein (5e6 cells per sample) prepared similarly are washed with 600 μL of PBSF and then labeled with a chosen concentration of the small-molecule-biotin complex (100 nM of DIG-PEG_3-biotin in our case) in 400 μL of PBSF for the libraries or 200 μL of PBSF for the reference samples for >3 h at 4 °C. The concentration of the label should be sufficient to observe a binding signal with the parent clone. Labeled cells are washed with 200 μL of PBSF and then incubated with a secondary label solution of 0.8 μL of SAPE (Invitrogen) and 4 μL of anti-cymc-FITC in 400 μL of PBSF for 8 min at 4 °C. Cells are washed with 200 μL PBSF, resuspended in either 800 μL of PBSF for the libraries or 400 μL of PBSF for the reference samples, and sorted.
11. Clones having binding signals higher than that of the parent reference sample are collected using FACS. Collected cells are recovered overnight in ~1 mL of low-pH SDCAA supplemented with Pen/Strep at 30 °C, pelleted by centrifugation at 1700×*g* for 4 min, resuspended in 2 mL of low-pH SDCAA supplemented with Pen/Strep, and grown for an additional 24 h at 30 °C. Cells (2e7) are resuspended in 2 mL of SGCAA and induced overnight at 18 °C.
12. To reduce noise from the first round of cell sorting, the sorted libraries are labeled and subjected to a second round of cell sorting using the same conditions and gates as in the first round. Collected cells are recovered and grown as described above.
13. One hundred million cells from the expression-sorted libraries and at least 2e7 cells from doubly sorted library are pelleted by

centrifugation at 1700×g for 4 min, resuspended in 1 mL of freezing solution (50 % YPD, 2.5 % glycerol), transferred to cryogenic vials, slow-frozen in an isopropanol bath, and stored at −80 °C until further use.

3.6 Next-Generation Library Sequencing

1. Library DNA is prepared as detailed previously [16]. Illumina adapter sequences and unique library barcodes are appended to each library pool through PCR amplification using population-specific HPLC-purified primers.
2. The library amplicons are verified on a 2 % agarose gel stained with SYBR Gold and then purified using an Agencourt AMPure XP bead-based purification kit. Each library amplicon is denatured using NaOH and then diluted to 6 pM. A sample of *PhiX* control DNA is prepared in the same manner as the library samples and added to the library DNA to create high enough sample diversity for the Illumina base-calling algorithm. The final DNA sample is prepared by pooling 300 μL of 6 pM *PhiX* control DNA (50 %), 102 μL of 6 pM expression-sorted library, and 33 μL of 6 pM sorted libraries each.
3. DNA is sequenced in paired-end mode on an Illumina MiSeq using a 300-cycle reagent kit and custom HPLC-purified primers.
4. Data from each next-generation sequencing library is demultiplexed using the unique library barcodes added during the amplification steps. For example, in our experiment, of a total 5,630,105 paired-end reads, 2,531,653 reads were mapped to library barcodes. For each library, paired-end reads are fused and filtered for quality (Phred ≥ 30).
5. The resulting full-length reads are aligned against the relevant segments of the template gene sequence using scripts from the software package Enrich [17].
6. For single mutations having ≥7 counts in the original input library, a relative enrichment ratio between the input library and each selected library is calculated [16, 18, 19]. This cutoff value is used to establish statistical significance in the final data set.
7. A pseudocount value (0.3 in our case) is added to the total reads for each selected library mutation, to allow calculation of enrichment values for mutations that disappeared completely during selection.

3.7 Selectivity Assays by Equilibrium Fluorescence Polarization Competition Assays

1. To verify binding and to measure binding dissociation constants, fluorescence polarization assays are using purified protein and fluorescent ligand (Fig. 5). Fluorescence polarization-based affinity measurements of designs and their evolved variants are performed as noted previously [20] using a small-molecule-fluorescent dye conjugate (in our case Alexa488-conjugated DIG; DIG-PEG$_3$-Alexa488).

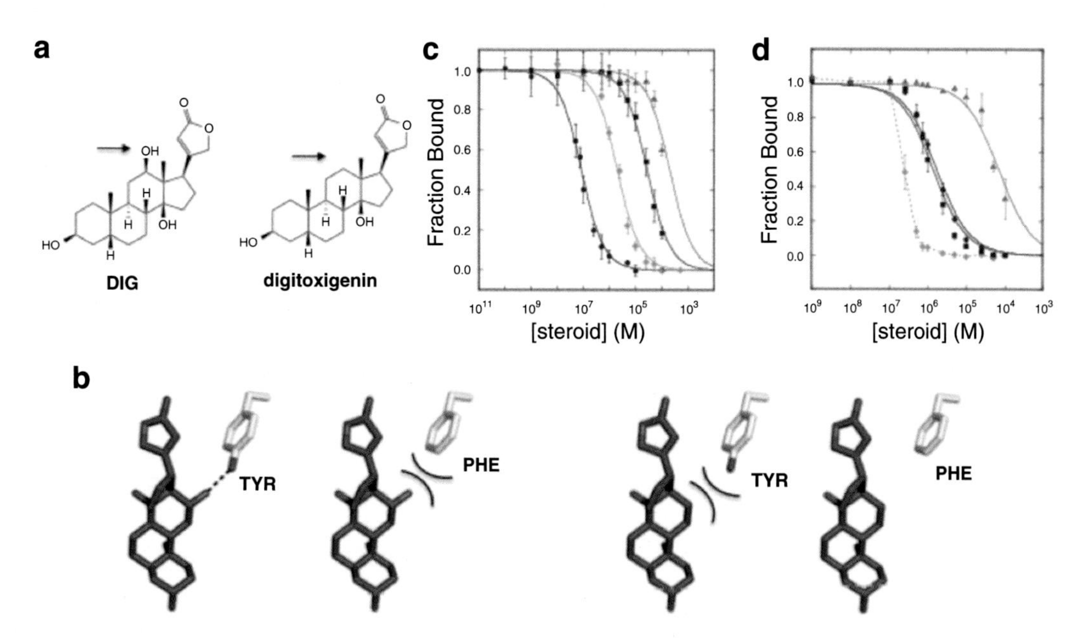

Fig. 5 Measuring and modulating selectivity of designed proteins guided by the computational model of binding. (**a**) The specificity of the designed binding protein can be modulated for congeneric ligands that differ in their chemical structure by as little as a hydroxyl group, as is the case with DIG and digitoxigenin. (**b**) Guided by the computational model of DIG10.3, in which tyrosine side chain groups were positioned to make hydrogen bonds with the DIG hydroxyl, a Tyr to Phe substitution was chosen, and (**c** and **d**) the selectivity of DIG10.3 and DIG10.3_Y110F was measured as described in the text. Robust specificity switching was observed (compare **c** and **d**), demonstrating the programmability of computationally designed ligand-binding proteins

2. In a typical experiment, the concentration of the conjugate is fixed near the K_d of the interaction being monitored, and the effect of the increasing concentrations of protein on the fluorescence anisotropy of the fluorescent dye is determined.
3. Fluorescence anisotropy (r) is measured in 96-well plate format at appropriate excitation and emission wavelengths (λ_{ex} = 485 nM and λ_{em} = 538 nM using a 515 nm emission cutoff filter, in our case). In all experiments, PBS (pH 7.4) is used as the buffer system and the temperature is 25 °C. For high-affinity complexes, it is important to use NBS-coated plates to improve the signal-to-noise aspect.
4. Equilibrium dissociation constants (K_d) are determined by fitting plots of the anisotropy averaged over a period of 20–40 min (equilibrium) after reaction initiation versus protein concentration to Eq. 1:

$$A = A_f + (A_b - A_f) \times \left(\frac{([L]_T + K_D + [R]_T) - \sqrt{(-[L]_T - K_D - [R]_T)^2 - 4[L]_T[R]_T}}{2[L_T]} \right) \quad (1)$$

where A is the experimentally measured anisotropy, A_f is the anisotropy of the free ligand, A_b is the anisotropy of the fully bound ligand, $[L]_T$ is the total ligand concentration, and $[R]_T$ is the total receptor concentration.

5. For ensuring assay robustness, reported K_d values should represent the average of at least three independent measurements with at least two separate batches of purified protein.

3.8 Fluorescence Polarization Equilibrium Competition Binding Assays

1. Fluorescence polarization equilibrium competition binding assays are used to determine the binding affinities of designed proteins and their variants for unlabeled ligands and congeneric compounds (for which selectivity measurements and modulation is desired; in our case, these were digoxigenin, digitoxigenin, progesterone, β-estradiol, and digoxin; Fig. 5a). During the computational design procedure, careful placement of interacting amino acid side chains allows for explicit design of selectivity (Fig. 5b). Selectivity can be switched by manipulation of these residues. In our case, we considered Tyr to Phe mutations as candidates to switch the specificity toward more hydrophobic steroids (Fig. 5b). The labeled small molecule (Subheading 3.5) is used, and the ability of different ligands to inhibit its binding to the designed protein variant is used to calculate their affinities for the protein.
2. In a typical experiment, the concentration of labeled small molecule is kept near or below the K_d of the interaction being monitored, the concentration of protein is fixed at a saturating value such that >95 % the labeled small molecule in the system is bound to protein, and the effects of increasing concentrations of unlabeled ligand on the fluorescence anisotropy of the fluorescent dye are determined as described above in Subheading 3.5.
3. If the ligands being considered are insoluble or sparingly soluble in aqueous buffers, stock solutions are typically made in organic solvents such as DMSO or methanol. For each ligand concentration, a negative control sample containing only the appropriate dilution of the corresponding organic solvent-only control solution (in aqueous assay buffer, PBS in our case) is measured. While we found that at all concentrations employed, methanol or DMSO solvents did not affect fluorescence anisotropy with our binding assay. However, correction for this effect must be made.
4. The concentration of total unlabeled ligand producing 50 % binding signal inhibition (I_{50}) is determined by fitting a plot of the anisotropy averaged over a period of 30 min to 3 h after reaction initiation versus unlabeled ligand concentration [20]. *See* **Note 7**.

5. For cases in which K_d for competitor is much smaller than K_d for the labeled small molecule, the data cannot be fit to the model and only qualitative conclusions can be reached (Fig. 5c, d).
6. The inhibition constant for each protein–ligand interaction, K_i, is calculated from the measured IC_{50} and the K_d of the protein-label interaction according to a model accounting for receptor-depletion conditions [20].
7. IC_{50} values, the concentrations of free unlabeled ligand producing 50 % binding signal inhibition, are calculated from the measured I_{50} values [20].
8. For assay robustness, reported I_{50} and subsequent K_i values should represent the average of at least three independent measurements from at least two batches of purified protein and a fresh unlabeled inhibitor stock prepared for each experiment.

4 Notes

1. In a typical experiment using DIG-BSA-biotin or DIG-RNase-biotin, 4e6 cells are resuspended in 50 μL of a premixed solution of PBSF containing a 1:100 dilution of anti-cmyc-FITC, 2.66 μM DIG-BSA-biotin or DIG-RNase-biotin, and 664 nM SAPE (to achieve a 1:4 streptavidin/biotin ratio). The use of carrier protein–ligand molecules offers a highly avid label for detection of weak binders. The avidity of the system (i.e., number of copies of the ligand on the carrier protein) can be tailored by changing the concentration of reagents in the carrier protein–ligand conjugation reaction.
2. In our case, competition assays with free digoxigenin were performed: between 750 μM and 1.5 mM of digoxigenin (Sigma Aldrich, St. Louis, MO) prepared as a stock solution in MeOH was added to each labeling reaction mixture, and binding of the resultant samples was determined as above. For "true" hits, the addition of excess free ligand should abolish the binding signal. Control experiments performed in a similar manner showed that the small amount of MeOH added does not affect the fluorescence or binding properties of SAPE.
3. It is best to restrict the library size such that each clone in the library can be oversampled by 10–100 in the transformed pool.
4. The stringency of the sort can be increased from round to round in order to hone in on one or a few binding clones by lowering the label concentration. However, it is important for the first round to be permissive to ensure that clones with low representation in the library pool are able to enrich if they have desirable binding properties.

5. Transformation of Kunkel libraries is typically not as efficient as is transformation of other library formats, so we found that preparing the library DNA in more efficient *E. coli* prior to transformation into yeast led to higher overall transformation efficiencies and a better chance of having complete clone coverage in the transformed library.
6. It is best to restrict the total library size so that each clone can be oversampled at 10–100 in both the transformed library and in the sequencing run (Illumina MiSeq runs currently yield up to 10^7 reads/run).
7. Note that for some experiments, due to the lack of solubility, limiting competitor ligand concentrations can make it impossible to collect data in the regime of complete inhibition. In these cases, data are fit by fixing the anisotropy at infinite steroid concentration to a value measured for other ligands for which this value could be determined experimentally.

Acknowledgment

SDK acknowledges support from the NSF (grant MCB1330760).

References

1. Rothlisberger D, Khersonsky O, Wollacott AM, Jiang L, DeChancie J, Betker J, Gallaher JL, Althoff EA, Zanghellini A, Dym O, Albeck S, Houk KN, Tawfik DS, Baker D (2008) Kemp elimination catalysts by computational enzyme design. Nature 453(7192):190–195. doi:10.1038/nature06879
2. Jiang L, Althoff EA, Clemente FR, Doyle L, Rothlisberger D, Zanghellini A, Gallaher JL, Betker JL, Tanaka F, Barbas CF 3rd, Hilvert D, Houk KN, Stoddard BL, Baker D (2008) De novo computational design of retro-aldol enzymes. Science 319(5868):1387–1391. doi:10.1126/science.1152692
3. Fleishman SJ, Whitehead TA, Ekiert DC, Dreyfus C, Corn JE, Strauch EM, Wilson IA, Baker D (2011) Computational design of proteins targeting the conserved stem region of influenza hemagglutinin. Science 332(6031):816–821. doi:10.1126/science.1202617
4. Tinberg CE, Khare SD, Dou J, Doyle L, Nelson JW, Schena A, Jankowski W, Kalodimos CG, Johnsson K, Stoddard BL, Baker D (2013) Computational design of ligand-binding proteins with high affinity and selectivity. Nature 501(7466):212–216. doi:10.1038/nature12443
5. Khare SD, Fleishman SJ (2013) Emerging themes in the computational design of novel enzymes and protein-protein interfaces. FEBS Lett 587(8):1147–1154. doi:10.1016/j.febslet.2012.12.009
6. Fleishman SJ, Baker D (2012) Role of the biomolecular energy gap in protein design, structure, and evolution. Cell 149(2):262–273. doi:10.1016/j.cell.2012.03.016
7. Griss R, Schena A, Reymond L, Patiny L, Werner D, Tinberg CE, Baker D, Johnsson K (2014) Bioluminescent sensor proteins for point-of-care therapeutic drug monitoring. Nat Chem Biol 10(7):598–603. doi:10.1038/nchembio.1554
8. Kuhlman B, Baker D (2000) Native protein sequences are close to optimal for their structures. Proc Natl Acad Sci U S A 97(19):10383–10388
9. Chao G, Lau WL, Hackel BJ, Sazinsky SL, Lippow SM, Wittrup KD (2006) Isolating and engineering human antibodies using yeast surface display. Nat Protoc 1(2):755–768
10. Mazor Y, Blarcom TV, Mabry R, Iverson BL, Georgiou G (2007) Isolation of engineered, full-length antibodies from libraries expressed in *Escherichia coli.* Nat Biotechnol 25(5):563–565
11. Nilsson B, Moks T, Jansson B, Abrahmsén L, Elmblad A, Holmgren E, Henrichson C, Jones TA, Uhlén M (1987) A synthetic IgG-binding

domain based on staphylococcal protein A. Protein Eng 1(2):107–113

12. Gietz RD, Schiestl RH (2007) High-efficiency yeast transformation using the LiAc/SS carrier DNA/PEG method. Nat Protoc 2(1):31–34
13. Schlehuber S, Beste G, Skerra A (2000) A novel type of receptor protein, based on the lipocalin scaffold, with specificity for digoxigenin. J Mol Biol 297(5):1105–1120
14. Kunkel TA (1985) Rapid and efficient site-specific mutagenesis without phenotypic selection. Proc Natl Acad Sci U S A 82(2): 488–492
15. Benatuil L, Perez JM, Belk J, Hsieh C-M (2010) An improved yeast transformation method for the generation of very large human antibody libraries. Protein Eng Des Sel 23(4):155–159
16. Whitehead TA, Chevalier A, Song Y, Dreyfus C, Fleishman SJ, De Mattos C, Myers CA, Kamisetty H, Blair P, Wilson IA, Baker D (2012) Optimization of affinity, specificity and function of designed influenza inhibitors using deep sequencing. Nat Biotechnol 30(6): 543–548
17. Fowler DM, Araya CL, Gerard W, Fields S (2011) Enrich: software for analysis of protein function by enrichment and depletion of variants. Bioinformatics 27(24):3430–3431
18. Fowler DM, Araya CL, Fleishman SJ, Kellogg EH, Stephany JJ, Baker D, Fields S (2010) High-resolution mapping of protein sequence-function relationships. Nat Methods 7(9):741–746
19. McLaughlin RN Jr, Poelwijk FJ, Raman A, Gosal WS, Ranganathan R (2012) The spatial architecture of protein function and adaptation. Nature 491(7422):138–142
20. Rossi AM, Taylor CW (2011) Analysis of protein-ligand interactions by fluorescence polarization. Nat Protoc 6(3):365–387

Chapter 10

Computational Design of Multinuclear Metalloproteins Using Unnatural Amino Acids

William A. Hansen, Jeremy H. Mills, and Sagar D. Khare

Abstract

Multinuclear metal ion clusters, coordinated by proteins, catalyze various critical biological redox reactions, including water oxidation in photosynthesis, and nitrogen fixation. Designed metalloproteins featuring synthetic metal clusters would aid in the design of bio-inspired catalysts for various applications in synthetic biology. The design of metal ion-binding sites in a protein chain requires geometrically constrained and accurate placement of several (between three and six) polar and/or charged amino acid side chains for every metal ion, making the design problem very challenging to address. Here, we describe a general computational method to redesign oligomeric interfaces of symmetric proteins for the purpose of creating novel multinuclear metalloproteins with tunable geometries, electrochemical environments, and metal cofactor stability via first and second-shell interactions.

The method requires a target symmetric organometallic cofactor whose coordinating ligands resemble the side chains of a natural or unnatural amino acid and a library of oligomeric protein structures featuring the same symmetry as the target cofactor. Geometric interface matches between target cofactor and scaffold are determined using a program that we call symmetric protein recursive ion-cofactor sampler (SyPRIS). First, the amino acid-bound organometallic cofactor model is built and symmetrically aligned to the axes of symmetry of each scaffold. Depending on the symmetry, rigid body and inverse rotameric degrees of freedom of the cofactor model are then simultaneously sampled to locate scaffold backbone constellations that are geometrically poised to incorporate the cofactor. Optionally, backbone remodeling of loops can be performed if no perfect matches are identified. Finally, the identities of spatially proximal neighbor residues of the cofactor are optimized using Rosetta Design. Selected designs can then be produced in the laboratory using genetically incorporated unnatural amino acid technology and tested experimentally for structure and catalytic activity.

Key words Metalloprotein, Metalloenzyme design, Multinuclear metal site, Unnatural amino acid, 2,2′-Bispyridine, Computational design

1 Introduction

Much progress has been made in the last two decades toward the de novo design of novel metalloproteins [1–9], where the guiding principle is simultaneous placement of two or more metal coordinating side chain groups from naturally occurring amino acid

Barry L. Stoddard (ed.), *Computational Design of Ligand Binding Proteins*, Methods in Molecular Biology, vol. 1414,
DOI 10.1007/978-1-4939-3569-7_10, © Springer Science+Business Media New York 2016

residues, cysteines, aspartate and glutamate, and histidine residues. However, successful design attempts have been largely dominated by *mononuclear* (a single metal ion per designed protein) insertions into a single type of scaffold—the geometrically well defined alpha helical bundles [3]. One of the challenges while designing a *multinuclear* (metal ion site composed of two or more metal ions) metalloproteins is the need to incorporate multiple side chain coordinating groups in close spatial proximity in a single protein—placing exacting constraints on design. Another challenge is the design of the electrostatic environment of the metal ions, which has a large impact on the stability of the highly charged cofactor and the associated catalytic activity.

Computational algorithms could, in principle, aid in addressing both challenges. We previously developed an algorithm that utilized the metal-chelating unnatural amino acid 2,2′-bispyridyl alanine (BPY) [10, 11] for designing mononuclear metal-binding sites [9]. The algorithm uses RosettaMatch [12] to combinatorially search, in a given protein scaffold (typically a single chain), for a constellation of backbone structures that can support the multiple (~3–6) side chain metal-chelating functional groups in the appropriate coordination geometry. The use of BPY simplified the combinatorial design problem as, unlike any natural amino acid side chain, the bipyridyl moiety contributes two metal ligands from the same amino acid side chain. Metalloproteins featuring BPY with His and Asp/Glu residues were designed, and their crystallographic structure demonstrated close agreement with the design model. However, this algorithm is limited by its combinatorial complexity and is not applicable, practically, to construct multinuclear metal-binding sites.

Here, we describe an approach to computationally design incorporation a symmetric multinuclear metallo-cofactor via integration into a similarly symmetric protein scaffold (Fig. 1). For this task, we have developed a matching algorithm, symmetric protein recursive ion-cofactor sampler (SyPRIS), and implemented it in Python. This algorithm allows expanding metalloprotein design to scaffolds other than alpha helical bundles, as well as gaining access to a greater variety of symmetric multinuclear cofactors such as iron-sulfur clusters and cubane complexes. We illustrate the method by describing the incorporation of the D_2 symmetric cobalt-oxygen cube-like cofactor (Co-cubane) [13–20]. This cofactor is a mimic of the water oxidation center in photosystem II and features four bipyridyl moieties coordinating four Co-ions, respectively. Though Co-cubane is used as an example, the method is generally applicable to incorporate all types of cofactors of either C or D symmetry within any complementary symmetric scaffold. Theozyme [21] matches generated from SyPRIS can be further designed with the enzyme design modules in the Rosetta macromolecular modeling software [12, 22–25] (Fig. 2).

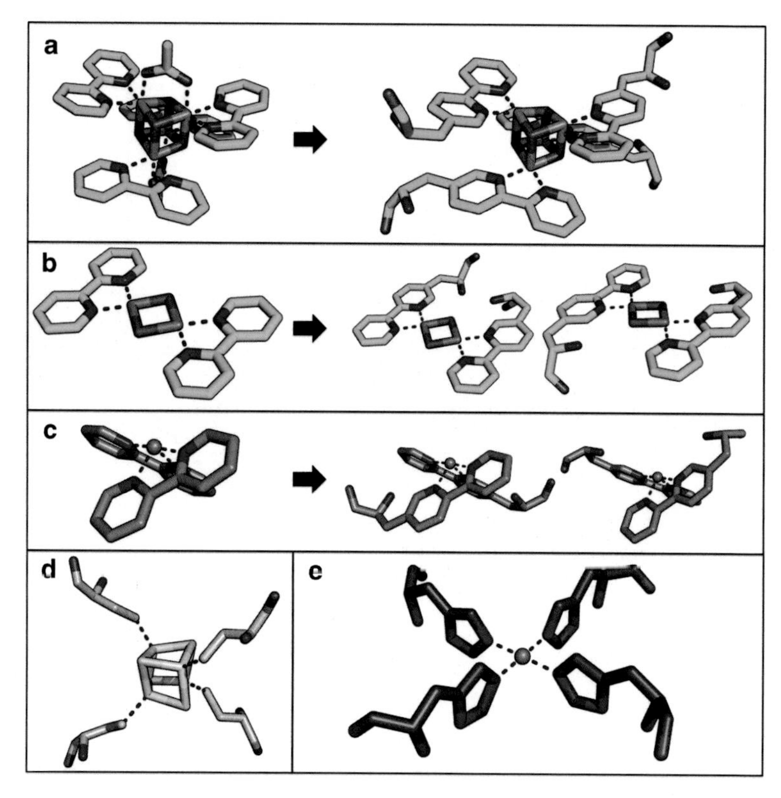

Fig. 1 Several target cofactors that this method was intended to implement using scaffolds of various symmetries. (**a**) $Co_4O_4(Ac)_2(bipyridine)_4$ converted from CCDC crystal structure to noncanonical amino acid-bound model featuring D_2 symmetry. (**b**) $Cu_2(OH)_2(bipyridine)_2$ converted to models featuring C_2 symmetry. (**c**) $CuOH(bipyridine)_2$ converted to models featuring C_2 symmetry. (**d**) $Fe_4S_4(Cys)_4$ cluster featuring D_2 symmetry. (**e**) $Cu(OH)_2(His)_4$ featuring C_4 symmetry

2 Methods

2.1 The General Pipeline for the Method (Fig. 3a) Includes the Following Steps (Also See Note 1)

1. Generate and standardize a symmetric scaffold library (Fig. 3b).
2. Prepare a target cofactor for symmetric insertion (Fig. 3c).
3. Use SyPRIS to identify inverse rotamer positions suitable for design (Fig. 3d).
4. Perform kinematic loop closure on residue matches that reside within a loop secondary structure (Figs. 3e, f).
5. Design the oligomeric interface with constraints (Fig. 3g).
6. Revert extraneous residue mutations to favor wild-type sequence.
7. Experimental validation through protein expression, purification, and crystallization (not discussed here).

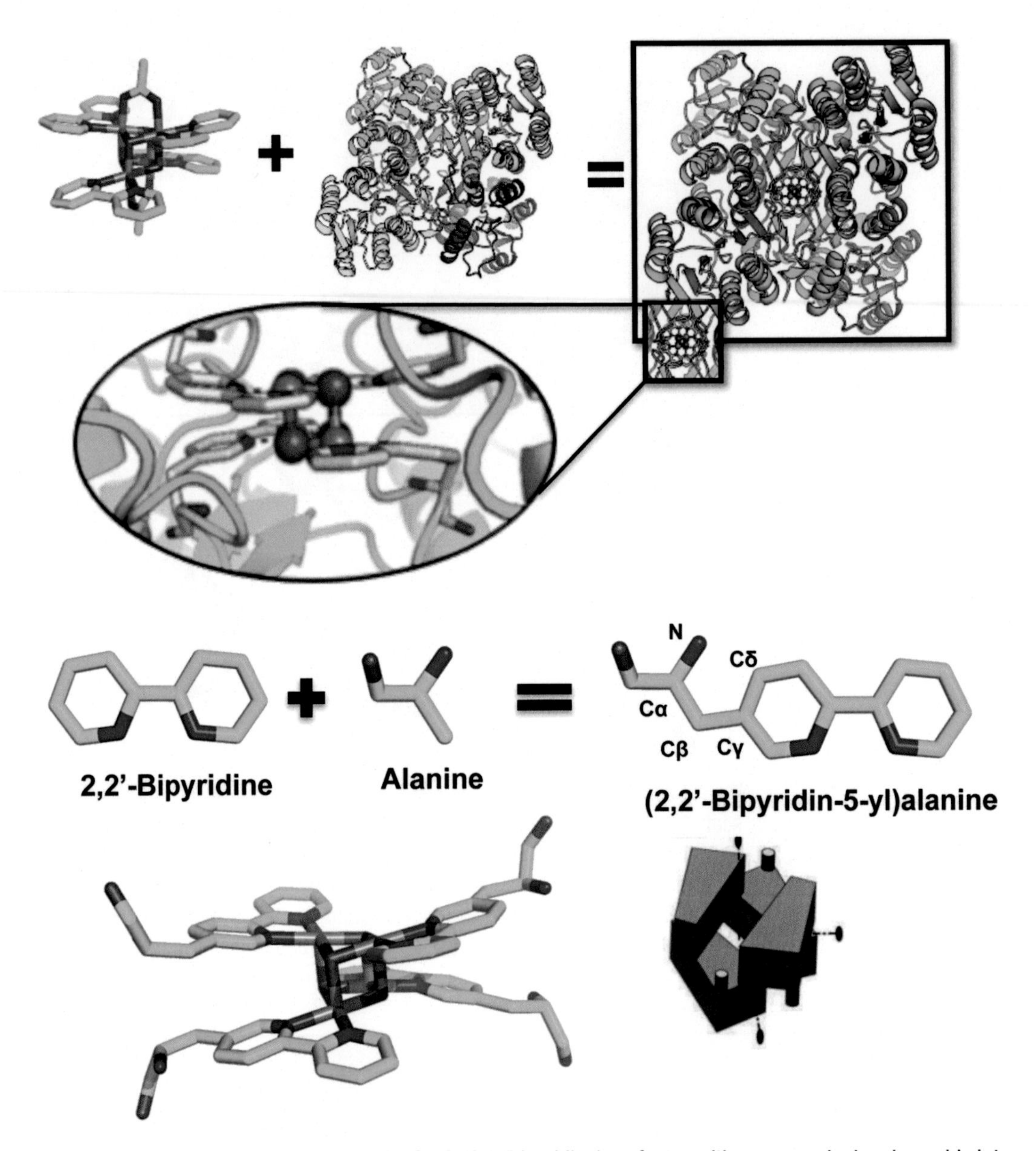

Fig. 2 Method overview, incorporation of a $Co_4O_4(Ac)_2(bipyridine)_4$ cofactor with noncanonical amino acids into a D_2 symmetric scaffold

2.2 Generate and Standardize Symmetric Scaffold Library

Potential protein scaffold candidates are selected from the RCSB protein databank to feature a given symmetry in the oligomeric protein, i.e., D_2, $C_{2,3,4...}$, etc. Search parameters include symmetry type, chain stoichiometry, expressibility in *E. coli*, 90 % sequence identity threshold, and <3.0 Å resolution (for structures determined by X-ray crystallography). From these constraints, a raw scaffold library is generated. More than 70 % of the scaffold files generated in this way contain asymmetries in the form of incomplete chains—due to missing electron density in the crystal structures.

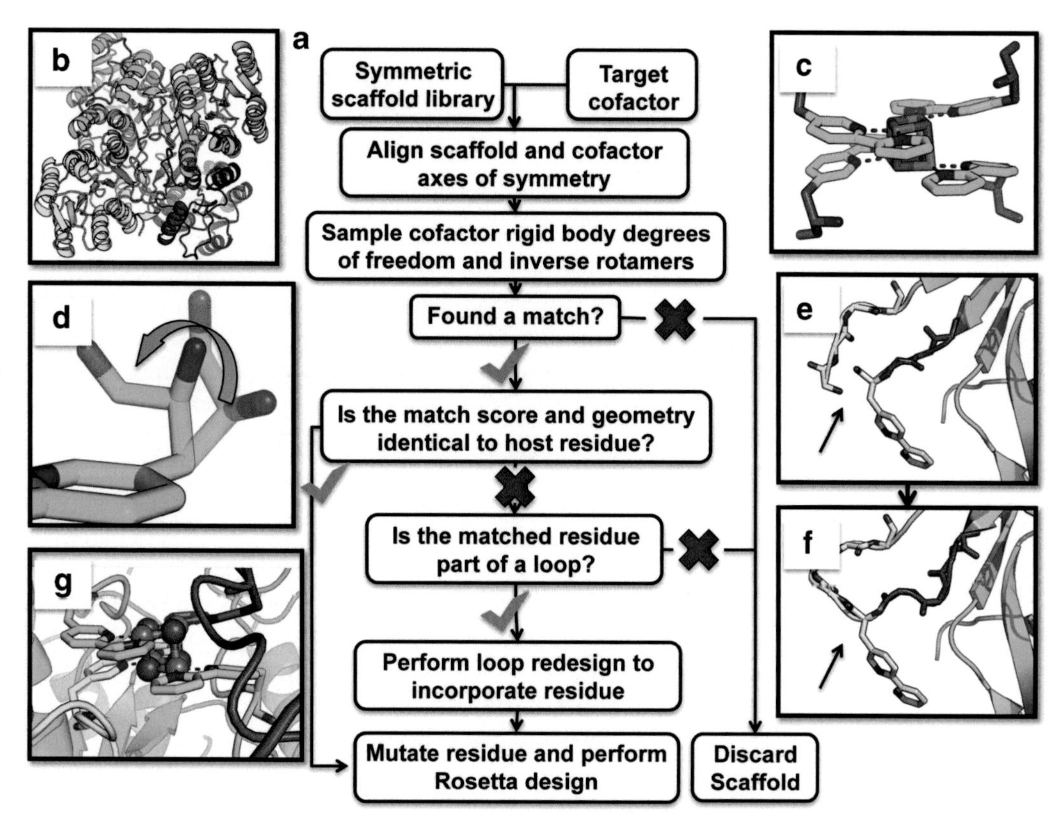

Fig. 3 (**a**) SyPRIS flow chart starting from generating scaffold library and ultimately ending in designable or discarded match. (**b**) An example scaffold, part of a library, will be considered by SyPRIS for the incorporation of a target cofactor. (**c**) A target cofactor, in this case an oxocobalt cubane coordinated by bipyridine ligands, has been modified with the appended magenta atoms creating a noncanonical amino acid. (**d**) The rotameric degrees of freedom for the atoms comprising the new backbone are sampled recursively with a chi distribution file (or exhaustively if desired) and compared to that of nearby backbone residues of the scaffold. (**e**) If the matched residue is part of a loop and the match was not geometrically identical, the loop is remodeled. (**f**) Three residues upstream and downstream of the translated backbone position are remodeled using Generalized KIC in Rosetta. (**g**) A fully designed oligomeric interface showing incorporated cofactor

In order to use the symmetry package of the Rosetta suite, all input files must be composed of chains that are equal in both residue length and residue type. To correct the intrinsic asymmetries, a hybrid Smith-Waterman local alignment is performed on all combinations of chains, removing residues absent from other chains, until a single converging monomeric sequence and all its symmetric partner protomers in the structures are found.

2.3 Target Cofactor

Cofactors of interest include organometallic compounds containing ligands that resemble either canonical amino acids or previously characterized noncanonical amino acids. PDB files are generated for cofactors of interest using their crystal structures and, where needed, the programs Mercury 3.5 and ConQuest 1.17 from the

Cambridge Crystallographic Database (CCDC). Small structural changes may be applied to the supplied atom positions to reduce asymmetries within the X-ray crystallographic models. If necessary, backbone atoms are appended to each symmetric ligand, and all dihedrals are set to a default 0.0° prior to matching. To identify dihedral positions acceptable for each cofactor, an ensemble is generated of all dihedral rotations while simultaneously performing internal atomic clash checks. Dihedral rotations that pass the clash check are stored and plotted against each subsequent dihedral rotation within a heat map. Preferred geometries are classified as regions of the heat map with the highest bin density at a determined threshold. These geometric constraints are then converted into a "chi distribution" file necessary for the symmetric protein recursive ion sampler (SyPRIS). A chi distribution file depicts the four atoms participating in a dihedral rotation, a range of values between which to sample, and the degree with which to iterate. A Rosetta parameter file, which stores information about the asymmetric unit of the multinuclear cluster (i.e., one Co-ion and one oxygen atom for the Co-cubane, one Fe and one S atom for an iron-sulfur cluster), is defined for integration within the Rosetta suite during design. Lastly, a Rosetta enzyme design constraints file, which adds an energy term favoring the coordination geometry between ligand and complex, is generated to more accurately determine the energy of the integrated cofactor.

2.4 Symmetric Protein Recursive Ion Sampler (SyPRIS)

With the scaffold set and cofactor model in place, the following steps are utilized in finding symmetric matches between the cofactor coordinated to an UAA and the protein scaffold.

2.4.1 Align Scaffold and Cofactor Axes of Symmetry

1. The axis of symmetry for the scaffold protein and each cofactor are determined by finding the eigenvector and eigenvalues—multiplying the coordinate matrix by its transpose matrix. Consequently, this creates unit vectors for each set of coordinates and supplies the principal rotational axes defined as the eigen minimum and maximum and their orthogonal cross product. In C-symmetry proteins, the eigen minimum and maximum can each be the target axis of symmetry. To correctly identify the axis of symmetry in a C-system, the midpoint of all symmetric Cα atoms is generated, and the average of all vectors connecting atoms to the origin becomes the symmetric axis.
2. Translate all Cartesian atoms of all files so that the axis of symmetry origin of the scaffold and each model lie on a theoretical (0, 0, 0) origin.
3. Align the axes of symmetry of the complex so that the eigen maximum and eigen minimum are aligned with that of the given scaffold (Fig. 4b). In C-symmetry, the eigen minimum of the cofactor is aligned to the midpoint average vector generated in **step 1**.

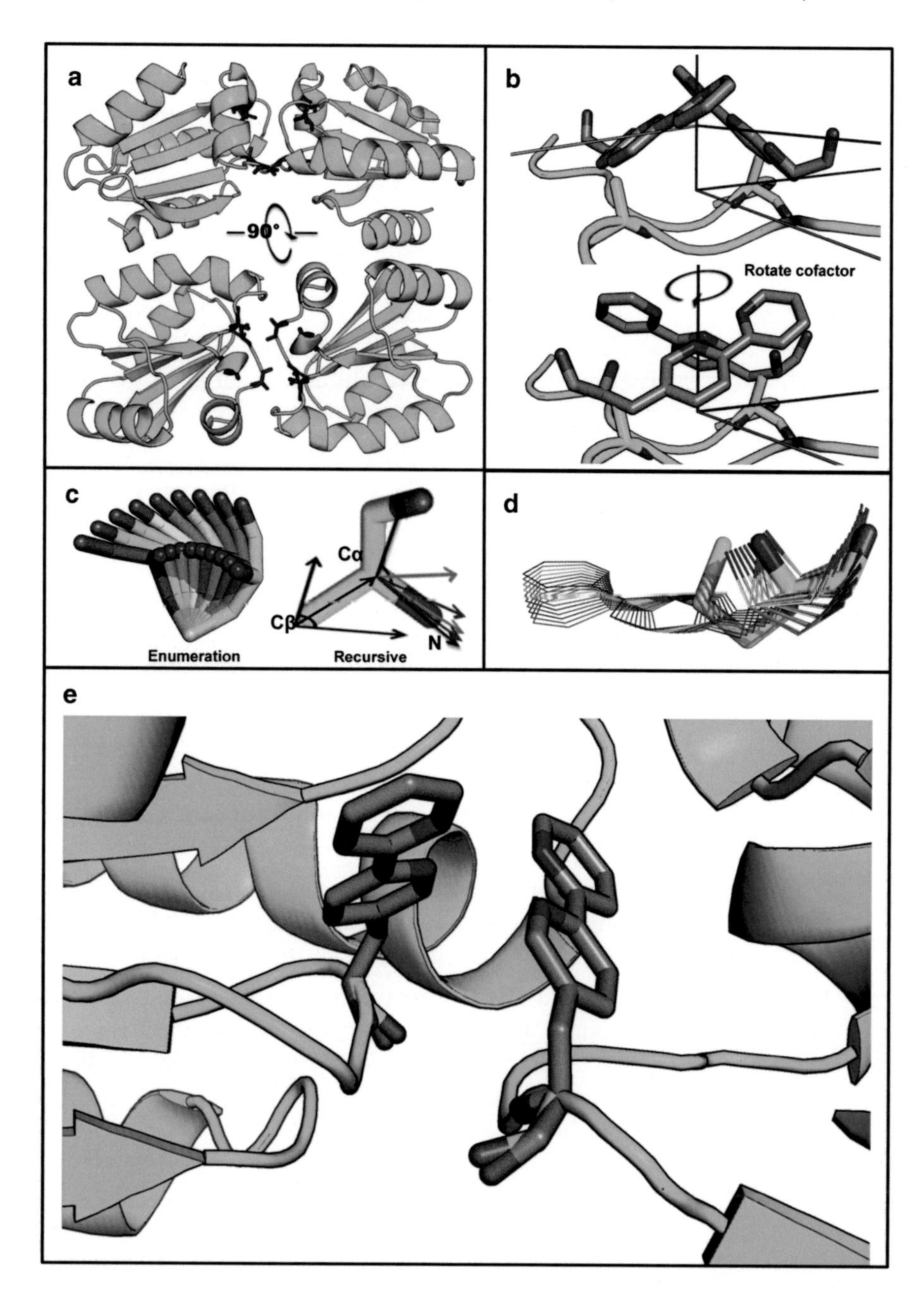

Fig. 4 (**a**) Residues that satisfy user-specified distance from symmetric axis highlighted in red sticks. (**b**) Rigid body rotation about symmetric axis to align symmetric axes. (**c**) Pictorial view of the enumerative exhaustive backbone sampling (*left*). Schematic view of the recursive atom placing algorithm for direct matching (*right*). (**d**) Ensemble of backbone positions generated via the recursive method. (**e**) A matched cofactor output from SyPRIS ready for Rosetta Design

4. If the input features C-symmetry, SyPRIS will locate the midpoint of the Cβ atoms of the cofactor and translate to the midpoint of each protein Cβ combination that is within ± <user input (default = 1.0)> Å of the cofactor Cβ radii (Fig 3a). The cofactor is then rotated about the plane of symmetry until the Cβ atoms of both the cofactor and protein are aligned (Fig 3b). Each rotational/translational position unique to a residue subset will store the lowest atom magnitude difference position as well as two other rotational positions clockwise and counterclockwise to the aligned atoms within a <user input (default = 1.0)> Å direct distance. The four unaligned positions will be stored to further generate an ensemble of positions and dihedrals starting from **step 6**, below.
5. If the input features d-symmetry, SyPRIS will perform 90° and 180° rotations of the cofactor about the vectors that correspond to each of the defined symmetric axes. Each rotational position will be further sampled in **step 6**.

2.4.2 Sample Inverse Rotamers

1. A cofactor to scaffold backbone clash check is performed by determining distances between all heavy atoms of the cofactor not included in the chi distribution file and the backbone heavy atoms of nearby residues (not including the residue making the match ± one residue position proximal in sequence). Any distances to heavy atoms <user input (default = 2.8 Å) are considered clashes and discarded.
2. For each unique cofactor rotation, cofactor backbone atoms (branches) are rotated within the range of values about the bonds defined by the atoms in the chi distribution file.
3. To score a given rotation, a vector is produced from the last stationary atom (LASA) to the first atom changing location (FACL). For example, while rotating about a chi1 bond of BPY UAA, the LASA is the alpha carbon, while the FACL would be the backbone nitrogen atom. The vector produced by the LASA and FACL of the cofactor is compared to that of the scaffold. The angle difference is calculated as an AngleLog:

$$\text{AngleLog} = \log\left(\Sigma\Delta\left[\left(\cos^{-1}\left(<xyz> \bullet <xyz'> / \|xyz\| \times \|xyz'\|\right) n / 20 \times n\right)\right]\right)$$

where n is the number of compared vectors and a value of zero is an average deviation of 20° across all n vectors. To further score a matched position, the magnitude of the cofactor FACL to the compared scaffold atom is calculated. The default threshold for AngleLog and atom magnitude is <user input (default = 0.0)> and <user input (default = 0.8)> Å, respectively.

4. Enumerative sampling. A predefined ensemble of inverse rotameric states is stored within one cofactor file. Each state is sampled exhaustively (Fig. 4c, left).
5. Recursive sampling. For any range of values tested in the chi distribution file, the best scoring rotation (as long as it meets the thresholds) is stored along with the best adjacent rotation. Recursive ½ angles are sampled within this range to minimize to the best solution. The algorithm to locate new half dihedrals:

$$\text{A)} \quad (\varphi_o + \varphi_n / 2)^n \quad \text{or} \quad \text{B)} \quad (\varphi_{n-1} + \varphi_n / 2)^n$$

where n is the number of half angles calculated as set by the user, φ_o is first dihedral (best scored), and $n = 1$ is the best scoring adjacent dihedral. SyPRIS starts with the algorithm in A. If two of the newly calculated half angles score better than the original dihedral, the B algorithm takes over for subsequent tests. Only the φ_0, φ_1, and φ_n ($n = \max$) FACL rotated branches will be stored to further sample a wider ensemble of positions (Fig. 4c, right). This algorithm occurs for each subsequent torsion angle at all stored positions (3^# of chis). Therefore, a cofactor with three chis featuring D_2 symmetry will store 27 positions (with tunable tolerance) at a given rotation. A C_2 cofactor with the same number of chis will store up to five times this many positions due to the rigid body rotational degrees of freedom (Fig. 4d).
6. For both the recursive and enumerative methods, final matches are determined by scoring the average AngleLog and RMSD over all FACL atom positions as defined in **step 8** (Fig. 4e).
7. A table for each protein is generated containing all the intrinsic properties of the ion cluster at a given match—model number and rotation about an axis. The table also includes the residue matched within the scaffold, the average AngleLog score, each individual AngleLog for all chains, the RMSD for all compared atoms, and the scaffold name. If an exact match is found (priority 1 designs), the scaffold will be mutated at the given residue position and passed to Rosetta Design. All other matches are subjects for the KIC procedure (priority 2 designs).

2.5 Kinematic Loop Closure (KIC)

This predesign method takes the tables generated by SyPRIS and locates the preferred residues for replacement with the ligand-like amino acid within the protein scaffold. The secondary structure of that residue with ± <user input (default = 3)> residues is determined based on Ramachandran preferred angles of phi and psi using a standard DSSP check. If the query within the scaffold is a loop region, the scaffold is accepted as designable; otherwise, if the region is helical or forms beta sheets, the scaffold is rejected. The scaffolds containing loops at match locations are then subjects of programs that:

1. Take the scaffold and corresponding model as arguments.
2. Translate the backbone coordinates of the matched residue on the scaffold to the location of the model to ensure exact match (generally changing atom positions by 0.5 Å across the entire residue).
3. Generate a coordinate constraint file (*see* **Note 2**) of the heavy atoms comprising the multinuclear cluster in the model corresponding to chain A for use during design. A coordinate constraint (CST) file contains coordinates that ensures that the metal cluster atoms do not change positions during design.
4. Generate two "loops" files (upstream and downstream of the matched residue) specific to each scaffold and matching residues necessary for performing KIC. The loop file contains information for which residue backbones will be sampled to make connection to another end point residue (i.e., remodeling the upstream or downstream loop about the ligand-like residue).
5. Utilizing a Rosetta-generalized KIC [26, 27], the four residues upstream and downstream are remodeled to accommodate the new position of the matched residue (step II). The remodeling includes sampling of backbone phi and psi angles while progressively closing the chain break. More details can be found in Kortemme et al.
6. A deterministic de novo loop is generated for each use of generalized KIC.
7. Generated loops are evaluated based on void formation, electrostatic repulsion, etc.

2.6 Rosetta Design

All redesigned loop scaffolds that pass are subject to four rounds of rotamer sampling followed by gradient-based minimization of side chain and backbone atoms. Design and repack shells are defined as residues with Cα atoms within 12 and 16 Å radii, respectively, about the matched residue. The design shell specifies that all residues within the shell excluding the metal cofactor and UAA will be allowed to mutate to other more favorably scoring residues. Residues within the repack shell sample their rotameric preferred side chain conformations while keeping their identity fixed. The talaris2013 symmetric score function with constraints is used to evaluate the states of the protein during design. The coordinate constraint file generated in **step 3** of Subheading 2.5 is used to force the ligand-like residue into a conformation conducive for

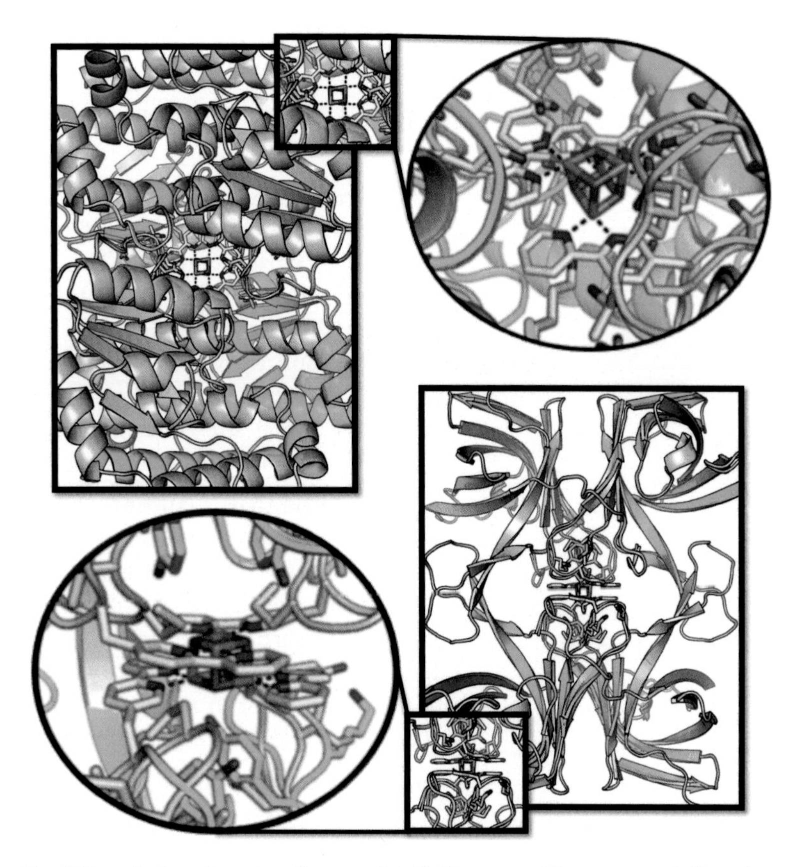

Fig. 5 Two designs incorporating a catalytic D_2 symmetric organometallic cofactor ($Co_4O_4(Ac)_2(bipyridine)_4$). The noncanonical amino acid bipyridine is incorporated on one chain, forming the cofactor upon oligomerization. The design protein (*green* and *white*) is compared to the wild-type scaffold (*wheat*). Mutation positions are represented by *sticks*

coordinating the ions of the cofactor. The symmetry definition file generated in stage 2 was used to copy any change made on the master unit to all slave units as defined by Rosetta symmetry. Backbone minimization is allowed for residues that are part of the UAA-containing loop and nearby residues. Heavy coordinate constraints are placed on the scaffold to only allow movement of backbone atoms if necessary due to redesigned loop clashes. Final designs are chosen by low backbone RMSD of the design shell, smallest change to void volume, and favorable energies of interaction of the design shell residues with the cofactor (*see* **Notes 3** and **4**). Lastly, reversions are made on extraneous residues (*see* **Note 5**) to favor the wild-type sequence, and the protein is ready for expression (Fig. 5).

3 Notes

1. All Python scripts and skeleton RosettaScripts XML files are attached.
2. The Rosetta force field, as other molecular mechanics force fields, does not accurately model interactions of protein functional groups with metal ions. Therefore, it is necessary to treat these interactions with restraints. The weights used in the restraints will be system dependent, but in the final models, one should end up with a metal site geometry similar to the one from the starting crystal structure with some small deviation. If the metal site is completely distorted, the weights of the restraints should be increased to keep the geometry fixed.
3. Another metric that is currently evaluated by human intuition in our protocol is that access of small ions/substrates to the metal site has not been blocked by new mutations introduced in the design protocol. Conformational changes upon substrate binding are not modeled, and system-dependent knowledge of the dynamics of the closure and opening of the active site should be kept in mind when either picking out scaffolds for design and evaluating designs by inspection.
4. Many substitutions can be introduced, but as a designer, one should also make sure that the initial protein scaffold can accommodate these changes in the absence of any substrate; otherwise, the enzyme will either not express or be unfolded. In particular, we paid special attention to the maintenance of the symmetric interface of the oligomer in question.
5. Chemical intuition is almost always required to evaluate the goodness of designs.

References

1. Ghosh D, Pecoraro VL (2004) Understanding metalloprotein folding using a de novo design strategy. Inorg Chem 43:7902–7915. doi:10.1021/ic048939z
2. Hellinga HW (1996) Metalloprotein design. Curr Opin Biotechnol 7:437–441. doi:10.1016/S0958-1669(96)80121-2
3. Peacock AFA (2013) Incorporating metals into de novo proteins. Curr Opin Chem Biol 17:934–939. doi:10.1016/j.cbpa.2013.10.015
4. Zastrow ML, Pecoraro VL (2013) Designing functional metalloproteins: from structural to catalytic metal sites. Coord Chem Rev 257:2565–2588. doi:10.1016/j.ccr.2013.02.007
5. Lu Y, Yeung N, Sieracki N, Marshall NM (2009) Design of functional metalloproteins. Nature 460:855–862. doi:10.1038/nature08304
6. Grzyb J, Xu F, Weiner L et al (2010) De novo design of a non-natural fold for an iron-sulfur protein: Alpha-helical coiled-coil with a four-iron four-sulfur cluster binding site in its central core. Biochim Biophys Acta Bioenerg 1797:406–413. doi:10.1016/j.bbabio.2009.12.012
7. DeGrado WF, Summa CM, Pavone V et al (1999) De novo design and structural characterization of proteins and metalloproteins. Annu Rev Biochem 68:779–819. doi:10.1146/annurev.biochem.68.1.779

8. Degrado WF, Summa CM, Pavone V et al (1999) De novo design and structural characterization of proteins. Biochemistry 68: 779–819
9. Mills JH, Khare SD, Bolduc JM et al (2013) Computational design of an unnatural amino acid dependent metalloprotein with atomic level accuracy. J Am Chem Soc 135:13393–13399. doi:10.1021/ja403503m
10. Liu CC, Schultz PG (2010) Adding new chemistries to the genetic code. Annu Rev Biochem 79:413–444. doi:10.1146/annurev.biochem.052308.105824
11. Imperiali B, Fisher SL (1991) (S)-u-amino-2,2′-bipyridine-6-propanoic acid: a versatile amino acid for de novo metalloprotein design. J Am Chem Soc 113:8527–8528. doi:10.1021/ja00022a053
12. Richter F, Leaver-Fay A, Khare SD et al (2011) De novo enzyme design using Rosetta3. PLoS One 6:1–12. doi:10.1371/journal.pone.0019230
13. Smith PF, Kaplan C, Sheats JE et al (2014) What determines catalyst functionality in molecular water oxidation? Dependence on ligands and metal nuclearity in cobalt clusters. Inorg Chem 53:2113–2121. doi:10.1021/ic402720p
14. Li X, Clatworthy EB, Masters AF, Maschmeyer T (2015) Molecular cobalt clusters as precursors of distinct active species in electrochemical, photochemical, and photoelectrochemical water oxidation reactions in phosphate electrolytes. Chemistry 21(46):16578–16584. doi:10.1002/chem.201502428
15. Dimitrou K, Brown AD, Christou G et al (2001) Mixed-valence, tetranuclear cobalt (iii, iv) complexes: preparation and properties of [Co4O4(O2CR)2(bpy)4]3+ salts. Chem Commun 4:1284–1285. doi:10.1039/b102008k
16. Evangelisti F, Guettinger R, More R et al (2013) Closer to photosystem II: A Co4O4 cubane catalyst with flexible ligand architecture. J Am Chem Soc 135(50):18734–18737. doi:10.1021/ja4098302
17. McCool NS, Robinson DM, Sheats JE, Dismukes GC (2011) A Co4O4 cubane water oxidation catalyst inspired by photosynthesis. J Am Chem Soc 133:11446–11449. doi:10.1021/ja203877y
18. Berardi S, La Ganga G, Natali M et al (2012) Photocatalytic water oxidation: tuning light-induced electron transfer by molecular Co4O4 cores. J Am Chem Soc 134:11104–11107. doi:10.1021/ja303951z
19. Chakrabarty R, Bora SJ, Das BK (2007) Synthesis, structure, spectral and electrochemical properties, and catalytic use of cobalt (III)–oxo cubane clusters. Polyhedron 46: 9450–9462
20. Najafpour MM, Rahimi F, Aro E-M et al (2012) Nano-sized manganese oxides as biomimetic catalysts for water oxidation in artificial photosynthesis: a review. J R Soc Interface 9:2383–2395. doi:10.1098/rsif.2012.0412
21. Tantillo DJ, Chen J, Houk KN (1998) Theozymes and compuzymes: theoretical models for biological catalysis. Curr Opin Chem Biol 2:743–750. doi:10.1016/S1367-5931(98)80112-9
22. Siegel JB, Zanghellini A, Lovick HM et al (2010) Computational design of an enzyme catalyst for a stereoselective bimolecular diels-alder reaction. Science 105:1–6
23. Röthlisberger D, Khersonsky O, Wollacott AM et al (2008) Kemp elimination catalysts by computational enzyme design. Nature 453: 190–195. doi:10.1038/nature06879
24. Jiang L, Althoff EA, Clemente FR et al (2008) De novo computational design of retro-aldol enzymes. Science 319:1387–1391. doi:10.1126/science.1152692
25. Bradley P, Misura KMS, Baker D (2005) Toward high-resolution de novo structure prediction for small proteins. Science 309:1868–1871. doi:10.1126/science.1113801
26. Mandell DJ, Kortemme T (2009) Backbone flexibility in computational protein design. Curr Opin Biotechnol 20:420–428. doi:10.1016/j.copbio.2009.07.006
27. Mandell DJ, Coutsias EA, Kortemme T (2009) Sub-angstrom accuracy in protein loop reconstruction by robotics-inspired conformational sampling. Nat Methods 6:551–552. doi:10.1038/nmeth0809-551

Chapter 11

De Novo Design of Metalloproteins and Metalloenzymes in a Three-Helix Bundle

Jefferson S. Plegaria and Vincent L. Pecoraro

Abstract

For more than two decades, de novo protein design has proven to be an effective methodology for modeling native proteins. De novo design involves the construction of metal-binding sites within simple and/or unrelated α helical peptide structures. The preparation of α_3D, a single polypeptide that folds into a native-like three-helix bundle structure, has significantly expanded available de novo designed scaffolds. Devoid of a metal-binding site (MBS), we incorporated a 3Cys and 3His motif in α_3D to construct a heavy metal and a transition metal center, respectively. These efforts produced excellent functional models for native metalloproteins/metalloregulatory proteins and metalloenzymes. Morever, these α_3D derivatives serve as a foundation for constructing redox active sites with either the same (e.g., 4Cys) or mixed (e.g., 2HisCys) ligands, a feat that could be achieved in this preassembled framework. Here, we describe the process of constructing MBSs in α_3D and our expression techniques.

Key words De novo protein design, Three-helix bundle, Metal-binding site, Metalloprotein, Metalloregulatory protein, Metalloenzyme, Protein expression

1 Introduction

De novo protein design offers a methodology for modeling the metal centers of metalloproteins and metalloenzymes [1–3]. This approach involves the construction of a desired metal-binding site(s) in a peptide scaffold with a sequence that is not found in nature, thus allowing scientists to uncover physical properties that may remain hidden from direct studies of native proteins. The most commonly used scaffolds have an α-helical fold and have previously been engineered to contain heme, nonheme iron, and zinc centers [4]. Much of our efforts have focused on building a 3Cys site in the TRI and Coil-Ser (CS) peptide system [3, 5]. This thiol-rich site is accomplished through the self-association of a single TRI or CS peptide into a three-stranded coiled tertiary structure (3SCC) (Fig. 1a) [6]. Our work with the 3SCC scaffolds has generated excellent spectroscopic, structural, and functional models for native

Barry L. Stoddard (ed.), *Computational Design of Ligand Binding Proteins*, Methods in Molecular Biology, vol. 1414,
DOI 10.1007/978-1-4939-3569-7_11, © Springer Science+Business Media New York 2016

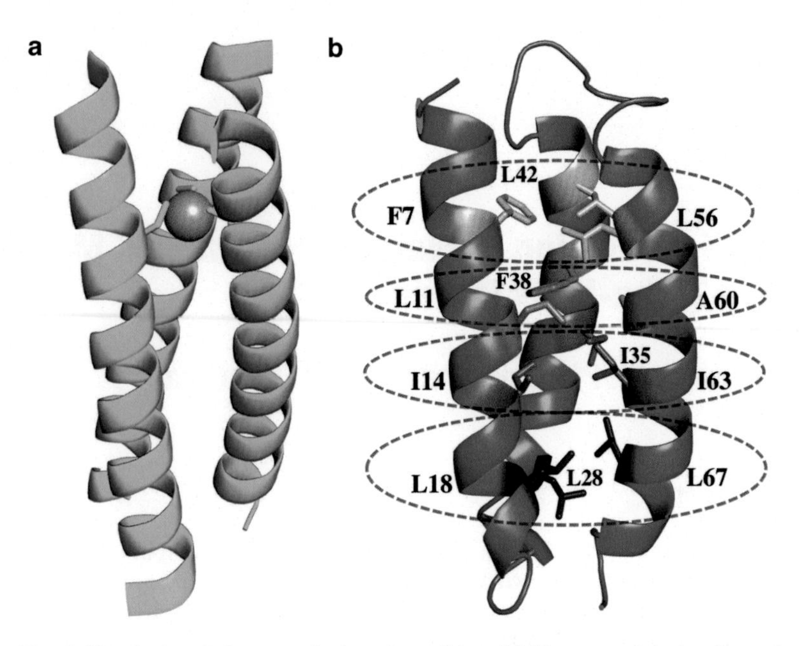

Fig. 1 Structures of de novo designed peptides. (**a**) X-ray crystal structure of As(III) bound CSL9C (PDB 2JGO), a three-stranded coiled coil scaffold. (**b**) Solution structure of α_3D. Apolar residues of α_3D divided into four layers, as indicated by varying shades of *gray*. The first layer comprises F7, L42, and L56 at the N-terminal end of the bundle. Subsequent layer contains L11, F38, and A60. The third layer has all isoleucine residues at the 14th, 35th, and 63rd positions. The C-terminal layer composes of L18, L28, and L67. These layers were predicted to provide a 3Cys metal-binding site

metalloregulatory proteins that bind toxic heavy metals, including arsenic, cadmium, mercury, and lead [3]. Moreover, in an effort to recapitulate the activity of metalloenzymes bound to a transition metal, TRI constructs with a 3His site had also been developed and shown to possess copper nitrite reductase activity [7, 8] and zinc carbonic anhydrase [9, 10].

DeGrado and coworkers expanded available de novo designed scaffolds through the preparation of a native-like peptide, α_3D [11] (Fig. 1b). This scaffold is a single polypeptide chain that preassembles into an antiparallel three-helix bundle, a major advancement in de novo protein design. Lacking a metal-binding site, our first approach aimed to introduce a 3Cys site in α_3D. Through the substitutions of apolar residues, as shown in Fig. 1b, four locations (categorized as layers) were identified that could accommodate this design. Based on Nuclear Magnetic Resonance (NMR) analysis on α_3D, the fourth layer, which is composed of L18, L28, and L67, was predicted to be the most amenable to mutations. Chakraborty et al. prepared α_3D**IV** (Fig. 2a), an α_3D derivative with a 3Cys site at the C-terminal end of the bundle [12]. The authors showed that α_3D**IV** binds heavy metals Cd, Hg, and Pb in

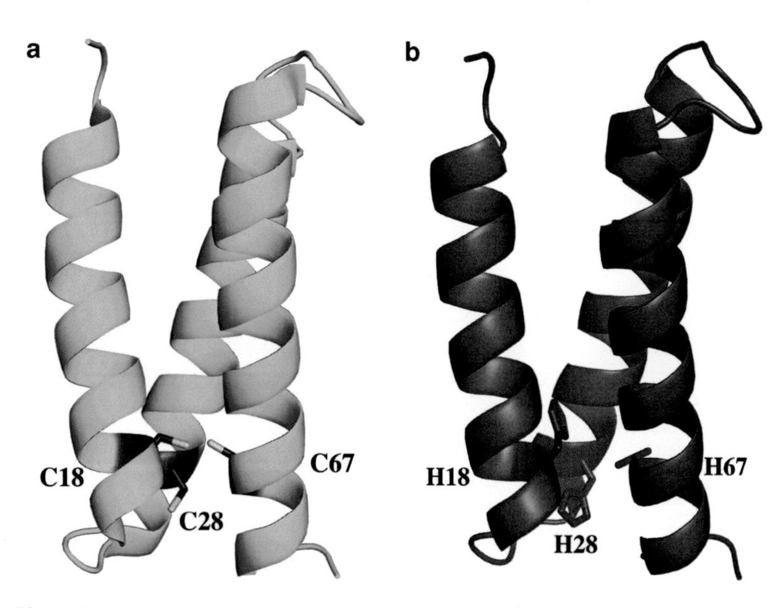

Fig. 2 Subsequent α_3D derivatives for heavy and transition metal binding. (**a**) Solution structure of α_3D**IV**, which exhibits a 3Cys site at positions 18, 28, and 67 that coordinates Cd, Hg, and Pb. (**b**) Model of a 3His α_3D derivative, α_3D**H$_3$**, which was demonstrated to bind Zn and perform the function of carbonic anhydrase. This model was constructed from the α_3D**IV** structure

Table 1
Amino acid sequence of α_3D constructs

Construct	Sequence	Molecular weight (Da)	PDB code
α_3D	MGSWAEFKQR LAAIKTR LQAL GGS EAELAAFEKE IAAFESE LQAY KGKG NPEVEALRKE AAAIRDE LQAYRHN	7977.2	2A3D
α_3DIV	MGSWAEFKQR LAAIKTR **C**QAL GGS EAE**C**AAFEKE IAAFESE LQAY KGKG NPEVEALRKE AAAIRDE **C**QAYRHN	7946.9	2MTQ
α_3D**H$_3$**[a]	MGSWAEFKQR LAAIKTR **H**QAL GGS EAE**H**AAFEKE IAAFESE LQAY KGKG NPEVEALRKE AAAIRDE **H**QAYR**V**N**GSGA**	8283.5	

Bolded residues indicate change from the sequence of α_3D
[a] *See* **Note 1**

the expected mode, serving as a spectroscopic and functional model for metalloregulatory proteins that contain an MS_3 center. The NMR structure of α_3D**IV** was also solved, which revealed that the overall fold of α_3D was not significantly perturbed after the removal of stabilizing Leu residues [13]. Subsequently, a 3His zinc metal site was also incorporated in the fourth layer, generating α_3D**H$_3$** [14] (Fig. 2b, Table 1).

Construct α_3**D**$\mathbf{H_3}$ was extended with a glycine-serine-glycine-alanine (GSGA) tail in an attempt to increase its overall stability after the incorporation of bulky His residues inside the core, without perturbing the overall framework of α_3D (*see* **Note 1**). This tail can also be modified to glycine-serine-glycine-cysteine (GSGC) with an A77C mutation. Both derivatives resulted in high expression yields of 100 mg/L, and from chemical denaturation studies, the GSGA construct increased the Gibbs free energy of unfolding (ΔG_U) of α_3**D**$\mathbf{H_3}$ to 3.1 from 2.5 kcal/mol compared to α_3**DIV**. Moreover, α_3**D**$\mathbf{H_3}$ was shown to bind Zn and perform the CO_2 hydrolysis associated with carbonic anhydrase. Overall, these efforts increased in scope the use of α_3D as a viable framework for modeling the metal centers of native proteins. They provide the opportunity to tackle redox active sites with either the same ligands (e.g., 4Cys) or mixed ligands (e.g., 2HisCysMet) [15], which can be achieved in this preassembled scaffold. This chapter presents our design and expression techniques in preparing α_3D derivatives.

2 Materials

Prepare all solutions using MQ or double distilled H_2O. Prepare and store solutions at room temperature, unless noted otherwise. Prepare all solutions on a sterile lab bench, cleaned with 10 % bleach and followed by 70 % ethanol. Autoclave all the necessary glassware.

2.1 Modeling Using PyMOL

Access to a computer console connected to the Internet that contains a more recent version of PyMOL (1.3–1.7) is required [16]. A payment is required to obtain a license for PyMOL (http://www.pymol.org), but a free version for students and educators is available (http://pymol.org/edu/?q=educational/). Once a computer is equipped with PyMOL, download the structure of α_3D (PDB code 2A3D) and/or α_3**DIV** (PDB code 2MTQ) from the RCSB Protein Data Bank (RCSB PDB) (http://www.rcsb.org/pdb/home/home.do) by entering the PDB code in the search box and downloading the PDB text file under the Download Files tab. A computer mouse with at least three customizable buttons is ideal for visualizing structures on PyMOL.

2.2 Transformation Components

1. A synthetic gene that contains the DNA sequence of the designed α_3D derivative cloned into pET-15b.
2. One-shot (50 μL) BL21(DE3) chemically competent *Escherichia coli* cells.
3. LB agar plates: Plates are prepared on a sterilized lab bench and under a flame provided by an isopropanol lamp. Suspend 4.0 g LB agar powder in 250 mL beaker containing 100 mL water

and autoclave using a liquid program. Allow the solution to cool to touch and then add 100 μL of a 100 mg/mL ampicillin (amp) solution. Pour LB agar solution in 100×15 mm petri dishes, allow to solidify, and store plates upside down in 4 °C.

4. SOC media, which can be prepared or commercially purchased. SOB media: Dissolve 0.20 g tryptone, 0.05 g yeast extract, and 0.005 g NaCl in 9.8 mL H_2O. Autoclave this solution using a liquid program and allow to cool to room temperature. Subsequently, add 100 μL of 1.0 M $MgCl_2$ and 100 μL of 1.0 M $MgSO_4$ to the SOB solution. A 1.0 mL SOC stock media is prepared by adding 20 μL of 20 % glucose (w/v) into 980 μL SOB media. Store leftover SOB and SOC media in 4 °C or −20 °C for short or long storage, respectively.

2.3 Protein Expression Components

1. An autoinduction media [17] is the preferred expression media (*see* **Note 2**) and prepared in 6 L batches (3×2 L solutions), which contains a rich media and a sugar solution. In a 4 L flask, suspend 48 g yeast extract powder and 24 g tryptone powder in 1.8 L H_2O. For a 6 L rich media, prepare the sugar solution by adding in a 2 L flask containing 600 mL H_2O 13.8 g KH_2PO_4 (monobasic), 62.0 g K_2HPO_4 (dibasic), 5.0 mL glycerol, 0.5 g glucose, and 2.0 g lactose. Autoclave the rich and sugar solutions using a short liquid program (*see* **Note 3**). The autoinduction media is prepared by aliquoting 0.2 L of the sugar solution into a 1.8 L of rich media (*see* **Note 4**).
2. LB media: Suspend 10 g tryptone powder, 5 g yeast extract powder, and 10 g NaCl in a 2 L flask containing 1.0 L H_2O. Autoclave using a liquid program.

2.4 Protein Purification Components

1. Lysis buffer: 1X PBS and 2 mM DTT. To prepare a 1 L 10X PBS buffer, dissolve in 800 mL H_2O 80 g NaCl, 2.0 g KCl, 14.4 Na_2HPO_4, and 2.4 KH_2PO_4. Adjust pH to 7.4 and autoclave using a liquid program. For a 100 mL 1X lysis buffer solution, add 10 mL of 10X PBS solution into 90 mL H_2O and dissolve 30.9 mg DTT. Prepare the lysis buffer solution fresh every expression.
2. A centrifuge for 1 L and 50 mL cell cultures.
3. A sonicator and a steel cup that can hold 100–200 mL volume.
4. A water bath set at 55 °C.
5. A pH electrode and lyophilizer.
6. A reverse-phase C18 HPLC. Solvents: The polar solvent is composed of H_2O and 0.1 % trifluoroacetic acid and the nonpolar solvent is comprised of 90 % acetonitrile, 10 % H_2O, and 0.1 % trifluoroacetic acid.

3 Methods

3.1 Design of α_3D Derivatives Using PyMOL

1. Run PyMOL and open the α_3D**IV** (PDB 2MTQ) or α_3D structure (PDB 2A3D).
2. Show structure as carton.
3. To model a new metal-binding site in the layers described in the introduction (Fig. 1), in the Menu tab, choose Wizard and then Mutagenesis.
4. In Mutagenesis option, select backbone-dependent rotamers and show residues as sticks.
5. Pick a residue to mutate and then the desired residue that can provide a metal-binding ligand such as S(Cys), N(His), or O(Asp or Glu).
6. Notice that several rotamers are possible. Choose the rotamer that is conducive to metal binding, that is, where the ligand is oriented toward the hydrophobic core.
7. Repeat according to the number of desired ligands (*see* **Note 5**).
8. Under the Wizard tab, use the Measurement option to determine the distances between the ligands. To obtain a qualitative sense of a suitable metal binding, these distances should be between 3.5 and 4.5 Å.

3.2 Transformation

Prior to the transformation experiment, prepare the amino acid sequence with the desired mutations. The gene for this sequence is placed between restriction sites BamHI and NcoI in the pET-15b vector (*see* **Note 6**).

1. Add 4–5 μL of 1 ng/μL of DNA to a tube of one-shot (50 μL) BL21(DE3) chemically competent *E. coli* cells thawed on ice for 10 min. Let stand for 10 min.
2. Heat shock in a 42 °C water bath for 30 s.
3. Cool on ice for 2 min.
4. Add 200 μL SOC and shake at 200 rpm in 37 °C for 30–50 min.
5. Prepare a diluted culture solution by adding 10 μL of BL21(DE3) cells into 90 μL fresh SOC.
6. Plate 100 μL culture on LB agar amp plate and incubate upside down overnight in 37 °C.
7. Save unused cells in 4 °C, which can be re-platted if the overnight plate does not show single colonies or is overgrown with no distinguishable single colonies.

3.3 Protein Expression Using Autoinduction Media

1. Pick single colonies from the overnight plate and inoculate 20 mL LB broth containing 20 μL of 100 mg/mL amp. Grow cultures overnight at 200 rpm and 37 °C.

2. Add 2 mL of 100 mg/mL amp to a 2 L autoinduction media. Inoculate with a 20 mL overnight culture.
3. Incubate overnight, 16–20 h, at 180 rpm and 25–30 °C.
4. Harvest cells by spinning down in 1 L centrifuge tubes at 8,000 × *g* and 4 °C. Re-suspend pelleted cells with 15–25 mL of lysis buffer pre-chilled on ice.
5. Transfer re-suspended cells in a steel cup chilled on an ice bucket.
6. Insert sonicator tip in steel cup, ~80 % submerged in re-suspended cells. Keep steel cup on ice.
7. Sonicate for total of 5 min, at 30 s on and 30 s off intervals. Repeat three times or until the solution has turned translucent.
8. Transfer to centrifuge tubes (50 mL) and spin-down at 17,000 × *g* and 4 °C for 30 min.
9. Transfer supernatant to 50 mL conical tubes and heat denature at 55 °C for 20–30 min. Transfer to the appropriate centrifuge tubes and spin-down at 17,000 × *g* and 4 °C for 30 min.
10. Pour supernatant in a beaker and acidify to pH 1.9 to precipitate salts and cellular debris. Transfer to the appropriate centrifuge tubes and spin-down at 17,000 × *g* and 4 °C for 30 min.
11. Place supernatant in 50 mL conical tubes and flash-freeze in liquid nitrogen for 10–15 min or until completely frozen. Lyophilize frozen protein for 2–3 days or until dry.

3.4 Protein Purification

1. Redissolve dry protein powder in H_2O (15–20 mL) and check pH (*see* **Note 7**).
2. Purify on a reversed-phase C18 HPLC using a flow rate of 20 mL/min and a linear gradient of polar solvent (0.1 % TFA in water) to nonpolar solvent (0.1 % TFA in 9:1 CH_3CN/H_2O) over 45 min.
3. Retention time of α_3D constructs is between 26 and 30 min.
4. The molecular weight is determined using an electrospray mode on a Micromass LCT Time-of-Flight mass ionization spectrometer. The MW accounts for 72 of the 73 amino acids as Met1 is cleaved posttranslation (*see* **Note 8**).

4 Notes

1. The sequence of α_3D**H**$_3$ was extended with a GSGA tail to increase the overall protein stability [14]. This extension also improved protein expression yield to ~100 mg/L compared from 50 mg/L compared to α_3D**IV**.

2. Autoinduction media and induction via IPTG work by the same mechanism, which involves the induction of gene expression by relieving the repression of the *lac* promoter. Using the latter induction technique, repression is relieved by the binding of IPTG (allolactose analog). However, in the case of autoinduction media, protein overexpression is controlled by the availability of sugar source instead of the addition of IPTG. Cell density relies on sugar source, as well as the expression of the designed gene without monitoring the cell density at 600 nm (OD_{600}). After exhausting the more metabolically available sugars, glucose and glycerol, the cells will use lactose as the sugar source. This natural switch will turn on the components of the *lac* operon, including our gene of interest that is downstream of the *lac* promoter region in pET-15b. Overall, compared to IPTG induction in LB media, the autoinduction technique eliminates 3–5 h waiting for the cell density to reach the proper OD_{600} and has also improved our protein yield by 20–50 mg/mL.
3. For the sugar solution, autoclave using a short liquid program to avoid sugar oxidation, which can make them less bioavailable. If this sugar oxidation is suspected, the solution can be prepared by dissolving all the components in autoclaved water and vacuum filter through a sterile 0.22 μm filter.
4. Prepare the autoreduction media several hours before inoculation to avoid contamination. Contamination was often observed when the media was prepared one to two days prior to expression.
5. Depending on your design goals, each metal-binding residue can be positioned on separate helices to form a triangular pocket. Or the two ligands can be placed on the same helix, spaced by two to three residues, to replicate a chelate-like motif and the third on a second helix, requiring only two of the three helices to form a metal-binding site.
6. Carefully indicate where to place the stop codon in the sequence. If a 73 amino acid sequence is desired, place the stop codon (TAA, TAG, or TGA) after Asn73. If a GSGA or GSGC tail is desired, place the stop codon after Ala/Cys77 (*see* **Note 1**). As described in the introduction and Note 1, the addition of a GSGA tail had a significant effect on the expression yield (100 mg/L) of α_3**DH**$_3$ and improved the ΔGU of α_3**DH**$_3$ by 0.6 kcal/mol compared to α_3D**IV**. The latter effect, which demonstrates an increase in stability, showed that the addition of these tails does not change or perturb the overall framework of α_3D. Moreover, an A77C mutation to generate a GSGC tail also had the same effect on the expression yield of α_3**DH**$_3$. Therefore, we expect that the addition of a GSGA or GSGC tail is essential in stabilizing α_3D derivatives that aim to modify layers 2 or 3 (see Fig. 1b).

7. The pH of the polar solvent is about 1.9. It is important to make sure that the crude protein solution matches this pH condition. When the pH conditions do not match, we have observed precipitation after the crude solution is mixed with the polar solvent. This precipitate can clog the HPLC tubings and lines and ultimately damage the solvent pump system.
8. Met1-containing species is observed in the mass spectrum of α_3**DIV** and α_3**DH**$_3$ but at a low amount compared to the Met1-cleaved species. Met1-containing species make up about <5 % of the total protein.

Acknowledgments

J.S.P. and V.L.P. would like to thank the National Institutes of Health (NIH) for financial support for this research (ES012236). J.S.P. thanks the Rackham Graduate School at the University of Michigan for a research fellowship.

References

1. DeGrado WF, Summa CM, Pavone V, Nastri F, Lombardi A (1999) *De novo* design and structural characterization of proteins and metalloproteins. Annu Rev Biochem 68:779–819. doi:10.1146/annurev.biochem.68.1.779
2. Lu Y, Berry SM, Pfister TD (2001) Engineering novel metalloproteins: design of metal-binding sites into native protein scaffolds. Chem Rev 101(10):3047–3080. doi:10.1021/cr0000574
3. Yu F, Cangelosi VM, Zastrow ML, Tegoni M, Plegaria JS, Tebo AG, Mocny CS, Ruckthong L, Qayyum H, Pecoraro VL (2014) Protein design: toward functional metalloenzymes. Chem Rev 114(7):3495–3578. doi:10.1021/cr400458x
4. Lu Y, Yeung N, Sieracki N, Marshall NM (2009) Design of functional metalloproteins. Nature 460(7257):855–862. doi:10.1038/nature08304
5. Peacock AF, Iranzo O, Pecoraro VL (2009) Harnessing nature's ability to control metal ion coordination geometry using *de novo* designed peptides. Dalton Trans 13:2271–2280. doi:10.1039/b818306f
6. Touw DS, Nordman CE, Stuckey JA, Pecoraro VL (2007) Identifying important structural characteristics of arsenic resistance proteins by using designed three-stranded coiled coils. Proc Natl Acad Sci U S A 104(29):11969–11974. doi:10.1073/pnas.0701979104
7. Tegoni M, Yu F, Bersellini M, Penner-Hahn JE, Pecoraro VL (2012) Designing a functional type 2 copper center that has nitrite reductase activity within alpha-helical coiled coils. Proc Natl Acad Sci U S A 109(52):21234–21239. doi:10.1073/pnas.1212893110
8. Yu F, Penner-Hahn JE, Pecoraro VL (2013) *De novo*-designed metallopeptides with type 2 copper centers: modulation of reduction potentials and nitrite reductase activities. J Am Chem Soc 135(48):18096–18107. doi:10.1021/ja406648n
9. Zastrow ML, Peacock AF, Stuckey JA, Pecoraro VL (2012) Hydrolytic catalysis and structural stabilization in a designed metalloprotein. Nat Chem 4(2):118–123. doi:10.1038/nchem.1201
10. Zastrow ML, Pecoraro VL (2013) Influence of active site location on catalytic activity in *de novo*-designed zinc metalloenzymes. J Am Chem Soc 135(15):5895–5903. doi:10.1021/ja401537t
11. Walsh ST, Cheng H, Bryson JW, Roder H, DeGrado WF (1999) Solution structure and dynamics of a *de novo* designed three-helix bundle protein. Proc Natl Acad Sci U S A 96(10):5486–5491. doi:10.1073/pnas.96.10.5486
12. Chakraborty S, Kravitz JY, Thulstrup PW, Hemmingsen L, DeGrado WF, Pecoraro VL (2011) Design of a three-helix bundle capable

of binding heavy metals in a triscysteine environment. Angew Chem Int Ed Engl 50(9): 2049–2053. doi:10.1002/anie.201006413

13. Plegaria JS, Pecoraro VL (2015) Sculpting metal-binding environments in *de novo* designed three-helix bundles. Isr J Chem 55(1):85–95. doi:10.1002/ijch.201400146
14. Cangelosi VM, Deb A, Penner-Hahn JE, Pecoraro VL (2014) A *de novo* designed metalloenzyme for the hydration of CO_2. Angew Chem Int Ed Engl 53(30):7900–7903. doi:10.1002/anie.201404925
15. Plegaria JS, Duca M, Tard C, Friedlander TJ, Deb A, Penner-Hahn JE, Pecoraro VL (2015) *De novo* design and characterization of copper metallopeptides inspired by native cupredoxins. Inorg Chem. doi:10.1021/acs.inorgchem.5b01330
16. The PyMOL Molecular Graphics System, Version 1.5.0.4 Schrödinger, LLC
17. Studier FW (2005) Protein production by auto-induction in high density shaking cultures. Protein Expr Purif 41(1):207–234. doi:10.1016/j.pep.2005.01.016

Chapter 12

Design of Light-Controlled Protein Conformations and Functions

Ryan S. Ritterson, Daniel Hoersch, Kyle A. Barlow, and Tanja Kortemme

Abstract

In recent years, interest in controlling protein function with light has increased. Light offers a number of unique advantages over other methods, including spatial and temporal control and high selectivity. Here, we describe a general protocol for engineering a protein to be controllable with light via reaction with an exogenously introduced photoisomerizable small molecule and illustrate our protocol with two examples from the literature: the engineering of the calcium affinity of the cell–cell adhesion protein cadherin, which is an example of a protein that switches from a native to a disrupted state (Ritterson et al. J Am Chem Soc (2013) 135:12516–12519), and the engineering of the opening and closing of the chaperonin Mm-cpn, an example of a switch between two functional states (Hoersch et al.: Nat Nanotechn (2013) 8:928–932). This protocol guides the user from considering which proteins may be most amenable to this type of engineering, to considerations of how and where to make the desired changes, to the assays required to test for functionality.

Key words Photoswitches, Computational protein design, Light-modulatable proteins, Protein engineering

1 Introduction

There has been considerable interest in light-based control of protein functions [1], and successful applications include light modulation of neuronal ion channels [2], light-switchable cell adhesion proteins [3], and light-controlled protein machines [4]. Light-based methods offer titratable, precise spatial, and temporal regulation that has been demonstrated in vitro [5], in cell culture [6, 7], and in whole animals [8]. Most examples of light-based control fall into one of two categories: (a) those that are genetically encoded using fusions with a light-sensitive protein borrowed from nature [6] and (b) those created via targeted insertion of amino acids into a protein sequence and subsequent reaction with them of an exogenously introduced photoisomerizable small molecule, typically azobenzene based [9]. Azobenzene and its derivatives undergo a reversible *cis–trans* isomerization upon illumination

Barry L. Stoddard (ed.), *Computational Design of Ligand Binding Proteins*, Methods in Molecular Biology, vol. 1414,
DOI 10.1007/978-1-4939-3569-7_12,

with either near-ultraviolet or visible light, leading to a change in end-to-end distance of ~18 Å in the *trans* state to a ~5–12 Å distribution in the *cis* state; this change in molecular shape can then be coupled to changes in protein function.

In this chapter, we describe two related methods for designing category (b) molecules (Fig. 1). The first method was applied to engineer light control of the group II chaperonin Mm-cpn [4] and illustrates the design of protein photoswitches that reversibly change between two known protein conformations. The described design method is useful when the target protein has two known functional conformations (e.g., chaperonins, nuclear hormone receptors, many small molecule binders), and the researcher would like to maintain both of them. (Note that, due to the two-state nature of azobenzene-based molecules, directing proteins into three or more conformations using light alone would require additional engineering.) The second method was applied to engineer light control of the cell adhesion protein cadherin [3] and designed a protein photoswitch that reversibility changes from a native to a disrupted conformation. Such designs are useful when a target protein has one functional conformation of interest to be disrupted, and the conformation in the disrupted state need not be specified in detail. We also include basic protocols for conjugating the small molecule to the protein, as well as for measuring the extent to which the protein is conjugated and switchable.

In general, the considerations for both design protocols share much in common. Most importantly, the target protein must be

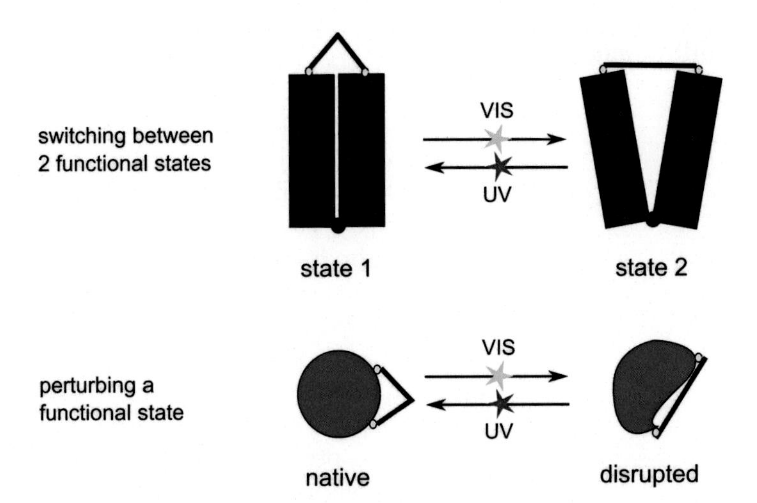

Fig. 1 Cartoon representation of two strategies for controlling protein function with light. In the first (*top*), the protein is switched between two defined conformations. In the second (*bottom*), the protein is switched between a functional state and a "disrupted" state, in which, for example, the conformation of an active or ligand-binding site is destabilized

suitable for cross-linking with a cysteine-reactive small molecule at defined sites, which requires elimination of all or most native cysteines and introduction of two nonnative cysteines to use as conjugation sites for cross-linking. In addition, the distance of the side-chain sulfur atoms of the engineered cysteines in the conformation(s) of the target to be maintained must match the end-to-end distance of the chosen azobenzene molecule in its *trans* and/or *cis* states, and the side-chain sulfur atoms of the engineered cysteines must be solvent accessible for cross-linking. The requirement of ensuring geometric compatibility necessitates experimentally determined structures or high-quality models for the protein conformations to be maintained. In addition, as initial characterizations of the photoswitches are carried out in vitro, protein targets should be amenable to protein purification and, ideally, be stable.

Where the protocols differ is in the detailed constraints necessary to satisfy. With two protein target states, one must find an interatomic distance between a pair of residues in the target such that in one state the distance is only compatible with one isomer of the chromophore and the second state only compatible with the other isomer of the small molecule; illumination should then lead to selective destabilization of one of the two protein states and hence reversible interconversion between them. Two states provide the advantage that the desired structures of both illumination states are known, increasing the probability that successful conjugation will produce a functional photoswitch. However, because of the additional geometric constraints, it is likely that only a small number of suitable cross-linking sites will be available, reducing the likelihood of finding one compatible with protein stability, structure, and function.

In comparison, with one functional state, one only needs to find a pair of attachment sites compatible with one isomer of the chromophore. Illumination should alter this distance and thus disrupt the conformation and function of the protein target. Because the geometric constraints need only be satisfied in one conformation, there are likely many more possible attachments sites. As the conformation of the illuminated state is not known, however, it is more likely that chromophore illumination may not result in the desired change in protein function. For example, if the local conformation of the protein is too flexible, the change in structure of the chromophore may be accommodated without a significant change in protein structure [9, 10].

2 Materials

1. Azobenzene chromophore suitable for cross-linking cysteine residues. For Mm-cpn we used azobenzene–dimaleimide (ABDM), and for cadherin we used 3,3′-bis(sulfonato)-4,4′-bis(chloroacetamido)azobenzene (BSBCA).

2. Structure(s) or high-quality models of protein of interest.
3. Academic license of the molecular modeling and design program Rosetta [11].
4. A protein structural viewer (such as PyMOL) capable of measuring distances, or script to compute distances from coordinates.
5. Supplies required to purify the target protein.
6. An assay for target functionality. Mm-cpn, 4 % native polyacrylamide gel (PAGE); cadherin, surface plasmon resonance (SPR).
7. UV–Vis spectrophotometer to determine the switching efficiency of the chromophore.
8. An assay to determine the efficiency of chromophore conjugation/cross-linking. Mm-cpn, 4–20 % SDS-PAGE gel; cadherin, mass spectrometer that is capable of detecting 1 Da changes in whole proteins, such as a Waters LCT Premier.
9. Illumination sources for ultraviolet and visible light to switch the chromophore between the *cis* and *trans* isomerization states. Mm-cpn and cadherin: high-power LEDs with emission wavelength of 365 nm (1 W, Advancemart) and 455 nm (3 W, SparkFun).

3 Methods

3.1 Computational Design of Protein Photoswitches

In this section, we detail the strategies and techniques to computationally design protein photoswitches with the goal of producing a ranked-ordered list of pairs of cysteine mutations to introduce into the target protein. At certain points in the method, the procedures bifurcate into parallel methods, depending on whether the target protein of interest has two conformations to maintain and switch between or a single conformation to disrupt.

3.1.1 Mutational Robustness

In order to preserve target structure and stability, locations must be identified within the target structure where a pair of cysteines can replace the native residues with minimal disruption to the overall fold. One strategy to identify the safest positions for these mutations is to estimate the folding free energy contribution of each native side chain. This energy is commonly estimated by independently mutating each residue to alanine, as alanine reduces the side chain to a single methyl group. This procedure, "alanine scanning," can be performed experimentally or computationally, which has provided rich data for the development of robustly tested computational protocols, including within Rosetta [12, 13]. Alanine scanning can be run within Rosetta using the RosettaScripts XML scripting interface. Detailed instructions are available within the RosettaScripts documentation [14].

We use computational alanine scanning within Rosetta to identify positions amenable to mutation to cysteine by only allowing mutations at positions where mutation to alanine is predicted not to destabilize the protein significantly (energy increase of less than 1 Rosetta energy units (approximately 1 kcal/mol); all positions that had a decrease in energy were accepted).

3.1.2 Distance Matching

1. To shift an equilibrium between two defined conformational states of a protein, select cross-linking positions for which the following criteria are satisfied:

   (a) The expected distance of the side-chain sulfur atoms of the engineered cysteines in one conformational state matches the end-to-end distance of the chosen chromophore in its *trans* state, and the other conformational state matches the length of the *cis* state (positive design) (*see* **Note 1**).

   (b) Each isomerization state is only compatible with one of the two conformational states of the protein and not the other (negative design).

2. To shift an equilibrium between one functional and one nonfunctional state (or ensemble) of a protein, select cross-linking positions for which the following criteria are satisfied:

   (a) The expected distance of the side-chain sulfur atoms of the engineered cysteines in the functional, known state matches the end-to-end distance of the chromophore in either the *trans* or *cis* states.

   (b) The expected distance of the side-chain sulfur atoms of the engineered cysteines after isomerization is incompatible with the functional state, leading to a disrupting conformation change in the protein. This can be accomplished, e.g., by disrupting secondary structure or distorting conformations of functional loops.

3.1.3 Solvent Accessibility

In order for the azobenzene chromophore to react efficiently with the target, the residues chosen as cross-linking sites must be solvent accessible. Using PyMOL [15] (or another method of the users' choice), identify all residues that are buried (have a solvent accessible surface area below a given threshold) and remove them from the list of possible mutations.

3.1.4 Steric Clashes (Visual Inspection)

Finally, residues to be cross-linked must be pointing toward each other, and the chromophore must be sterically compatible with the protein structure (e.g., a line drawn between the C_β atoms should not penetrate the protein). Generally, this process is easily done by visual inspection; in our experience, the vast majority of potential cross-link pairs are obviously sterically incompatible, leaving only a few pairs for consideration.

3.1.5 Protocol for the Structure-Based Design of a Photoswitchable Mm-cpn

The following two sections describe the specific parameters of the general protocol that we used for engineering photoswitchable Mm-cpn (Subheading 3.1.5) or cadherin (Subheading 3.1.6):

1. Using the PDB structures of Mm-cpn in the open and closed conformation (identifiers 3IYE and 3IYF, cryo-EM structures, and 3KFB and 3KFK, X-ray structures), calculate the expected distances between sulfur atoms for every possible pair of cysteine mutations in neighboring subunits of Mm-cpn as well as the expected accessible surface area of the sulfur. To create the models of the cysteine mutants and to do the calculations, use PyMOL or software of your choice.
2. Screen for residue pairs with an expected sulfur distance of 5–14 Å in the closed and 16.6–19.5 Å in the open state and a minimum expected surface accessible area for the sulfur atoms of 8 $Å^2$ (10 % of the maximum surface-exposed area of the sulfur atom in a deprotonated cysteine residue).
3. Keep residue pairs which satisfy the selection criteria in both sets of structures (3IYE/3IYF and 3KFB/3KFK).
4. Exclude residue pairs with an intra-monomer (Mm-cpn is a homooligomer of 16 subunits) distance smaller than 19.5 Å to avoid off-target cross-linking.
5. Visually inspect the list of possible cross-linking sites for residue pairs for which there is enough unoccupied space between the attachment sites to accommodate the chromophore ABDM, and choose promising cross-linking sites for in vitro testing.

3.1.6 Protocol for the Structure-Based Design of Photoswitchable Cadherin

1. Using Rosetta and the PDB structures of cadherin (identifiers 1FF5, 1EDH, 2O72, and 1Q1P) with the methodology described in Subheading 3.1.1, computationally mutate all residues in the protein to alanine, and record the predicted change in protein stability. Eliminate all residues with predicted change in stability >1 Rosetta energy unit.
2. Eliminate all residues that directly bind calcium ions.
3. For the residues that remain, compute the $C_β$–$C_β$ distance between all possible remaining pairs. Eliminate all pairs whose distances do not fall into the range 17–20 Å. After this step, the number of potential cross-linking pairs was reduced to approximately 1500.
4. Eliminate all pairs that do not have at least one cross-linking site ($C_β$ atom) within 20 Å of a calcium ion.
5. Eliminate all pairs that do not have solvent accessible surface area (SASA) of both cross-linking sites >30 $Å^2$. After this step, the number of potential pairs was reduced to approximately 300.

6. Visually inspect the remaining pairs by drawing a line between the cross-linking sites. Eliminate all pairs whose line intersects with the protein structure, by using the surface representation in PyMOL. This step reduced the number of possible pairs to approximately 30.
7. Using experimenter judgement, select a subset of pairs that meet the criteria of the study. We chose ten pairs based on a desire to have a diverse set of potential cross-linking sites.

3.2 Protein Engineering

After selecting the potential cross-linking sites via computational design, the next stage is to express and test the selected pairs experimentally. This method describes the removal of native cysteines and the addition of the cross-linking cysteines.

3.2.1 Elimination of Native Cysteines

The chromophores we used (ABDM and BSBCA) are cysteine reactive. As native cysteines in the protein target may also be reactive and give undesired side products, native cysteines should be removed, if possible, prior to mutation to cysteine residues at the cross-linking sites (*see* **Note 2**).

1. Using a suitable cloning method (e.g., Gibson assembly, site-directed mutagenesis), mutate a single native cysteine residue in the protein-coding sequence to an alternate amino acid (*see* **Note 3**).
2. Express and purify the mutated protein using a method appropriate for the specific protein, and then test the change in stability of the protein after the mutation using a method of choice (*see* **Note 4**).
3. If the protein has a specific function to be maintained, test changes in protein functionality after the mutation using a method appropriate to the specific function.
4. If **steps 1–3** result in a satisfactory outcome, repeat **steps 1–3** for an additional cysteine residue, continuing until all possible cysteines have been removed.
5. If **steps 1–3** do not result in a stable or functional protein, replace that cysteine by a different residue (repeat **steps 1–3**) or maintain the cysteine and repeat **steps 1–3** for the next cysteine in series (*see* **Note 5**).

3.2.2 Addition of Cross-Linking Cysteines

After all possible native cysteines have been removed, mutations to nonnative cysteines can be made.

1. Using a suitable cloning method (e.g., Gibson assembly, site-directed mutagenesis), mutate both native residues at the targeting cross-linking sites in the protein-coding sequence to cysteine.

2. Express and purify the mutated protein using a method of choice appropriate for the specific protein, and then test the change in stability of the protein after the mutation using a method of choice (*see* **Note 6**).
3. If the protein has a specific function to be maintained, test changes in protein functionality after the mutation using a method appropriate to the specific function.
4. If **steps 1–3** result in a satisfactory outcome, keep the potential pair. If not, eliminate it from future consideration.

3.3 Conjugating Protein with Small Molecule

In this section, we detail how to cross-link the azobenzene-based chromophore to the target protein, with the goal of optimizing the percentage of cross-linked and folded protein. We describe details of the important parameters controlling the outcome of the reaction, including how they may change based on the particular chromophore chosen.

3.3.1 Choice of Chromophore and Reactive Group (ABDM vs. BSBCA)

1. Chromophore absorption spectrum. Azobenzene cross-linkers with have been recently developed to enable the user to choose between a wide variety of wavelengths to switch the isomer equilibrium of the chromophore [9, 16–18]. The two chromophores used in the methods described in this chapter have the following absorption properties: The absorption peak of the π–π* transition of the *trans* state of ABDM is at 342 nm, and the long wavelength n–π* band of the *cis* state used for selective *cis–trans* isomerization is at 440 nm [19]. For BSBCA, the π–π* band is shifted to 363 nm and the *cis* n–π* band is at 450 nm [20].
2. Reactive groups. Maleimide, the reactive group of ABDM, reacts fast and specifically with thiols at a pH between 6.5 and 7.5 but is unstable in water. Proteins can be cross-linked at incubation times of less than 1 h at RT and at fairly low concentrations of protein and cross-linker (e.g., *see* Subheading 3.3.3). This strategy may be advisable for the conjugation of sensitive target proteins or for the conjugation of metastable protein states.

   Chloroacetamide, the reactive group of BSBCA, is also specific toward thiols, but is stable in water, and its reactivity is considerably lower than that of maleimides. This makes incubation times of several hours, high chromophore concentrations, optimized buffer conditions, and elevated incubation temperatures necessary to achieve satisfactory conjugation efficiency (*see* Subheading 3.3.4).
3. Chromophore solubility and bistability of azobenzene isomerization states. ABDM is not very soluble in aqueous solutions in its *trans* isomerization state. Therefore, it is advisable to cross-link a protein with ABDM in the *cis* state after pre-illumination with UV light. An advantage of ABDM, however, is the

high bistability of its two isomers. The *cis* isomerization state is stable for several hours due to a low rate of the thermal *cis* to *trans* isomerization [4, 19].

BSBCA is designed to be highly soluble in water due to the addition of sulfonate groups to the aromatic rings of the azobenzene. The rate of thermal *cis to trans* isomerization at room temperature is approximately 20 min, though this can be considerably longer when conjugated to protein [3, 20].

3.3.2 Reaction Conditions

See Burns et al. [20] for a comprehensive overview of conjugation reaction conditions with BSBCA. For ABDM refer to [4, 19, 21] or the protocol below.

3.3.3 Protocol for Conjugating Mm-cpn with ABDM

1. Dilute purified Mm-cpn to 500 μl at a concentration of 0.25 μM Mm-cpn (complex concentration) in Buffer A (20 mM HEPES pH 7.4, 50 mM KCl, 5 mM $MgCl_2$, 10 % glycerol).
2. Bias the conformational equilibrium of Mm-cpn toward the closed state via addition of $ADPAlF_X$ (a phosphate analogue which binds to hydrolyzed ATP after phosphate release) by adding 1 mM ATP, 6 mM $Al(NO_3)_3$, and 25 mM NaF to the solution (buffer A+, pH 7.0), and incubate the sample for 20 min at 43 °C [22].
3. Dissolve ABDM in dimethylformamide (DMF) to a concentration of 1.2 mM. Prior to cross-linking, illuminate ABDM for 1.5 min using the UV LED. UV illumination results in an accumulation of ~75 % *cis* isomer in the solution (estimated by analyzing the absorption spectrum of the sample [19]).
4. Add ABDM to the Mm-cpn solution at a ratio of 1 μl ABDM solution per 50 μl protein solution and shield the sample from background illumination. Quench the reaction after 40 min incubation time by adding dithiothreitol (DTT) to a concentration of 2 mM.

3.3.4 Protocol for Conjugating Cadherin with BSBCA

1. Dilute purified cadherin (protocol described in Ritterson et al. [3]) to a final concentration of 160 μM in 25 mM Tris–HCl pH 8.5, 400 mM NaCl, 1 mM EDTA, 3 mM KCl, 3 mM Tris(2-carboxyethyl)phosphine (TCEP), 500 μM BSBCA.
2. Place reactions at 25 ° C in the dark for 72 h.
3. Desalt excess chromophore using a HiPrep 26/10 (GE) column (*see* **Note 7**) into 25 mM Tris–HCl pH 8.5, 400 mM NaCl, 1 mM EDTA, 3 mM KCl, 3 mM TCEP.

3.4 Measuring Protein Conjugatability

In this section, we describe methodologies for measuring the fraction of total protein conjugated, with the goal of providing the researcher insight into which parameters of the reaction may need optimization and information about which cysteine pairs conjugate most completely. As in other sections, the method splits into parallel methods, based on the target structure of interest (cadherin monomer versus Mm-cpn chaperonin protein complex). Generally, a wide range of potential methods are possible, and the particular method chosen will depend on the protein target of interest.

3.4.1 Measuring Cross-Linking Ratio for ABDM-Mm-cpn with an SDS-PAGE Gel

1. Remove 20 μl of the sample cross-linked in Subheading 3.3.3, and analyze it on a 4–20 % gradient SDS-PAGE gel. Formation of covalently linked Mm-cpn multimers after subunit cross-linking by ABDM leads to multimer bands which can be easily distinguished from the 60 kDa band of the Mm-cpn monomer (*see* Fig. 2 in ref. [4]).
2. Estimate the cross-linking stoichiometry of the sample defined as the fraction of cross-links to possible cross-linking sites in the protein ensemble by calculating the sum of the relative intensities of the multimer bands (band intensity divided by the sum of the band intensities for all multimers) weighted by their ratio of cross-links to subunits using, e.g., the ImageJ [23] software package.

3.4.2 Estimating Conjugatability for Cadherin Using Mass Spectrometry

This example method for photoswitchable cadherin assumes one has the results of a conjugation reaction on hand (from Subheading 3.3.4) and wishes to know to what extent the reaction completed. Buffers are provided in the original work [3].

1. Estimate the protein concentration using A_{280}.
2. Dilute protein to an estimated 1 μM concentration in pure water (*see* **Note 8**).
3. Inject the conjugated sample into the mass spectrometer, observing a peak at 23,813 Da for unconjugated protein and 24,266 Da for conjugated (*see* **Note 9**).
4. Estimate the fraction of protein conjugated by calculating the peak areas for each subspecies and dividing the area of the conjugated peak by the sum of the areas of all subspecies. Potential conjugatabilities range widely, from 0 to 100 % depending on the cysteine pair (*see* **Note 10**).

3.5 Measuring Chromophore Switchability/Rate of Thermal Cis–Trans Back Reaction

In this section, we describe a method to determine the extent to which the chromophore in a cross-linked system undergoes isomerization upon illumination, without describing whether that isomerization causes a functional change in protein structure or state. We also provide a method to measure the half-life of the *cis* isomerization state, so that a researcher may determine which photoswitchable candidates are most promising to test in functional assays.

3.5.1 Illumination Techniques

The *cis–trans* isomer equilibrium of azobenzene-based chromophores can typically be switched in the direction of the *cis* state by exciting the π–π* band of the *trans* state in the near UV. To switch the equilibrium in the direction of the *trans* state, excite the n–π* band of the *cis* isomer with blue (or green) light [9]. We recommend the use of high-power LEDs for illumination as they are widely available, relatively inexpensive, portable, and intense enough to isomerize protein in bulk within seconds to minutes. Lasers can also be used for illumination, particularly in microscope and other applications where high spatial precision is desirable. Keep isomerized protein in the dark to the extent possible to prevent undesirable isomerization due to ambient light.

3.5.2 UV–Vis Spectroscopy

Prior to assessing the extent to which illumination modulates protein function, we recommend determining whether the chromophore conjugated to the protein is photoisomerizable by illuminating *trans*-relaxed protein with UV light. The *trans* states of the azobenzene-based chromophores used in this protocol have a characteristic near-UV absorption peak of the π–π* transition, and, upon illumination at that wavelength, the peak amplitude decreases as the small molecule isomerizes into the *cis* state.

1. Measure extinction coefficient $\varepsilon_{\text{trans}}$ of the unconjugated protein for the peak wavelength of the π–π* transition of the chromophore (ABDM: 342 nm; BSBCA 363 nm) using protein at a known concentration (if the protein has no cofactor bound that absorbs light at that wavelength, ε should be approximately zero) (*see* **Note 11**).
2. Compute $\varepsilon_{\text{trans}}$ for the conjugated protein using the sum of the extinction coefficients of the free chromophore in the *trans* state and the unconjugated protein.
3. Measure the absorption spectrum of the conjugated protein prior to illumination (*see* **Note 12**).
4. Illuminate the protein at the absorbance maximum (ABDM: 342 nm; BSBCA 363 nm) using a method of choice, and remeasure the absorption spectrum of the protein every 2 min of illumination time. Cease illumination once the absorption of the π–π* band reaches a minimum.
5. Estimate the fraction of protein that photoswitches using the following equation:

$$\text{Frac} = \frac{\varepsilon_{\text{peak},trans} - \varepsilon_{\text{peak, mix}}}{\varepsilon_{\text{peak},trans} - R * \varepsilon_{\text{peak},trans}}$$

where R is the *cis–trans* extinction coefficient ratio for the π–π* band of the chromophore; *peak* refers to the wavelength of the chromophore-specific absorption maximum; ε*trans* is the measured extinction coefficient for the thermodynamically equilibrated, 100 % *trans* state; and ε_{mix} is the measured extinction coefficient for the photostable, UV-illuminated state containing a mix of *cis* and *trans* (**Notes 12** and **13**).

3.5.3 Measuring Bistability/Relaxation Rate

6. Measure the absorbance of the conjugated protein at the peak wavelength of the π–π* transition of the chromophore (ABDM: 342 nm; BSBCA 363 nm) prior to illumination.
7. Using the same methods as in Subheading 3.5.2, illuminate the protein to photostability.
8. Measure the absorbance of the conjugated protein at the peak wavelength of the π–π* transition of the chromophore (ABDM: 342 nm; BSBCA 363 nm) immediately following illumination (time zero, t_0).
9. Keep the conjugated protein in the dark. Every 5 min (or a time of the experimenter's choosing), remeasure the absorbance of the sample.
10. Repeat **step 3** until the protein relaxes back to the unilluminated state. The half-life ($t_{1/2}$ of the illuminated state) is the time point at which the absorbance of the sample is halfway between the absorbances measured in **steps 1** and **2**.

3.6 Structural/Functional Assay

The particular method chosen for assaying whether photoswitching induces a structural or functional change will depend on the target protein. Here, we provide an example of a native gel assay used to determine changes in conformation upon illumination.

3.6.1 Native Gel Assay to Probe the Light-Induced Conformational Switching of ABDM-Mm- cpn

1. Use the cross-linked samples from Subheading 3.3.3.
2. To switch azobenzene between the *cis* and the *trans* isomerization states, expose the cross-linked Mm-cpn sample to alternating illumination for 20s with the blue LED (*cis* → *trans* isomerization) or for 90s with the UV LED (*trans*→ *cis* isomerization). For this the sample is pipetted in a 200 μl PCR tube without a cap and placed in a PCR tube rack. Illuminate from the top by placing the LED as close to sample as possible (in our case in a distance of ~1 cm) to maximize light exposure (*see* **Note 14**).
3. Illuminate the sample alternately with blue and UV light. After each illumination step, remove 20 μl of the sample for structural characterization.
4. Load the samples on a 4 % native PAGE gel and run it for 30 min at 160 V.
5. Stain and destain the gel with Coomassie blue and take a picture of the gel. You can observe the light-induced switching between the closed and open conformations of Mm-cpn via a

clear distinct band shift between both conformations on the gel (*see* Fig. 2 in ref. [4]).

4 Notes

1. Keep in mind that the end-to end distance distribution of the *cis* isomerization state is significantly broader than the one of the *trans* state (i.e., more rigid due to the planar extended π electron system) [9].
2. Deeply buried cysteines in the native protein may not be reactive and could be maintained. However, proteins often have some flexibility and can transiently expose buried positions. As a result, we recommend attempting to mutate all native cysteines and adding back those that cannot be mutated to an alternate residue without compromising protein stability or function.
3. We recommend using serine as a replacement for surface-exposed cysteine and alanine for buried cysteines. In theory, all cysteines could be removed in one step. Sequential mutation, although time consuming, allows one to identify any particularly troublesome cysteines that may have to be added back later.
4. There are a multitude of protein expression and purification methods, and the choice of a particular method is outside the scope of this chapter.
5. It is possible that surface-exposed cysteines distant from the intended cross-linking sites, even if they are labeled with chromophore, will not cross-link the protein and thus may not affect function. However, the presence of those additional labeled cysteines complicates measurement of protein concentration, conjugatability, and switchability.
6. When purifying and handling cysteine-containing proteins, maintain reducing agent (e.g., DTT, TCEP) wherever possible to avoid oxidation/disulfide bond formation of cysteine residues. Note, however, that thiol-based reducing agents can interfere with chromophore conjugation.
7. BSBCA tends to migrate slowly in common chromatography media and can be difficult to elute, especially in the presence of salt. It can be removed by washing the column thoroughly and repeatedly with pure water.
8. The presence of salts leads to adduct formation and the appearance of side peaks in the instrument, obfuscating the results. Cadherin is stable for hours in pure water without any salt; the stability of other proteins may vary.

9. For BSBCA, the conjugated, cross-linked protein will appear at +453 Da relative to unconjugated protein. In our hands, we never observed single-linked protein or protein conjugated to two single-linked chromophores. This is likely due to the much faster intramolecular reaction rate of the single-linked protein to the remaining cysteine compared to side reactions. If native cysteines in the protein were required to be maintained, however, reaction to them by chromophore will result in the appearance of additional peaks.
10. We assume the ionizability of the cross-linked protein is the same as the uncross-linked for the purposes of computing cross-linked fraction.
11. Different chromophores have different extinction coefficients. If protein concentration is to be measured using A_{280}, ε_{280} for the chromophore can be measured using pure chromophore of known concentration, and the conjugated protein concentration can be calculated using $\varepsilon_{280,conjugate} = \varepsilon_{280,chromophore} + \varepsilon_{280,unconjugated\ protein}$. This assumes the extinction coefficient of the small molecule does not change during conjugation; this assumption can be checked by comparing band intensities of unconjugated and unconjugated proteins at the same nominal concentrations with an alternate assay (e.g., Bradford or SDS-PAGE).
12. A pure population of *trans* protein can be obtained by first illuminating protein with visible light, followed by keeping protein in the dark for an extended period of time (e.g., overnight).
13. Computing R relies on knowing ε_{peak} for the *cis* chromophore, which may be difficult to obtain, as *cis* chromophore may be difficult to isolate for measurement. A previous study used an *R* value of 0.541 for computing protein concentrations, based on measurements of BSBCA chromophore isomers separated by HPLC [3].
14. UV light is absorbed by conventional glass and plastic.

Acknowledgments

Work on light-switchable protein functions in our group was supported by grants from the Program for Breakthrough Biomedical Research and the Sandler Family Foundation (to T. K.), the National Institutes of Health (PN2EY016525, PI Wah Chiu), the National Science Foundation (NSF CBET-1134127 to T.K.), and a Deutsche Forschungsgemeinschaft postdoctoral fellowship (HO 4429/2-1 to D.H.). We particularly thank A. Woolley (U Toronto) for the advice, discussions, and gifts of ABDM and BSBCA.

References

1. Krauss U, Drepper T, Jaeger KE (2011) Enlightened enzymes: strategies to create novel photoresponsive proteins. Chemistry 17(9):2552–2560. doi:10.1002/chem.201002716
2. Banghart M, Borges K, Isacoff E, Trauner D, Kramer RH (2004) Light-activated ion channels for remote control of neuronal firing. Nat Neurosci 7(12):1381–1386. doi:10.1038/nn1356
3. Ritterson RS, Kuchenbecker KM, Michalik M, Kortemme T (2013) Design of a photoswitchable cadherin. J Am Chem Soc 135(34):12516–12519. doi:10.1021/ja404992r
4. Hoersch D, Roh SH, Chiu W, Kortemme T (2013) Reprogramming an ATP-driven protein machine into a light-gated nanocage. Nat Nanotechnol 8(12):928–932. doi:10.1038/nnano.2013.242
5. Woolley GA, Jaikaran ASI, Berezovski M, Calarco JP, Krylov SN, Smart OS, Kumita JR (2006) Reversible photocontrol of DNA binding by a designed GCN4-bZIP protein. Biochemistry 45(19):6075–6084. doi:10.1021/bi060142r
6. Levskaya A, Weiner OD, Lim WA, Voigt CA (2009) Spatiotemporal control of cell signalling using a light-switchable protein interaction. Nature 461(7266):997–1001. doi:10.1038/nature08446
7. Zhang F, Muller KM, Woolley GA, Arndt KM (2012) Light-controlled gene switches in mammalian cells. Methods Mol Biol 813:195–210. doi:10.1007/978-1-61779-412-4_12
8. Wyart C, del Bene F, Warp E, Scott EK, Trauner D, Baier H, Isacoff EY (2009) Optogenetic dissection of a behavioural module in the vertebrate spinal cord. Nature 461(7262):407–410. doi:10.1038/nature08323
9. Beharry AA, Woolley GA (2011) Azobenzene photoswitches for biomolecules. Chem Soc Rev 40(8):4422–4437. doi:10.1039/c1cs15023e
10. Ali AM, Woolley GA (2013) The effect of azobenzene cross-linker position on the degree of helical peptide photo-control. Org Biomol Chem 11(32):5325–5331. doi:10.1039/c3ob40684a
11. Rosetta Commons (2015) Rosette license and download. https://www.rosettacommons.org/software/license-and-download. Accessed 5/31/2015
12. Kortemme T, Baker D (2002) A simple physical model for binding energy hot spots in protein-protein complexes. Proc Natl Acad Sci U S A 99(22):14116–14121. doi:10.1073/pnas.202485799
13. Kortemme T, Kim DE, Baker D (2004) Computational alanine scanning of protein-protein interfaces. Science STKE 2004(219):pl2. doi:10.1126/stke.2192004pl2
14. Rosetta Commons (2015) Rosetta documentation. https://www.rosettacommons.org/docs. Accessed 27 June 2015
15. Schrodinger LLC (2010) The PyMOL molecular graphics system, Version 1.3r1
16. Samanta S, McCormick TM, Schmidt SK, Seferos DS, Woolley GA (2013) Robust visible light photoswitching with ortho-thiol substituted azobenzenes. Chem Commun (Camb) 49(87):10314–10316. doi:10.1039/c3cc46045b
17. Samanta S, Babalhavaeji A, Dong MX, Woolley GA (2013) Photoswitching of ortho-substituted azonium ions by red light in whole blood. Angew Chem Int Ed Engl 52(52):14127–14130. doi:10.1002/anie.201306352
18. Beharry AA, Sadovski O, Woolley GA (2011) Azobenzene photoswitching without ultraviolet light. J Am Chem Soc 133(49):19684–19687. doi:10.1021/ja209239m
19. Umeki N, Yoshizawa T, Sugimoto Y, Mitsui T, Wakabayashi K, Maruta S (2004) Incorporation of an azobenzene derivative into the energy transducing site of skeletal muscle myosin results in photo-induced conformational changes. J Biochem 136(6):839–846. doi:10.1093/jb/mvh194
20. Burns DC, Zhang F, Woolley GA (2007) Synthesis of 3,3′-bis(sulfonato)-4,4′-bis(chloroacetamido)azobenzene and cysteine cross-linking for photo-control of protein conformation and activity. Nat Protoc 2(2):251–258. doi:10.1038/nprot.2007.21
21. Schierling B, Noel AJ, Wende W, le Hien T, Volkov E, Kubareva E, Oretskaya T, Kokkinidis M, Rompp A, Spengler B, Pingoud A (2010) Controlling the enzymatic activity of a restriction enzyme by light. Proc Natl Acad Sci U S A 107(4):1361–1366. doi:10.1073/pnas.0909444107
22. Douglas NR, Reissmann S, Zhang J, Chen B, Jakana J, Kumar R, Chiu W, Frydman J (2011) Dual action of ATP hydrolysis couples lid closure to substrate release into the group II chaperonin chamber. Cell 144(2):240–252. doi:10.1016/j.cell.2010.12.017
23. Abràmoff MD, Magalhães PJ, Ram SJ (2004) Image processing with ImageJ. Biophoton Int 11(7):36–42

Chapter 13

Computational Introduction of Catalytic Activity into Proteins

Steve J. Bertolani, Dylan Alexander Carlin, and Justin B. Siegel

Abstract

Recently, there have been several successful cases of introducing catalytic activity into proteins. One method that has been used successfully to achieve this is the theozyme placement and enzyme design algorithms implemented in Rosetta Molecular Modeling Suite. Here, we illustrate how to use this software to recapitulate the placement of catalytic residues and ligand into a protein using a theozyme, protein scaffold, and catalytic constraints as input.

Key words Enzyme design, De novo enzyme design, Rosetta, Theozyme

1 Introduction

The design of enzyme catalysts with catalytic proficiencies rivaling natural enzymes remains a major challenge in biochemistry. Successful design of active enzymes would be the ultimate proof of our understanding of enzymatic catalysis. The field of computational enzyme design has had successes in the past decade, with the successful computational design of enzyme catalysts that perform Kemp elimination [1], retro-aldol condensation [2], and Diels-Alder cyclization [3]. These examples used the Rosetta Molecular Modeling Suite [4] to introduce activity into protein scaffolds.

The introduction of new chemistry into a protein using the Rosetta Molecular Modeling Suite consists of the following steps:

1. **Theozyme generation**: A theozyme is a geometric description of the transition state of a reaction that is stabilized by interactions from protein residues [5]. There are several ways to generate theozymes: they may be calculated using QM methods, by direct observation of a crystal structure with an inhibitor bound, or by chemical intuition. The goal of a theozyme is to

Barry L. Stoddard (ed.), *Computational Design of Ligand Binding Proteins*, Methods in Molecular Biology, vol. 1414,
DOI 10.1007/978-1-4939-3569-7_13, © Springer Science+Business Media New York 2016

define the geometry of the transition state relative to the amino acids in the protein performing the chemistry. This includes selecting a low-energy mechanism for the reaction of interest and testing various amino acids to stabilize and lower the reaction energy. The geometry of the theozyme is written in a constraint file, which defines the residues and atoms involved and distances, angles, and dihedrals between them.

2. **Scaffold selection**: In order to place the theozyme into a protein, the engineer must choose a set of proteins to use as scaffolds. The proteins in this set will depend on the project goals. For example, if the end goal will be to introduce function into a protein that will be used in a thermophilic environment, then it would be best to select a subset of the proteins from the PDB that are already thermophilic proteins. It is recommended to work with crystal structures at high resolution (<2.0 Å).
3. **Match**: With both a set of protein scaffolds and a theozyme in hand, the next step is to find a protein in which the amino acids of the theozyme can be introduced and the ligand built off in a geometric orientation that satisfies all of the constraints, while not sterically overlapping with the protein backbone. RosettaMatch [6] is one software package that can perform this search. It requires a geometric description of the catalytic residues geometric orientation relative to the transition state (i.e., a constraint file) and a set of "scaffold" PDB files and positions within the PDB to search for potential placement of the catalytic amino acids (i.e., a position file). First, all the amino acids in the positions defined in the position file are converted into alanine. Next, the algorithms searches residue by residue, attempting to find a set of positions that allow both the catalytic residues and the theozyme to be introduced in an orientation that is within the geometric parameters defined in the constraint file. After running RosettaMatch, the result will be a set of scaffolds containing mutations and a ligand matching the geometry in the constraint file (i.e., matches).
4. **Enzyme design**: The next step is optimizing additional molecular interactions at the protein–theozyme interface by introducing mutations predicted to stabilize the conformations of residues involved in the theozyme (including the ligand). The enzyme design protocol in Rosetta [7] starts with a "match" from the matcher containing the theozyme grafted onto the protein backbone and designs the local region for complementarity to the ligand and interactions that stabilize the catalytic residues. This step also refines the ligand placement with finer sampling than is performed in the matching step.
5. **Manual refinement of designs**: The final computational step is an interactive assessment of the designs using real-time eval-

uation with Rosetta's energy function. Foldit [8] is used to enable the researcher to refine the automated designs using chemical intuition or external knowledge in order to optimize designs before experimental characterization.

6. **Experimental validation and characterization**: Designed sequences should be predicted to have overall low energy and be in close agreement with the defined constraints. Designs may be selected for testing by a combination of visual assessment and energetic scores, as well as other considerations such as viability of expression. In order to evaluate whether the introduced function is truly rate enhancing, we suggest following the guidelines of Wolfenden and calculating the k_{cat}/k_{uncat} [9, 10]. The result, if all goes well, is a novel enzyme that performs the desired function.

2 Materials

Here, we will provide an example of the design of an enzyme by recapitulating the active site of a glycosyl hydrolase through placing a p-nitrophenyl-beta-d-glucopyranoside (pNPG) substrate and catalytic amino acids into the native enzyme scaffold using the protein scaffold, theozyme, and chemical constraints as inputs. The protein selected in this case natively performs the chemistry desired. But this illustrates how the entire process works and could be readily adopted for introducing function into protein scaffolds that do not natively carry out the desired chemistry. The materials needed are:

1. A 3D model of the desired substrate, pNPG.
2. A constraint file describing the theozyme. In this case, we will use an experimentally derived geometry from the crystal structure 2JIE [11], which contains a transition state inhibitor (2-deoxy-2-fluoro-alpha-d-glucopyranose) that closely mimics the shape of the desired substrate, pNPG (Fig. 1).
3. A protein scaffold into which we can match the theozyme described in the constraint file. In this tutorial, we will use 2JIE itself for this walkthrough as discussed earlier.
4. A computer with Rosetta installed and Internet connection. For this tutorial, Rosetta has been installed from source into the home directory. The path to Rosetta binaries is ~/Rosetta/main/source/bin (*see* **Note 1**).
5. FoldIt. In order to load designs with transition state models, this must be compiled from source. Instructions for doing so can be found at *https://wiki.rosettacommons.org/index.php/Foldit_Getting_Started*.

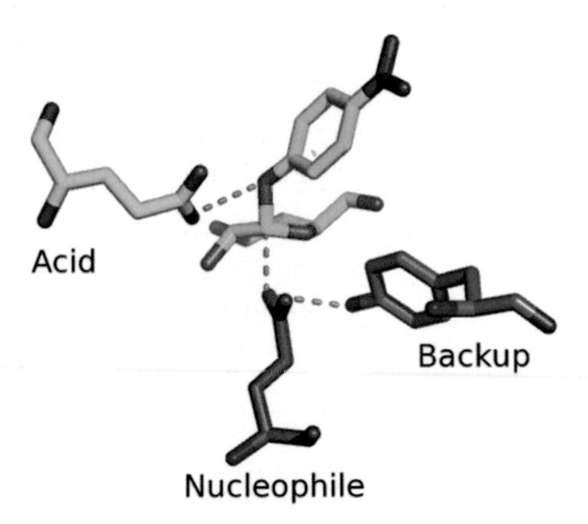

Fig. 1 The theozyme used in this example, showing the modeled pNPG, two glutamates, and one tyrosine residue in the proper orientation for hydrolysis

6. PyMOL version 1.5.0.5 or greater. Any molecular modeling program may be used, but the commands in this example are tested with the executable version of PyMOL, built upon version 1.5.0.5 © Schrodinger LLC [12].

3 Methods

3.1 Theozyme Generation

Entries beginning with > indicate the line is to be executed at a command prompt. Entries beginning with PyMOL> indicate the line is to be executed inside of the PyMOL command prompt. All of the files and commands used in this tutorial are available at https://github.com/SiegelLab/matcher_files (*see* **Note 2**).

1. Begin with a 3D conformer of pNPG as a Sybyl Mol 2 file. The model reaction here will be modeled as an S_n2-like reaction where the substrate transitions into a pentacoordinate transition state in which the anomeric carbon is approximately sp^2 hybridized. Therefore, the three atoms bonded to the anomeric carbon should be planar (Fig. 2). Using a program such as Spartan [13] constraints can be implemented using this chemical information where distances and dihedrals are locked at the proposed relative geometries for the transition state. The remainder of the molecular can undergo molecular mechanics minimization. The resulting model should be saved as LG1.mol2.

   In order for Rosetta to understand how to treat the ligand, we must convert the mol2 formatted file into a Rosetta params file. This file contains the atom type (e.g., primary carbon, hydroxyl, acid, etc.) specifications in order for Rosetta to calculate the molecular energy of the system with the ligand in terms of the energy function being used.

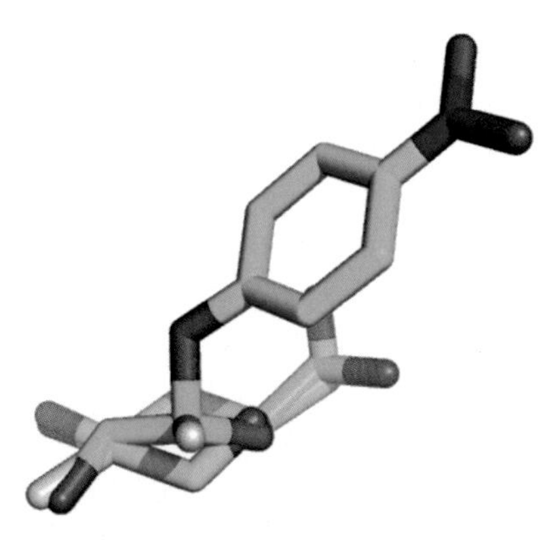

Fig. 2 The crystal structure ligand (*cyan*) versus the model of the transition state of pNPG (*green*). Only the hydrogen mentioned (part of the sp^2 hybridization of the anomeric carbon) is shown

2. To generate a parameters file for pNPG, run the Python script molfile_to_params.py using the 3D conformer of pNPG as the input:

```
>python ~/Rosetta/main/source/src/python/apps/public/
molfile_to_params.py -n LG1 LG1.mol2.
```

   The script will write a parameters file, LG1.params, and a PDB of the ligand called LG1_0001.pdb, which defines Rosetta atom types for each atom in the ligand. The params file defines a "neighbor" atom for the molecule, which is the center of mass atom of the ligand, by default. In the params file, this neighbor atom is denoted on the line that starts with "NBR_ ATOM." This atom is used when overlaying alternate conformations and may be a problem if the atom chosen moves in the alternate conformations. It is recommended that the user open the LG1_0001.pdb file, turn on the atom names, and verify the neighbor atom chosen, and the atoms it is directly bonded are the most relevant atoms for the chemistry being carried out.

3. Convert the theozyme into a Rosetta enzyme design/matcher constraint file. Using the atom names in LG1_0001.pdb and the standard Rosetta atom names for canonical amino acids (located in the source at *~/Rosetta/main/database/chemical /residue_type_sets/fa_standard/residue_types/l-caa*), write the constraint file defining the theozyme geometry. This has been previously described, and we refer to the literature [7] and the online documentation found at https://www.rosettacommons.org/docs/wiki/rosetta_basics/file_types/match-cstfile-format. In addition, there are several resources available at the Meiler lab research page http://www.meilerlab.org/index.php/jobs/resources -> Rosetta Resources -> Enzyme Design.

   Briefly, we wish to define three distinct constraints.

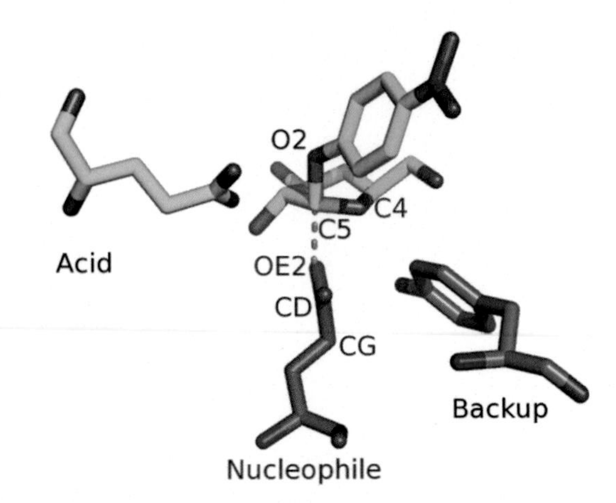

Fig. 3 Constraint Block 1 that describes the nucleophilic attach of the GLU353 in the native crystal structure

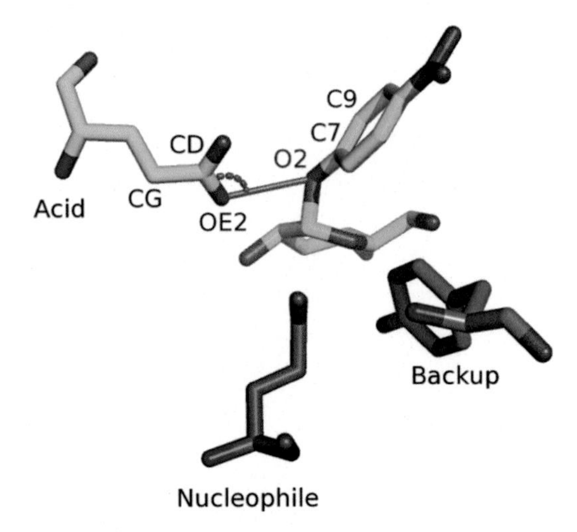

Fig. 4 Constraint Block 2 which describes the protonation of the leaving group. In the renumbered crystal structure 2jie, this is performed by residue GLU164

1. Nucleophile: The nucleophilic Glu353 must be constrained in accordance with an S_n2-like transition state geometry such that the carboxylate oxygen of Glu353, the anomeric carbon of the substrate, and the leaving group oxygen are collinear (Fig. 3).
2. Acid: The carboxylate of the acid–base, Glu164, must be constrained such that the carboxylate oxygen, the proton, and the leaving group oxygen are collinear (Fig. 4).
3. Backup: Finally, Tyr295 must be constrained to be within hydrogen bonding distance from the nucleophilic oxygen of Glu353 in order to maintain the Glu353 in the correct orientation for nucleophilic attack (Fig. 5).

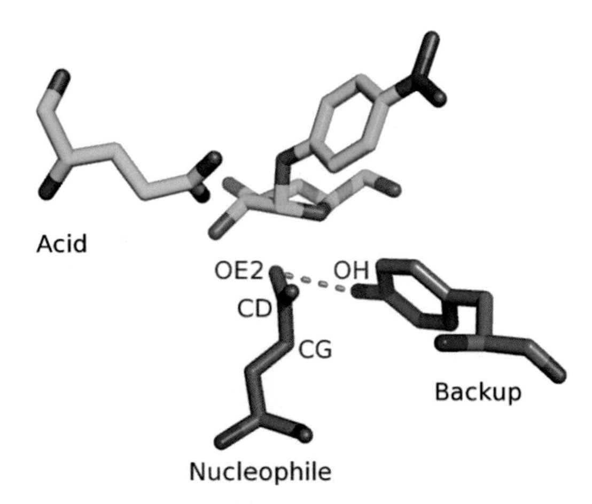

Fig. 5 Constraint Block 3 which describes the TYR interaction with the nucleophile, GLU

For this tutorial, it is sufficient to measure the distances, angles, and dihedrals from the crystal structure of 2JIE in complex with the transition state inhibitor and write these down in LG1.enzdes.cst for constraints #1 and #3. The inhibitor lacks a leaving group, and thus the researcher is left to make constraint #2 using knowledge of organic chemistry and idealized geometries. Here, we have idealized the angles and dihedrals to be round numbers in accordance with fundamental organic chemistry principles. The three constraints are as follows:

```
# GLU nucleophile to LG1
```

The following describe the geometry desired in the matcher and for the enzyme design:

```
TEMPLATE:: ATOM_MAP: 1 atom_name: C5 O2 C4
TEMPLATE:: ATOM_MAP: 1 residue3: LG1
TEMPLATE:: ATOM_MAP: 2 atom_name: OE2 CD CG
TEMPLATE:: ATOM_MAP: 2 residue1: E
CONSTRAINT:: distanceAB: 2.0 0.3 500.0 1 0
CONSTRAINT:: angle_A: 180.0 20.0 500.0 360 0
CONSTRAINT:: angle_B: 120.0 20.0 500.0 360 0
CONSTRAINT:: torsion_B: -180.0 30.0 500.0 360 0
CONSTRAINT:: torsion_AB: 180.0 30.0 500.0 360 0
CONSTRAINT:: torsion_A: -42.0 30.0 500.0 360 0
```

The last column describes the number of additional samples RosettaMatch will discretely test. When this column is set to 0, RosettaMatch will try the default value (2.0). If no matches are found, increase the degrees of freedom (DOFs) by increasing the values in this column (*see* **Note 3**).

```
# GLU acid to LG1 leaving group oxygen:
   TEMPLATE:: ATOM_MAP: 1 atom_name: O2 C7 C9
   TEMPLATE:: ATOM_MAP: 1 residue3: LG1
   TEMPLATE:: ATOM_MAP: 2 atom_name: OE2 CD CG
   TEMPLATE:: ATOM_MAP: 2 residue1: E
   CONSTRAINT:: distanceAB: 3.0 0.5 500.0 0
   CONSTRAINT:: angle_B: 120.0 25.0 500.0 360
   CONSTRAINT:: torsion_B: 180.0 35.0 500.0 180
   ALGORITHM_INFO:: match
   SECONDARY_MATCH: DOWNSTREAM
   ALGORITHM_INFO::END
# TYR backup to GLU nucleophile (Constraint block 1):
   CST::BEGIN
   TEMPLATE:: ATOM_MAP: 1 atom_name: OE2 CD CG
   TEMPLATE:: ATOM_MAP: 1 residue1: E
   TEMPLATE:: ATOM_MAP: 2 atom_type: OH
   TEMPLATE:: ATOM_MAP: 2 residue1: Y
   CONSTRAINT:: distanceAB: 3.0 0.5 500.0 0
   ALGORITHM_INFO:: match
   SECONDARY_MATCH: UPSTREAM_CST 1
   ALGORITHM_INFO::END
```

The resulting file in which these constraints are placed should be called LG1.enzdes.cst. This file serves as both the RosettaMatch file and the Rosetta enzyme design constraint file.

4. Generate conformations.
   This step allows the researcher to control over which degrees of freedom are sampled of the ligand. There are many ways to achieve this including Spartan [13], Omega [14], Gaussian [15], and Confab [16]. These programs evaluate a number of different conformations while taking in consideration the energy, intramolecular interactions, and steric interactions. It is highly recommended to use one of these programs to generate a representative set of conformations. However, for simplicity we shall generate our conformations by hand. What follows describes how to hand generate rotations about the oxygen atom of the leaving group, thus rotating the *p*-nitrophenyl group. It is important to note that when generating the conformations by hand, the energy of the ligand is not taken into consideration. This is critical since the intramolecular energy of the ligand is essentially not considered in the current implementation of the Rosetta Molecular Modeling Suite. Therefore, having low-energy conformations is of critical importance since Rosetta will not distinguish binding of a high-

energy to low-energy ligand conformation. The method described below assumes the conformation software does not change the atom names; if that is not the case, refer to the Notes section for a different method (*see* **Note 4**).

5. Open the LG1_0001.pdb file in PyMOL. Changing mouse mode from 3 button selection to 3 button editing, the *p*-nitrophenyl group can be rotated by holding Ctrl on the keyboard and right-clicking on the bond between the oxygen and the carbon on the ring, closer to the carbon side. This allows manipulation of the angle. Once the bond has been rotated, each conformation can be saved into separate files. Alternatively, for the file provided, use the following PyMOL commands:

   PyMOL>show labels, LG1

   ```
   PyMOL>get dihedral n. C5, n. O2,n. C7, n. C8 # -46.454 degrees.
   PyMOL>set_dihedral n. C5, n. O2,n. C7, n. C8, 73.55
   PyMOL>save LG1_rot2.pdb
   PyMOL>set_dihedral n. C5, n. O2,n. C7, n. C8, 193.55
   PyMOL>save LG1_rot3.pdb
   ```

6. Rename LG1_rot2.pdb to LG1.conf.pdb and concatenate the contents of LG1_rot3 on to the end in the command prompt, working in the same directory these files were created.

   ```
   >mv LG1_rot2.pdb LG1.conf.pdb; cat LG1_rot3.pdb >>
   ```

   LG1.conf.pdb.

   This creates a file called LG1.conf.pdb that contains rotations about the leaving group oxygen.

7. Add conformations to params file.
   To load the conformations into Rosetta, the conformation file must be identified in the params file. This will allow the transition state to adopt the generated ensemble of conformations during docking and design (*see* **Note 5**).

8. Add the following line to the bottom of the LG1.params file

   ```
   PDB_ROTAMERS conf.lib.pdb.
   ```

3.2 Scaffold Selection

3.2.1 Download the Crystal Structures to Use as Scaffolds

1. Open PyMOL.
2. PyMOL>fetch 2jie.

   This downloads the 2jie.pdb crystal structure into the directory from which PyMOL is executed from. Alternatively, just download the PDB directly from the website.

3. Renumber the PDB. It is important to renumber before creating the positions file. Many Rosetta applications will internally renumber proteins to start at 1. Renumbering thus avoids potential mismatches of residues.

4. >python ~/Rosetta/tools/renumber_pdb.py -pdb 2jie.pdb –o 2jie.renumbered.pdb -a 1.
5. Create a position file. In order for the matcher to run on a given scaffold and to limit sampling to only residues that are buried or in pockets, one may select a subset of all residue positions (*see* **Note 6**) and declare these in a space-delimited positions file (positions.pos). For this example, the renumbered crystal structure of 2jie was opened in PyMOL, the ligand selected, and any residues within 8Å of the ligand were selected. The 8 Å cutoff defines the location of the pocket as being comprised of these residues. However, all residues or any arbitrary set can be selected as potential positions to use during the transition state placement stage depending on the researchers goals and hypothesis of what an optimal catalytic site would be going into the modeling (*see* **Notes 8** and **9**).
6. The following assumes there are no other HETATM records in the PDB file other than waters (which get removed using the commands below). The following set of commands is to be executed in the PyMOL command prompt (*see* **Note 7**):

```
PyMOL>load 2jie.renumber.pdb
PyMOL>remove solvent
PyMOL>sele hetatm, HETATM
PyMOL>sele pos, hetatm expand 8 and n. CA
PyMOL>myfh = open("positions.pos","w")
PyMOL>iterate pos, myfh.write("%s " %resi)
PyMOL>myfh.close()
```

This loads the crystal structure, removes waters and other solvent molecules, selects the ligand in the crystal structure, expands the selection by 8 Å around the ligand, opens a file called postions.pos, and writes to it the list of residues by index with a space in between. The positions.pos file should read as one line with the following residues:

"15 16 19 119 163 164 220 293 294 295 352 353 354 355 356 399 400 404 405 406 407 415."

3.3 Match

1. Check the Constraint File –optional, requires additional modification to the constraint file (*see* **Note 10** and **11**). Although optional, this step is highly indicative of the accuracy of the constraint file. However, it does require the specification of all 6 DOFs for each constraint block. The Rosetta CstfileTo TheozymePDB app creates a model of the theozyme based off of the constraint file (this is the inverse problem of specifying a constraint file from the theozyme). The reader is highly encouraged to consult the online matcher documentation for more details and to remove any extra degrees of freedom to be sampled in the LG1.

enzdes.csts file when running this app. *Warning*: The LG1.enzdes.cst file as provided in the GitHub repository will not run without defining all of the degrees of freedom.

2. >~/Rosetta/main/source/bin/CstfileToTheozymePDB.default.linuxgccrelease -database ~/Rosetta/main/database -extra_res_fa LG1.params -match:geometric_constraint_file LG1.enzdes.csts.

This will create a number of PDB files that may be opened in PyMOL. These files may be used to verify the constraint file is specifying the distances, angles, and dihedrals as intended. The files should approximately match the theozyme from which the constraints were generated.

3. Run the matcher on the scaffold. Repeat for all scaffolds of interest; to do this just change the –s 2jie.renumber.pdb to another scaffold. Each scaffold should be renumbered and have its own unique position file. More information on each option used here may be found at https://www.rosettacommons.org/docs/latest/full-options-list.

4. >~/Rosetta/main/source/bin/match.default.linuxgccrelease (*add all of the following flags after this binary command; do not include the lines starting with #*).

   #File I/O

   -match:geometric_constraint_file LG1.enzdes.cst

   -s 2jie.renumbered.pdb

   -extra_res_fa LG1.params

   #Extra side chain rotamer samples

   -ex1 -ex2 -ex3 -ex1aro -ex2aro -use_input_sc #(*see* **Note 3**)

   #Matcher Options

   -match:lig_name LG1

   -match:scaffold_active_site_residues positions.pos

   -bump_tolerance 0.4

   -consolidate_matches T

   -output_matches_per_group 1

   -match_grouper SameSequenceGrouper

   #Other

   -ignore_unrecognized_res T

   -database ~/Rosetta/main/database

   -mute protocols.idealize

This will create a series of files starting with UM_ that describes the matches found, in the order of the constraint block (Fig. 6). For example,

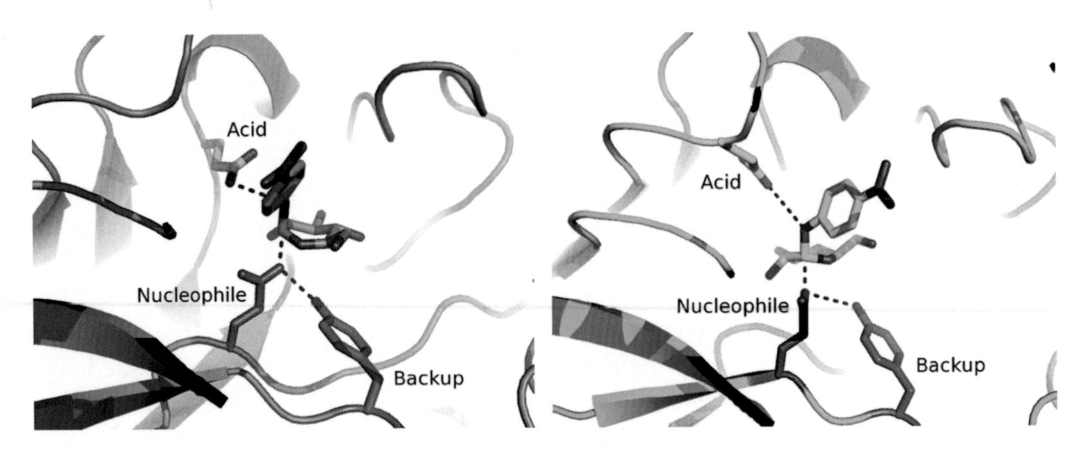

Fig. 6 Two examples of matches—pre optimization. Note that in these two examples, the Tyr is positioned at the same place, but the glutamic acids are on different loops (comparing the *left image* to the *right image*) resulting in a different orientation of the substrate

UM_13_E353E164Y295_2jie_LG1_1.pdb is the 13th hit found by the matcher, and it placed the first constraint block residue at position 353 in the protein and matched the second constraint block to position 164 and the third constraint block to the 295 tyrosine position. The LG1_1 refers to the first (of 3) ligand rotamers that were used in sampling.

3.4 Enzyme Design

At this point, for every scaffold there may be many different UM* hit files. In order to optimize the position of the ligand, as well as to backup the mutations made with additional interactions, an enzyme design run should be carried out. This samples different mutations around the space where the theozyme was inserted into the scaffold protein. This has previously been described in literature [7]; we have provided the flags to run the binary as well as the sampling script in the online GitHub repository (*see* **Note 12**). Further details on the format and movers called in the design_on.xml can be found at https://www.rosettacommons.org/docs/wiki/scripting_documentation/RosettaScripts/RosettaScripts. This should be run for every UM* file that comes out of the RosettaMatch simulation. Rosetta simulations stochastically sample the design space, and the results may differ from run to run.

1. >~/Rosetta/main/source/bin/rosetta_scripts.default.linuxgccrelease

```
@enzflags_parser
-parser:protocol design_on.xml
-s UM_13_E353E164Y295_2jie_LG1_1.pdb
-nstruct 10
-database ~/Rosetta/main/database
-run::preserve_header
```

```
-jd2::enzdes_out
-enzdes::cstfile LG1.enzdes.cst
-extra_res_fa LG1.params
```

This will create 10 new structures in which the geometric position of the ligand has been optimized and the mutations to stabilize the active site have been introduced. Each simulation also creates a score file and a tab-delimited file that contains the individual terms of the score function for each of the ten models. Each match has its own score file (when ran in different folders), and all of the score files may be concatenated into one combined score file for filtering. This can be achieved with the following Bash commands:

```
>find . –name "score.sc" > myscorefiles.
>while read x; do; cat $x >> combo.scores; done < myscorefiles.
```

This combines all of the scores into a new file called combo. score.

Rosetta carries out a Monte-Carlo simulation and is not deterministic of a low-energy solution. Ten simulations are not sufficient to thoroughly sample the degrees of freedom with side chain, backbone, and rigid body movement. However, ten provide a general indication of whether or not a low-energy solution is possible while minimizing computational time.

2. From the designs, a subset are chosen for further refinement (*see* **Note 13**). In this example, all of the score files for different position matches are combined into one file. Using the score terms as filters, a subset of designs can be selected for further refinement. This step may be done in a spreadsheet or using command line tools. There are no rules or established "best practices" for filtering and selecting a subset of designs. However, the following is one series of filters we commonly employ. First, select all of the matches that have 0 constraint energy (i.e., the constraint file specifications are fully satisfied). Then sort on interface energy (the term in the score file that ends with interf_E_1_2) and select the lowest 5 from this set to visually inspect. The lowest five found are 2 at position UM_18 (interface energies: -8.56, -7.56), 1 at position UM_31 (interface energy: -7.32), 1 at position UM_20 (interface energy: -7.15), and 1 at position UM_13 (interface energy: -6.49).
3. At this point, further refinement of the selected low-energy designs should be carried out. This can either include running a larger design simulation, which may be equivalent to the one above but with an –nstruct of 100 or 1000. However, at the very end, it is always critical to visually assess and evaluate the details of the designed interface. To do this, Foldit provides an

excellent way to understand why a mutation was chosen and the ability to assess if other mutations may also be favorable in real time. We will illustrate this with UM_13 in further detail.

3.5 Foldit

At this point, we have identified several designs from potentially several scaffolds that all appear to score well (*see* **Note 14**). As a final computational step, Foldit is utilized to interact with each design and make a manual and visual assessment of each design.

1. Choose a representative set of enzyme design outputs to manually refine in Foldit. In this case, we will select the structure UM_13_E353E164Y295_2jie_LG1_1_0001.pdb, which is one of the five best found by removing models that have a nonzero constraint score and sorting based on the interface score found in the score file. The first key step is to compare the design to the input scaffold to identify the mutations introduced. Mutations may be found by opening up the designed structure and the original crystal structure in PyMOL and visually identifying them. Alternatively, a Perl script located on the GitHub repository may be used to identify mutations given the crystal structure (>perl mutation_id.pl 2jie.renumbered.pdb UM_13_E353E164Y295_2jie_LG1_1_0001.pdb). In this case, the mutations are H119N,Y166A, and E399S.
2. Copy the output design (also renaming to a simpler name, such as lowE.pdb, is recommended), the parameters file for pNPG (LG1.params), the conformers file for pNPG (LG1.conf.pdb), and the ligand constraints file (LG1.enzdes.cst) in a working directory for loading into Foldit.
3. Start Foldit and enter any puzzle. Once in a puzzle (e.g., the introduction level 1-1 One Small Clash), activate the open dialog with Control-Alt-Shift-A and choose all four files.
4. Activate the Selection Interface by choosing Menu > Selection Interface and edit the viewing settings (recommended settings: Cartoon Thin structure and Score/Hydro+CPK coloring). Visually identify the ligand and zoom by hovering the moue over the ligand and pressing Shift-Q (no click necessary).
5. With the protein system loaded into Foldit, relax the global structure by alternating between shake (S) and wiggling the side chains (E). The enzyme design protocol only relaxed the environment local to the ligand. This relax step will lower the overall protein score as well as remove many of the clashing interactions that are seen in Foldit (*see* **Notes 15–17**).
6. Systematically evaluate each of the designed mutations introduced during design (*see* **Note 18**), potentially reverting residues chosen by the enzyme design algorithm back to the native residues from the crystal structure. Additional mutations may

also be incorporated at this stage based on chemical intuition and the Rosetta energy function. The end product should be a list of sequences to order and experimentally characterize.

3.6 Examples

Here we will briefly illustrate how to carry out by analysis of a design by looking at each of the three mutations in the UM_13 design:

1. **H119N**: This mutation exchanges a hydrogen bond to the ligand for two hydrogen bonds to neighboring protein residues. The per-residue score terms for any residue (including the ligand) can be viewed by hovering the cursor over it and pressing Tab. This opens an Info Panel which changes as we test reverting the asparagine 119 mutation back to the histidine found in the crystal structure. Make the mutation (click to select the residue then press M [mutate to] and N [Asparagine]), and select a sphere of 20–30 residues around the Asn by Control-Alt-Shift-drag (a residue counter in the top left corner counts the number of residues selected). Once the sphere is selected, re-optimize the local region by using the shake (S), wiggle the side chains (E), and wiggle the backbone and side chains together (W). When the backbone is allowed to move, discontinuities in the energy function may make large changes that are not realistic. Therefore, the timer and cycle number (in parentheses after the timer, top left corner) are used to limit the movement. The protein is near convergence when the cycle count advances rapidly. On a modern computer, a few seconds is likely sufficient to re-optimize the position of the side chain.

   After minimizing the structure, this reversion is predicted to maintain the total ligand score and only slightly increase the residue's score. Histidine has a much better Lennard-Jones attractive score term [4] than the Asparagine does, indicating that it is better packed in the area. What gives rise to the slightly higher total residue score is the Dunbrack energy [4]. The histidine has a much higher Dunbrack energy term, but if it is in fact well packed, this residue may be locked into the conformation found, regardless of the Dunbrack score. Given the ligand score stays the same and reverting the mutation seems to have little energetic effect, a conservative design would revert the mutation to the native crystal structure residue.

   Verdict: revert to native (H).

2. **Y166A**: At first, this mutation may seem desirable as reverting it to native in the enzyme design output structure has a significant energy penalty on the total score. However, a "local wiggle" as described above rapidly decreases the system energy to an input-like score, and, furthermore, the ligand score decreases slightly.

Verdict: order a sequence with this mutation (Y166A) and one without (Y166Y).

3. **E399S**: Reverting this mutation to a W is predicted to increase both the total protein energy and the ligand energy. The increase in energy comes from strong repulsion between this large amino acid and the ligand and surroundings. The local refinement within Foldit is unable to find an alternative conformation where this is a favorable interaction.

   Verdict: order this mutant.

4 Notes

1. Anytime a Rosetta app is used, the user may need to change the ending to match his or her build environment. For example, when the binary rosetta_scripts.default.linuxgccrelease is referred to, if the Clang compiler was used, the user would need to call the rosetta_scripts.default.linuxclangrelease binary.
2. The authors want to note the existence of an excellent review of this process on another substrate. This is located at https://github.com/RosettaCommons/teaching_resources/tree/master/OtherTeachingResources/whole_classes/tutorial_20121128jbei/tutorial_20121128jbei/enzyme_design.
3. One can restrict or remove the extra chi sampling for hydrogens or other DOFs. This can be controlled on the command line with –exX commands (where X is 1–4) and with additional commands in the LG1.enzdes.cst file in the algorithm blocks (see matcher documentation).
4. It is also possible to feed in a mol2 formatted file into molfile_to_params.py that already has conformations made. This may be the case if using another piece of software to generate the conformations. In this case, the researcher will not need to add the PDB_ROTAMER conf.lib.pdb line into the params file. See the help options from the molfile_to_params script for more information.
5. The time and memory for RosettaMatch rise dramatically as the degrees of freedom increase (i.e., the last column of the constraint lines). If the runtime lasts too long, try reducing the DOFs.
6. There are several methods available to automatically generate the positions files; see the matcher documentation for descriptions of these methods.
7. Ideally, PyMOL will be started from the command prompt in the directory the researcher will be working in. If PyMOL has been opened from an application icon or task bar, then upon

execution of some commands such as "fetch 2jie", it may be difficult to locate the actual file.

8. Many aspects may be involved in scaffold selection, such as the ability to express the protein in *E. coli* or another target organism of interest.
9. If no positions.pos file is included, the matcher will check every single residue in the protein, which may be a waste of computational resources and time. In addition, this will lead to many matches on the surface of proteins which may be undesirable.
10. Having too many constraints specified may hinder the ability to find matches. Only include as many constraints as are absolutely necessary.
11. The first constraint defined must have all 6 DOFs/constraint lines defined.
12. There are essentially unlimited enzyme design protocols available due to the flexibility of the RosettaScripts app. Only one was suggested here; however, we highly recommend adjusting the steps in this script to implement new design protocols to potentially produce better designs. Currently, there is no single protocol that has been conclusively demonstrated to be optimal.
13. Additional criteria for selecting which designs to visually check may be the number of mutations made. The more mutations made to a given protein, the more risk there is that the protein will not express or the backbone may shift.
14. Foldit may use a different energy function than what was used during design. This can lead to discrepancies between the predicted effect of a mutation in Foldit and enzyme design in Rosetta.
15. By selecting the ligand and clicking the left or right arrow keys, different ligand conformations are accessible, and the way Rosetta is sampling ligand conformations may be visually verified. If an undesired region of the ligand is moving during conformation sampling, adjusting the neighbor atom of the params file should fix this.
16. Undo (Cmd-Z) and Redo (Cmd-Y) work for most operations. The protein structure may be reset to the input by opening the Undo panel (press U) and choosing "Reset Puzzle."
17. It is recommended to sparsely use W (wiggle backbone and sidechains) or T (wiggle backbone). As a rule of thumb, if a full "register shift" in the backbone movement can be seen, it is advised to revert that change. The more the backbone moves, the less chance the model will be accurate.

18. Mutations from or to the following residues tend to disrupt the structure more so than other mutations: GLY, PRO, and CYS. In the case of GLY, if ALA can fit almost as well, that should be chosen over GLY.

Acknowledgments

The authors would like to thank UC Davis, Sloan Foundation (BR2014-012), ARPA-E (DE-AR0000429), and CDFA (SCB14037) for funding.

References

1. Röthlisberger D, Khersonsky O, Wollacott AM, Jiang L, DeChancie J, Betker J, Gallaher JL, Althoff EA, Zanghellini A, Dym O, Albeck S, Houk KN, Tawfik DS, Baker D (2008) Kemp elimination catalysts by computational enzyme design. Nature 453(7192):190–195
2. Jiang L, Althoff EA, Clemente FR, Doyle L, Röthlisberger D, Zanghellini A, Gallaher JL, Betker JL, Tanaka F, Barbas CF, Hilvert D, Houk KN, Stoddard BL, Baker D (2008) De novo computational design of retro-aldol enzymes. Science 319(5868):1387–1391
3. Siegel JB, Zanghellini A, Lovick HM, Kiss G, Lambert AR, Clair JLS, Gallaher JL, Hilvert D, Gelb MH, Stoddard BL, Houk KN, Michael FE, Baker D (2010) Computational design of an enzyme catalyst for a stereoselective bimolecular Diels-Alder reaction. Science 329(5989):309–313
4. Leaver-Fay A, Tyka M, Lewis SM, Lange OF, Thompson J, Jacak R, Kaufman K, Renfrew PD, Smith CA, Sheffler W, Davis IW, Cooper S, Treuille A, Mandell DJ, Richter F, Ban Y-EA, Fleishman SJ, Corn JE, Kim DE, Lyskov S, Berrondo M, Mentzer S, Popović Z, Havranek JJ, Karanicolas J, Das R, Meiler J, Kortemme T, Gray JJ, Kuhlman B, Baker D, Bradley P (2011) ROSETTA3: an object-oriented software suite for the simulation and design of macromolecules. Methods Enzymol 487:545–574. doi:10.1016/B978-0-12-381270-4.00019-6
5. Tantillo DJ, Chen J, Houk KN (1998) Theozymes and compuzymes: theoretical models for biological catalysis. Curr Opin Chem Biol 2(6):743–750
6. Zanghellini A, Jiang L, Wollacott AM, Cheng G, Meiler J, Althoff EA, Röthlisberger D, Baker D (2006) New algorithms and an in silico benchmark for computational enzyme design. Protein Sci 15(12):2785–2794
7. Richter F, Leaver-Fay A, Khare SD, Bjelic S, Baker D (2011) De novo enzyme design using Rosetta3. PLoS ONE 6(5):e19230. doi:10.1371/journal.pone.0019230
8. Eiben CB, Siegel JB, Bale JB, Cooper S, Khatib F, Shen BW, Players F, Stoddard BL, Popovic Z, Baker D (2012) Increased Diels-Alderase activity through backbone remodeling guided by Foldit players. Nat Biotechnol 30(2): 190–192
9. Radzicka A, Wolfenden R (1995) A proficient enzyme. Science 267(5194):90–93
10. Mak WS, Siegel JB (2014) Computational enzyme design: transitioning from catalytic proteins to enzymes. Curr Opin Struct Biol 27:87–94
11. Isorna P, Polaina J, Latorre-García L, Cañada FJ, González B, Sanz-Aparicio J (2007) Crystal structures of Paenibacillus polymyxa β-glucosidase B complexes reveal the molecular basis of substrate specificity and give new insights into the catalytic machinery of family I glycosidases. J Mol Biol 371(5):1204–1218. doi:10.1016/j.jmb.2007.05.082
12. The PyMOL Molecular Graphics System, Version 1.5.0.5 Schrodinger, LLC
13. Spartan '08, Wavefunction Inc., Irvine, CA
14. Hawkins PCD, Skillman AG, Warren GL, Ellingson BA, Stahl MT (2010) Conformer generation with OMEGA: algorithm and validation using high quality structures from the Protein Databank and Cambridge Structural Database. J Chem Inf Model 50(4):572–584
15. Frisch MJ, Trucks GW, Schlegel HB, Scuseria GE, Robb MA, Cheeseman JR, Scalmani G, Barone V, Mennucci B, Petersson GA, Nakatsuji H, Caricato M, Li X, Hratchian HP, Izmaylov AF, Bloino J, Zheng G, Sonnenberg

JL, Hada M, Ehara M, Toyota K, Fukuda R, Hasegawa J, Ishida M, Nakajima T, Honda Y, Kitao O, Nakai H, Vreven T, Montgomery JA Jr, Peralta JE, Ogliaro F, Bearpark MJ, Heyd J, Brothers EN, Kudin KN, Staroverov VN, Kobayashi R, Normand J, Raghavachari K, Rendell AP, Burant JC, Iyengar SS, Tomasi J, Cossi M, Rega N, Millam NJ, Klene M, Knox JE, Cross JB, Bakken V, Adamo C, Jaramillo J, Gomperts R, Stratmann RE, Yazyev O, Austin AJ, Cammi R, Pomelli C, Ochterski JW, Martin RL, Morokuma K, Zakrzewski VG, Voth GA, Salvador P, Dannenberg JJ, Dapprich S, Daniels AD, Farkas Ö, Foresman JB, Ortiz JV, Cioslowski J, Fox DJ (2009) Gaussian 09. Gaussian, Inc., Wallingford, CT

16. O'Boyle NM, Vandermeersch T, Flynn CJ, Maguire AR, Hutchison GR (2011) Confab-Systematic generation of diverse low-energy conformers. J Cheminform 3:8

Chapter 14

Generating High-Accuracy Peptide-Binding Data in High Throughput with Yeast Surface Display and SORTCERY

Lothar "Luther" Reich, Sanjib Dutta, and Amy E. Keating

Abstract

Library methods are widely used to study protein–protein interactions, and high-throughput screening or selection followed by sequencing can identify a large number of peptide ligands for a protein target. In this chapter, we describe a procedure called "SORTCERY" that can rank the affinities of library members for a target with high accuracy. SORTCERY follows a three-step protocol. First, fluorescence-activated cell sorting (FACS) is used to sort a library of yeast-displayed peptide ligands according to their affinities for a target. Second, all sorted pools are deep sequenced. Third, the resulting data are analyzed to create a ranking. We demonstrate an application of SORTCERY to the problem of ranking peptide ligands for the anti-apoptotic regulator Bcl-x_L.

Key words Yeast surface display, Deep sequencing, High-throughput assay, Protein–protein interaction, Bcl-2 family

1 Introduction

High-throughput analysis of functional mutations in proteins, peptides, or DNA by deep sequencing is emerging as a powerful technique. Properties such as protein stability, enzymatic activity, and peptide ligand or DNA binding have been studied [1–16]. The general approach involves screening a library of mutants or performing a selection for a desired function. Library sequences in pre- and post-selected pools are then identified by next-generation sequencing, and computational routines are used to extract information about how sequence relates to function.

Many selection or screening processes have been employed for these types of studies, including in vitro assays, phage display, yeast surface display in combination with fluorescence-activated cell sorting (FACS), and in vivo assays. Some studies have used the observed frequencies of mutant variants in selected pools to infer sequence–function relationships [1–5]. As an alternative measure, enrichment scores have been calculated from the ratio of pre- and

Barry L. Stoddard (ed.), *Computational Design of Ligand Binding Proteins*, Methods in Molecular Biology, vol. 1414,
DOI 10.1007/978-1-4939-3569-7_14,

post-selection frequencies [6–14]. The effects of mutations in particular sequence positions have been investigated, either by experimentally screening single-mutant libraries or by assuming positional independence during computational post-processing. Position weight matrices have been built that score binding, stability, and function using this approach, sometimes with correction for non-specific binding or consideration of enrichment changes over multiple selection rounds [5, 12, 13]. Analyzing single-residue substitutions benefits from enhanced statistical power, because it is easy to saturate a single-position sequence space. But important context-dependent effects may be neglected in this type of analysis.

In this chapter, we introduce a high-accuracy alternative to enrichment-based methods for probing mutational effects on the affinity of peptide ligands. Our protocol "SORTCERY" comprises the three steps of selection, deep sequencing, and computational analysis (Fig. 1a). The selection process involves two-color cell sorting of a yeast surface-displayed library based on the expression levels of displayed peptides and levels of binding to a target (Fig. 1b). Our sorting protocol builds on reports that two-color FACS can accurately distinguish between binders of different affinities [15–19] and agrees with a theoretical model describing the expected signals for clones expressing peptides with a range of binding strengths [20]. This model can guide sorting of a library into pools according to binding affinity, and the pools can then be deep sequenced to obtain information about individual library member affinities. SORTCERY extracts information from deep sequenced library pools using computational routines that rank observed mutant sequences according to binding strength.

Applying SORTCERY to study helical peptide affinities for the apoptosis-regulating protein Bcl-x_L, we obtained extremely accurate rankings for ~1000 sequences over a range of dissociation constants from 0.1 to 60 nM (Fig. 2a). Our study is described in Ref. [20], and the reader is referred to that paper for in-depth exposition of the theory underlying SORTCERY, the results when applied to Bcl-x_L, and further discussion of strengths and limitations of this method. A special variant of our approach is described here (Fig. 2b, *see* **Note 9**) that can potentially be used to analyze much larger libraries.

2 Materials

2.1 Cell Culture Media

1. SD + CAA/SG + CAA: Dissolve 5 g casamino acids, 1.7 g yeast nitrogen base, 5.3 g ammonium sulfate, 10.2 g Na_2HPO_4–$7H_2O$, and 8.6 g NaH_2PO4-H_2O in 700 ml water and autoclave for 15 min at 22 psi and 120 °C. For growth media (SD + CAA), dissolve 50 g glucose in 50 ml water then sterilize

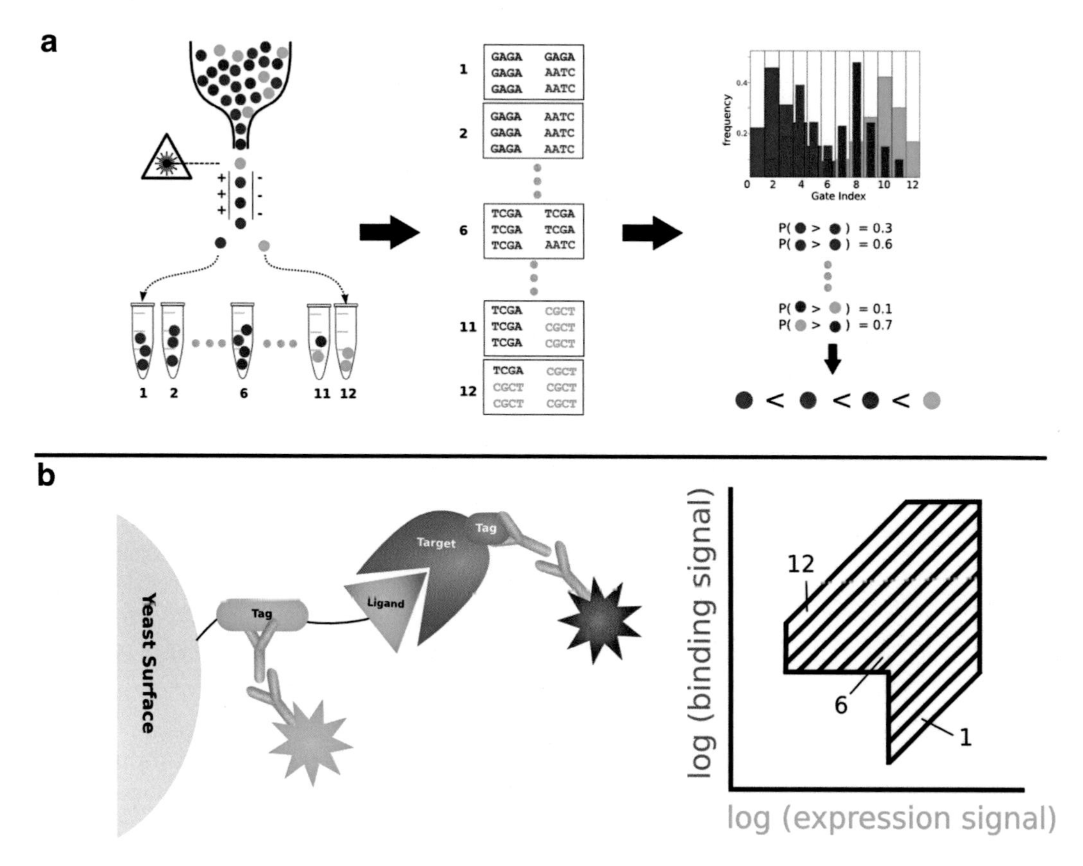

Fig. 1 (**a**) SORTCERY combines experimental and computational protocols to rank peptide ligands according to their affinity for a target. Yeast-displayed peptides are sorted into pools that include ligands of similar affinity using FACS. Deep sequencing information is generated for each sample, and the distribution of each sequence over the FACS gates is determined. Pairwise comparison of distributions permits calculation of the probability that one peptide binds more strongly than another, for each pair of peptides. A global rank order of affinities is computed from the probabilities. (**b**) SORTCERY's yeast-display and gate-setting schemes. Peptide expression and target binding are detected via tags that are recognized by pairs of primary and fluorescently labeled secondary antibodies. Two-color cell sorting is based on these two signals. Gates are set to optimally separate binders of different affinities and to exclude non-binders and non-expressing cells

with a 0.2 μm filter. Add 40 ml of this 50 % glucose solution to the autoclaved media and fill up to 1 l with sterile water. For induction media (SG + CAA), dissolve 20 g galactose in 100 ml water then sterilize with a 0.2 μm filter. Add 100 ml of this 20 % galactose solution to the autoclaved media and fill up to 1 l with sterile water.

2.2 Fluorescence-Activated Cell Sorting

1. Low protein binding 0.45 μm filter plates or bottle-top filters.
2. BSS pH 8.0: 50 mM Tris, 100 mM NaCl, 1 mg/ml BSA.
3. Primary antibody mixture: anti-HA (Roche) 1:100 dilution and anti-Myc (Sigma) 1:100 dilution in BSS.

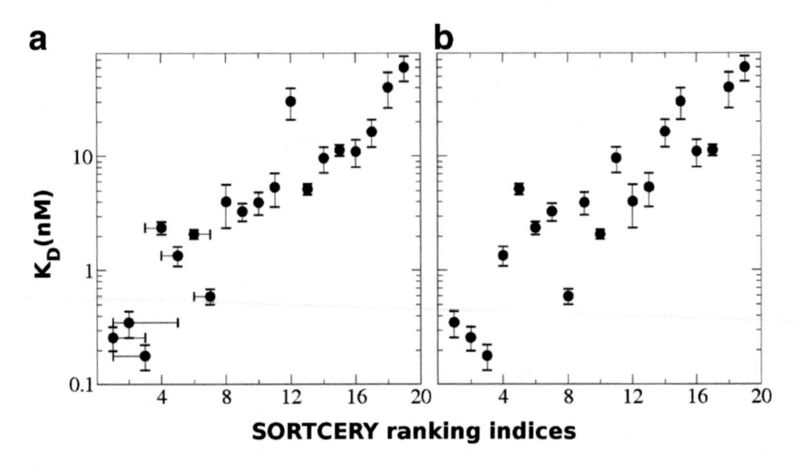

Fig. 2 (**a**) Individually measured dissociation constants vs. SORTCERY ranking indices for 19 sequences from a ranking of ~1000 sequences. Clones have been reindexed from 1 to 19. Error bars for rank indices are 95 % bootstrap confidence intervals: error bars for dissociation constants indicate standard deviations for four individual measurements. (**b**) Ranking indices for the same 19 clones as determined by convoluted SORTCERY (*see* **Note 9**). Figure panel (**a**) is adopted with publisher's permission from Fig. 4 in Ref. [20]

4. Secondary antibody mixture: APC-labeled anti-mouse (BD Biosciences) 1:40 dilution and PE-labeled anti-rabbit (Sigma) 1:100 dilution in BSS.

2.3 Deep Sequencing Sample Preparation (*See* Note 1)

1. Zymoprep Yeast Plasmid Miniprep I (Zymo Research).
2. Isopropanol.
3. High-Fidelity DNA Polymerase (e.g., Phusion).
4. Thermocycler.
5. Gel equipment.
6. PCR purification and gel extraction kits (QiaGen).
7. MmeI (New England Biolabs): MmeI restriction enzyme, NEB CutSmart Buffer, 1 mM SAM.
8. T4 Ligase.
9. Primers and oligos.

3 Methods

3.1 Cell Growth and Induction of Yeast Surface Display Library (*See* Note 2)

1. Dilute cells to OD_{600} of 0.05 in SD + CAA and grow for 8 h at 30 °C.
2. Dilute cells to OD_{600} of 0.005 in SD + CAA and grow to OD of 0.1–0.4 at 30 °C.
3. Dilute cells to OD_{600} of 0.025 in SG + CAA and grow to OD of 0.2–0.5 at 30 °C for induction of peptide expression.

3.2 Gate Setting

1. SORTCERY uses a two-color FACS setup to monitor expression (*Fe*) and binding (*Fb*) signals on a log/log or biexponential scale. On a log(*Fb*) vs. log(*Fe*) plot, points of equal binding strength lie on a line with a slope of 1 [20]. Subdivide the log/log plot accordingly into areas (gates) of different affinities by dissecting it with lines of slopes of 1 (red lines in Fig. 3). The number, position, and spacing of the lines will affect the performance of the procedure. We recommend an equal spacing between lines as this will result in optimal resolution between binders of different affinities. The number of lines (and the resulting gates) depends on the required resolution. This can be determined by measuring the FACS profiles of several yeast-displayed standards (*see* **Note 3**). Lines should be positioned such that the gates cover an area from the strongest binders to the baseline binding signal. FACS profiles of standards can help determine whether the experimental setup will generate samples with quality appropriate for a SORTCERY sort (*see* **Note 4**).
2. Gate boundaries should be set to exclude cells without significant expression signal and to prevent cells in the binding baseline from being captured in gates for higher affinities. Cutoffs

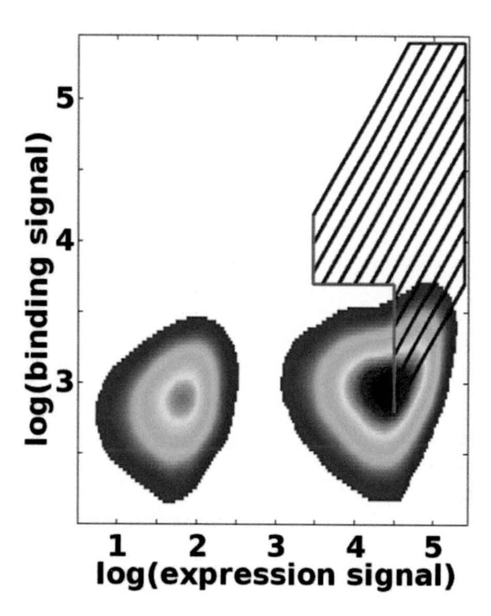

Fig. 3 Gate setting for an affinity sort with 12 gates. The *red, diagonal lines* subdivide the axis of affinity into different intervals and thus insure that each gate corresponds to a unique range of dissociation constants. The *green, lower left borders* exclude non-binding cells from higher-affinity gates and exclude non-expressing cells from all gates. The depicted FACS profile of a non-binder illustrates this. The *blue, upper-right borders* exclude cells with the maximum possible expression or binding signal, because affinities cannot be accurately estimated from such signals. This figure is adopted with the publisher's permission from supplemental Fig. 3 in Ref. [20]

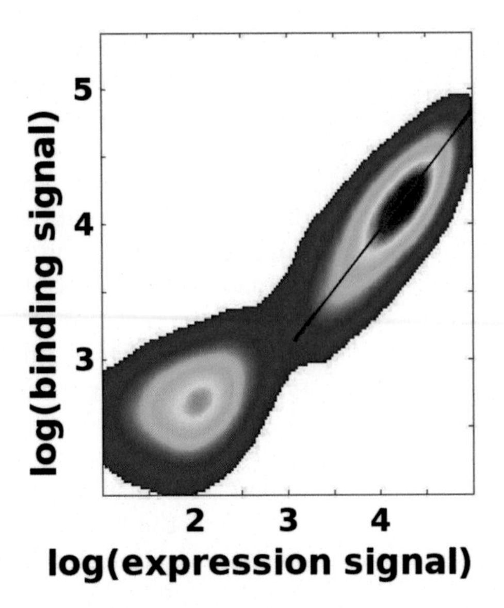

Fig. 4 FACS profile for a BH3 peptide ligand binding to Bcl-x_L. The *red line* indicates the orientation of the first principle component for the profile of the expressing cells. This figure is adopted with publisher's permission from Fig. 3 in Ref. [20]

can be established by monitoring the FACS profile of a non-binding yeast clone and noting: (1) the position of non-expressing cells (blob in the lower left corner of Fig. 3) and (2) the binding baseline (lower right area in Fig. 3). Determine appropriate cutoffs and set gate lower-edge boundaries accordingly (*see* example: green edges in Fig. 3).

3. Cell sorters assign maximum signal values to any signal intensity above their scale of measurement. Such signals have, therefore, not been accurately determined. Exclude the maximum expression and binding signal areas from the gates by setting gate boundaries accordingly (*see* example: blue edges in Fig. 3) (Fig. 4).

3.3 Cell Sorting

1. Filter grown and induced yeast cells (Subheading 3.1) and wash twice with BSS.
2. Incubate cells with target molecule in BSS for 2 h at 21 °C (*see* **Notes 5** and **6**). Shake gently during incubation.
3. Filter cells and wash twice with BSS.
4. Incubate with mixture of primary antibodies (20 μl per 10^6 cells, *see* **Notes 7** and **8**) at 4 °C.
5. Filter cells and wash twice with BSS.
6. Incubate with mixture of secondary antibodies at 4 °C.
7. Filter cells and wash twice with BSS. Resuspend cells in BSS for sorting.
8. Sort cells into each individual gate and retain sorted pools for deep sequencing analysis (*see* **Notes 9** and **10**). Note the number

of cells sorted into each pool. Also determine the library distribution across all gates by recording how many cells hit each gate during a set time interval, e.g., a minute. This information is important for the deep sequencing analysis (Subheading 3.5, **step 4**).

3.4 Deep Sequencing Sample Preparation

3.4.1 DNA Extraction

1. If >80,000 cells are sorted, spin cells down, aspirate supernatant, and add 150 μl of solution 1 from the Zymoprep kit + 2 μl Zymolyase. For smaller numbers of cells, directly add 50 μl of solution 1 per 100 μl cell suspension + 2 μl Zymolyase per 150 μl total volume.
2. Incubate at 37 °C for 1 h on a shaker.
3. Successively add 150 μl of solutions 2 and 3 per 150 μl incubation volume and vortex after each addition. Spin down precipitate, and retain supernatant.
4. Add 1 volume isopropanol and 0.1 volume 3 M NaOAc to each volume of DNA extract. Store at −20 °C overnight.
5. Spin at 14,000 × *g* at 4 °C for 10 min. Carefully remove supernatant. Resuspend DNA pellet in 20 μl sterile water (pellet may not be visible for small numbers of sorted cells).

3.4.2 DNA Amplification and Adapter Attachment

Most of this section is based on the excellent preparation protocol in Ref. [21].

1. For each sorted sample, separately, amplify DNA sequences encoding the peptide ligands out of plasmids by PCR. The 5′ end of the forward primer needs to contain a binding site for the MmeI restriction enzyme: 5′ GGGACCACCACCTCCGAC 3′ (*see* **Note 11**). The 5′ end of the reverse primer has to consist of a part of the Illumina adapter sequence: 5′ CGGTCTCGGCATTCCTGC 3′ (*see* **Notes 12** and **13**).
2. Purify PCR products with the Qiagen PCR purification kit. Elute in 30 μl sterile water.
3. Digest each PCR product with the MmeI restriction enzyme. Incubate the digestion mixture for 1 h at 37 °C, then heat inactivate for 20 min at 80 °C (*see* **Note 14**).

Digestion reagents	
PCR product	12.5 μl
1 mM SAM	2.5 μl
NEB CutSmart buffer	5 μl
MmeI	5 μl per 8.6 pmol PCR product
Sterile water	Fill up to 50 μl

4. Prepare double-stranded adapters by annealing single-stranded oligos. The forward strand should contain the standard

Illumina read binding site [22], a unique barcode for multiplexing (*see* **Note 15**) and a 3′ TC, resultung in the sequence: 5′ ACACTCTTTCCCTACACGACGCTCTTCCGATCTbarcode TC 3′. The reverse complement strand should be 5′ phosphorylated and lack the 5′ GA 3′ that would be complementary to the TC of the forward strand.

5. Ligate each digestion product with an adapter containing a unique barcode. Ligate for 30 min at 20 °C, then heat inactivate for 10 min at 65 °C.
6. Run the products of the ligation reaction on a gel. Gel-purify the bands of correct size with the QIAquick gel purification kit. Elute in 30 μl sterile water.
7. PCR-amplify the ligation product. Primers should contain overhangs that complete the Illumina adapter sequences.

   Forward Primer: 5′ AATGATACGGCGACCACCGAGATCTACACTCTTTCCCTACACGACGCT 3′.

   Reverse Primer: 5′ CAAGCAGAAGACGGCATACGAGATCGGTCTCGGCATTCCTGCATCTT 3′.

   15 PCR cycles should be sufficient using Phusion polymerase.
8. Purify PCR products with the Qiagen PCR purification kit. Elute in 30 μl sterile water.
9. Combine samples and perform a multiplexed deep sequencing run on an Illumina sequencer with the standard forward Illumina read primer: 5′ ACACTCTTTCCCTACACGACGCTCTTCCGATCT 3′. If a reverse read is also to be carried out, use a custom primer (*see* **Note 16**).

3.5 Computational Analysis

1. Filter the Illumina data by only considering sequences with a high Phred score for the mutated positions and a low number of read errors in unmutated positions (*see* **Note 17**). If a reverse read has been performed that overlaps the forward read, compare complementary mutant codons and choose the version with the higher Phred score.
2. Assign each Illumina read to its sorted pool/gate by barcode identification.
3. Count the copies of each unique sequence across all pools. Discard sequences with low copy numbers when summing up counts from all gates. Calculate the number of sorted cells that each unique sequence likely originated from. Dividing the number of cells that were sorted into a pool by the number of sequence reads for this sample provides a rough estimate of the cells per read. As a rule of thumb, require at least 100 sorted cells for each observed sequence.
4. If a convoluted sort strategy was used, *see* **Note 18**. Otherwise, calculate the distribution over the gates for each unique sequence.

$$f_{xj} = \frac{z_x \frac{n_{xj}}{\sum_i n_{xi}}}{\sum_y z_y \frac{n_{yj}}{\sum_i n_{yi}}}$$

Here, *fxj* is the normalized frequency of sequence j in gate *x*, *nxj* is the number of reads of sequence j in deep sequencing data set *x* (which corresponds to gate *x*), and *zx* is the number of cells that hit gate *x* when measuring the distribution of cells across all gates.

5. Calculate all possible pairwise probabilities that a peptide *A* is a stronger binder than a peptide *B* and vice versa:

$$p(A > B) = \sum_x f_{xA} \sum_{y<x} f_{yB}$$

Note that gate indices *x* and *y* are assigned from lowest to highest affinity gates, i.e., in the equation the sum over y runs over all gates corresponding to lower affinities than that of gate *x*. Assign these probabilities as weights to the edges of a directed graph. The vertices of the graph represent peptides and the directed edge running from vertex B to vertex A indicates the assumption that peptide *A* is a stronger binder than peptide *B* (Fig. 5a).

6. Find the maximum linear subgraph by first applying the method described in Ref. [23]. To do this, randomly choose a peptide/vertex A. For each other peptide/vertex B, compare the edge weights of the two edges that connect it to *A*. If $p(A > B) > p(B > A)$, then *B* is considered a worse binder than *A*; if $p(B > A) > p(A > B)$, then *B* is considered a better binder than *A*. Group all peptides according to whether they are better or worse binders than *A*. Then, within each group, repeat the procedure of randomly choosing one peptide and evaluating all others with respect to it, continuing to subdivide the groups until an ordering from best to worst binder has been constructed. Determine the likelihood score for this ordering by summing up the logarithms of the edge weights for all directed edges that agree with the ordering (Fig. 5b). Repeat the procedure of constructing an ordering several times and retain the one with the best score. Further refine this ordering by inserting each individual peptide into all possible positions and keeping the new position if a better score is obtained. Run the routine several times, alternately starting with the best and the worst binding peptide. Finally, run a Monte-Carlo search in which moves correspond to exchanging the positions of two peptides in the ordering. The final result represents an affinity ranking of all peptides.

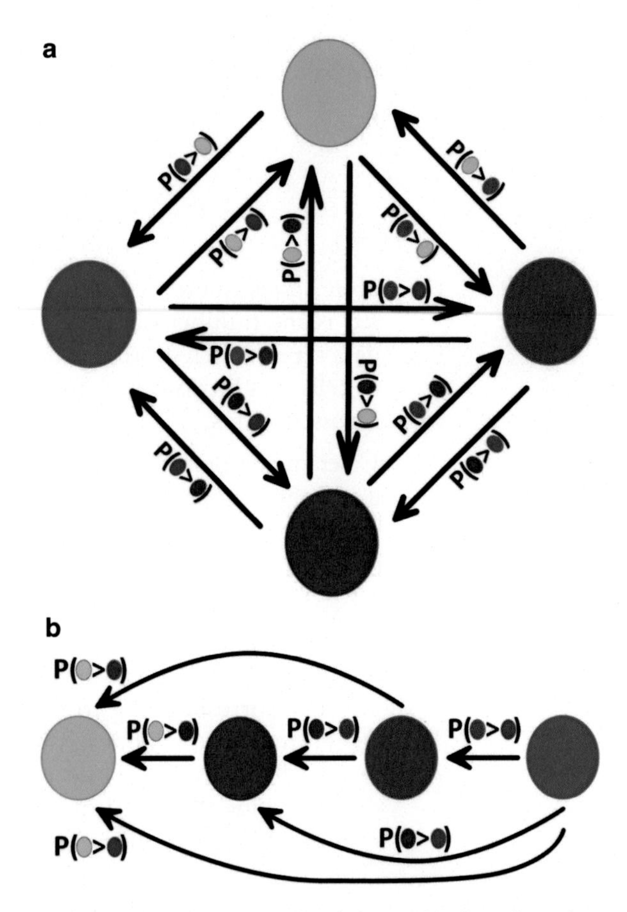

Fig. 5 (**a**) A directed graph representing four peptide ligands and assumptions about their relative binding strengths. Each edge is weighted by the probability that the ligand at its tail is a weaker binder than the ligand at its head. (**b**) A linear subgraph of (**a**). Note that no conflicting assumptions about binding strengths exist

4 Notes

1. We fine-tuned the protocols described in Subheading 3.4 using material from the specified suppliers. We have not tested corresponding products from other suppliers, and it is possible that these will also work for deep sequencing sample preparation. Experimenters may need to adjust protocols according to the specific products they use.
2. This growth protocol has been optimized for EBY100 cells that have been transformed with a pCTCON2 plasmid [17]. The experimenter may have to choose other parameters for a different setup. In the authors' experience, cell densities may have an impact on the quality of FACS profiles. Low-quality FACS profiles can lead to suboptimal sorts with respect to affinity. Users of the procedure should strictly monitor cell

densities. The first growth step in this protocol ensures that samples contain mostly live and healthy cells for correct measurements of ODs. It may be possible to skip this step if cells are not grown up from frozen stocks or plates.

3. The number and position of gates can be chosen based on a set of standards. Record the FACS profiles of several yeast-displayed standards in a same-day experiment at a target concentration chosen based on anticipated affinities. Construct a set of gates to be tested for adequate resolution. Determine for each FACS profile how many cells would have hit each gate. This provides a distribution over the gates for each standard. Then, simulate an experiment by drawing random samples with a size of ten cells for each standard. (Note that clones should be sampled more often than this during an actual SORTCERY sort. However, real samples may experience additional experimental noise during preparation for deep sequencing. Thus, we find 10 cells in this procedure provide useful information.) Use the random sample for each standard X and gate i to calculate the normalized frequency, fiX, with which the standard would be observed in the gate. Calculate the probability that standard X is a better binder than standard Y based on the random samples, using the formula given in Subheading 3.5, **step 5**. Compare the result to the actual affinities of the standards. Repeat this many times to determine the range of values the probability can take. Sufficient resolution, i.e., a sufficient number and appropriate placement of gates, will be indicated by mostly high probabilities for the correct ordering of standards.
4. Record several FACS profiles for standards. Consider data for expressing cells that have binding signals mostly above the baseline. Use a cutoff line with a slope of −1 to separate expressing from non-expressing cells; using other cutoffs may bias the analysis. Adjust the retained data by subtracting the average binding and expression signals from each data point. Calculate the covariance matrix of the data. Determine the first principal component by calculating the matrix's eigenvectors and eigenvalues. The vector with the largest corresponding eigenvalue indicates the orientation of the first principle component. Determine the first principle component's slope, i.e., the slope of the vector. High-quality FACS profiles should result in a value close to 1 (Fig. 4). Reduction in quality can have many different experimental origins, such as inappropriate growth protocols (*see* **Notes 1** and **2**), excess dissociation of target molecule during washing steps (*see* **Note 8**), or nonspecific binding to tube walls (*see* **Note 5**).
5. BSA is used as a blocking agent to prevent nonspecific binding to the cells and, more importantly, the test tube walls.

Adsorption to the tube walls may lead to significant depletion of target molecules and distortion of FACS profiles.

6. The number of target molecules should be in excess of the number of surface-displayed peptides. For example, our yeast strain expresses about 30,000 peptides per cell [24]. If 10^6 cells are incubated in 700 μl of 1 nM target molecule solution, then at most ~10 % of the target molecules are bound. Adjust your incubation volume accordingly. Choose the concentration of target molecule appropriately to investigate a specific range of affinities (*see* **Note 3**).
7. We have used an HA tag for detection of expression and a Myc tag for detection of binding. However, other tags may work with our protocol and may be preferred by the experimenter. Required antibody concentrations may depend on the exact choice. Always test whether the antibodies provide high-quality FACS profiles (*see* **Note 3**).
8. Swift application of antibodies is crucial because washing steps can disturb the equilibrium between free and bound target molecules. We have found that fully prepared samples are relatively stable, possibly because the antibodies cross-link the bound target molecules and thereby dramatically decrease dissociation.
9. Because gate setting requires a significant amount of time, gates should be drawn prior to sample preparation. Adjust PMT voltages so that the library's FACS profile largely covers the preset gates. Adjustments may be guided by a set of standards.
10. If the number of chosen gates exceeds the number of sample tubes that the cell sorter can sort into at the same time, gates have to be sampled successively. This may waste a huge number of labeled cells, because cells that hit unselected gates will be discarded. The experimenter can adopt an alternative, convoluted sorting strategy instead that permits sorting into all gates simultaneously. In this approach, cells from different gates are sorted into the same sample tubes. Successive sorts that combine different sets of gates can be carried out, which enables back-calculation of the number of cells in each gate for each clone in the subsequent analysis (*see* **Note 17**). For N gates, prepare *N* unique combinations of gates. A gate must not be paired with any other gate more than once in these combinations. Sort orthogonal sets of combinations successively. For example, if 12 gates are chosen and the sorter can only sort into four sample tubes at the same time, the following set of combinations would be appropriate: {1,2,3}, {4,5,6}, {7,8,9}, {10,11,12}, {1,4,7}, {2,5,10}, {3,8,11}, {6,9,12}, {1,5,8}, {2,4,11}, {3,9,10}, and {6,7,12}. Note that any pair of two gate indices appears together at most once. This set of combinations could be processed in three successive sorts col-

lecting four pools of cells (each pool derived from three gates, all pools sorted into individual sample tubes) at a time: first {1,2,3}, {4,5,6}, {7,8,9}, {10,11,12}, then {1,4,7}, {2,5,10}, {3,8,11}, {6,9,12}, and then {1,5,8}, {2,4,11}, {3,9,10}, {6,7,12}.

11. MmeI recognizes the sequence 5′ TCCRAC 3′. Additional nucleotides 5′ of the binding site can improve binding (e.g., 5′ GGGACCACCACC 3′ in **step 1**, Subheading 3.4.2). MmeI cuts 20 nucleotides 3′ of its binding sequence.
12. Use high-fidelity polymerase and as few PCR cycles as possible in order to reduce errors and amplification bias. 25 cycles generally suffice with the Phusion Polymerase standard protocol.
13. High salt content from the DNA extraction step may prove inhibitory to sufficient amplification. 5 μl DNA extract in a 100 μl reaction mixture generally provides enough dilution to obtain satisfactory results.
14. Excess MmeI may block digestion. MmeI activity is also curbed by high amounts of salt. Excess salt may enter the reaction mixture via the PCR product from the PCR purification step. In addition, MmeI has a very low turnover and stoichiometric amounts of MmeI are required for sufficient digestion. Experimenters need to take special care to use the exact amounts of PCR product and MmeI indicated in Subheading 2.
15. Diverse barcodes at the beginning of a deep sequencing read are required to ensure proper calibration of the base-calling algorithm. Barcodes need to be at least five nucleotides long, and deep sequencing runs should be multiplexed with at least 20 different barcodes. Barcode sequences should vary such that all bases appear in each position with roughly the same frequency.
16. Sequencing a library can be a difficult task for Illumina sequencers, because current base-calling algorithms expect significant sequence variety for all positions of a sample, whereas library samples generally contain regions of constant sequence. Spiking PhiX genome into the sample may help alleviate problems, as may running a reference lane with PhiX genome on the same flow cell.
17. MmeI sometimes cuts 19 or 21 bases 3′ of its binding site. Furthermore, the TC 3′ of the barcode may be missing in some reads. A small fraction of undigested but ligated sample may also be observed.
18. Analyze deep sequencing from convoluted sorts (*see* **Note 9**) in the following way: For each sequence j calculate its frequency in each pool x as

$$g_{xj} = \frac{n_{xj}}{\sum_i n_{xi}}$$

with nxj being the number of reads for sequence j in pool x. Then calculate the corrected number of cells in pool x that contained sequence j as

$$m_{xj} = g_{xj} \sum_y z_y$$

where zy is the number of cells that hit gate y considering the distribution of cells across all gates, and the index y runs over all those gates that are part of pool x. Solve a linear equation system of the form

$$\overrightarrow{M_j} = \overrightarrow{D_j Q_j}$$

for the elements of vector Qj. The xth entry of the vector Mj is mxi. The entry $dxyj$ in the xth row and yth column of matrix Dj is 1 if gate y is part of pool x and zero otherwise. The entry qyj in vector Qj is the time-corrected number of cells in gate y. Normalize vector Qj to obtain the frequencies that are required for **step 5**.

Acknowledgments

The authors thank Vincent Xue for preparing Fig. 1. The authors express their gratitude to the Swanson Biotechnology Center Flow Cytometry Facility and the MIT BioMicro Center for technical support.

This protocol was developed with support from NIGMS under award GM096466. It was also funded by grant no. RE 3111/1-1 of the German Merit Foundation to LR.

Figures 2a, 3, and 4 were reprinted from Publication "SORTCERY—a high-throughput method to affinity rank peptide ligands;" Reich L, Dutta S, Keating AE, J Mol Biol (2015) 427: 2135–2150 with permission from Elsevier.

References

1. Hietpas RT, Jensen JD, Bolon DNA (2011) Experimental illumination of a fitness landscape. Proc Natl Acad Sci U S A 108:7896–7901
2. DeKosky BJ, Ippolito GC, Deschner RP, Lavinder JJ, Wine Y, Rawlings BM et al (2013) High-throughput sequencing of the paired human immunoglobulin heavy and light chain repertoire. Nat Biotechnol 31:166–169
3. Ernst A, Gfeller D, Kan Z, Seshagiri S, Kim PM, Baderet GD et al (2010) Coevolution of

PDZ domain-ligand interactions analyzed by high-throughput phage display and deep sequencing. Mol Biosyst 6:1782–1790

4. DeBartolo J, Dutta S, Reich L, Keating AE (2012) Predictive Bcl-2 family binding models rooted in experiment or structure. J Mol Biol 422:124–144
5. Jolma A, Kivioja T, Toivonen J, Cheng L, Wei G, Enge M (2010) Multiplexed massively parallel SELEX for characterization of human transcription factor binding specificities. Genome Res 861:861–873
6. Reynolds KA, McLaughlin RN, Ranganathan R (2011) Hot spots for allosteric regulation on protein surfaces. Cell 147:1564–1575
7. McLaughlin RN Jr, Poelwijk FJ, Raman A, Gosal WS, Ranganathan R (2012) The spatial architecture of protein function and adaptation. Nature 491:138–142
8. Fowler DM, Araya CL, Fleishman SJ, Kellogg EH, Stephany JJ, Baker D et al (2010) High-resolution mapping of protein sequence-function relationships. Nat Methods 7:741–746
9. Whitehead TA, Chevalier A, Song Y, Dreyfus C, Fleishman SJ, DeMattos C et al (2012) Optimization of affinity, specificity and function of designed influenza inhibitors using deep sequencing. Nat Biotechnol 30:543–548
10. Zhu J, Larman HB, Gao G, Somwar R, Zijuan Zhang Z, Lasersonet U et al (2013) Protein interaction discovery using parallel analysis of translated ORFs (PLATO). Nat Biotechnol 31:331–333
11. Tinberg CE, Khare SD, Dou J, Doyle L, Nelson JW, Schena A et al (2013) Computational design of ligand-binding proteins with high affinity and selectivity. Nature 501:212–218
12. Araya CL, Fowler DM, Chen W, Muniez I, Kelly JW, Fields S (2012) A fundamental protein property, thermodynamic stability, revealed solely from large-scale measurements of protein function. Proc Natl Acad Sci U S A 109:16858–16863
13. Starita LM, Pruneda JN, Russell SL, Fowler DM, Kim HJ, Hiatt JB et al (2013) Activity-enhancing mutations in an E3 ubiquitin ligase identified by high-throughput mutagenesis. Proc Natl Acad Sci USA 110(14):E1263–E1272
14. Melamed D, Young DL, Gamble CE, Miller CR, Fields S (2013) Deep mutational scanning of an RRM domain of the Saccharomyces cerevisiae poly(A)-binding protein. RNA 19:1537–1551
15. Kinney JB, Murugana A, Callan CG Jr, Cox EC (2010) Using deep sequencing to characterize the biophysical mechanism of a transcriptional regulatory sequence. Proc Natl Acad Sci U S A 107:9158–9163
16. Sharon E, Kalma Y, Sharp A, Raveh-Sadka T, Levo M, Zeevi D et al (2012) Inferring gene regulatory logic from high-throughput measurements of thousands of systematically designed promoters. Nat Biotechnol 30:521–530
17. Chao G, Lau W, Hackel BJ, Sazinsky SL, Lippow SM, Wittrup KD (2006) Isolating and engineering human antibodies using yeast surface display. Nat Protoc 1:755–768
18. Liang JC, Chang AL, Kennedy AB, Smolke CD (2012) A high-throughput, quantitative cell-based screen for efficient tailoring of RNA device activity. Nucleic Acids Res 40: 138–142
19. Dutta S, Koide A, Koide S (2008) High-throughput analysis of the protein sequence stability landscape using a quantitative yeast surface two-hybrid system and fragment reconstitution. J Mol Biol 382:721–733
20. Reich L, Dutta S, Keating AE (2015) SORTCERY – a high-throughput method to affinity rank peptide ligands. J Mol Biol 427: 2135–2150
21. Hietpas R, Roscoe B, Jiang L, Bolon DNA (2012) Fitness analyses of all possible point mutations for regions of genes in yeast. Nat Protoc 7:1382–1396
22. Illumina (2015) Illumina Adapter Sequences, Document # 1000000002694 v00. Available on the Illumina web site. http://support.illumina.com/downloads/illumina-customer-sequence-letter.html. Accessed 13 Feb 2016.
23. Ailon N, Charikar M, Newman A (2008) Aggregating inconsistent information: ranking and clustering. JACM 55: article 23
24. Boder ET, Wittrup KD (1997) Yeast surface display for screening combinatorial polypeptide libraries. Nat Biotechnol 15:553–557

Chapter 15

Design of Specific Peptide–Protein Recognition

Fan Zheng and Gevorg Grigoryan

Abstract

Selective targeting of protein–protein interactions in the cell is of great interest in biological research. Computational structure-based design of peptides to bind protein interaction interfaces could provide a potential means of generating such reagents. However, to avoid perturbing off-target interactions, methods that explicitly account for interaction specificity are needed. Further, as peptides often retain considerable flexibility upon association, their binding reaction is computationally demanding to model—a stark limitation for structure-based design. Here we present a protocol for designing peptides that selectively target a given peptide-binding domain, relative to a pre-specified set of possibly related domains. We recently used the method to design peptides that discriminate with high selectivity between two closely related PDZ domains. The framework accounts for the flexibility of the peptide in the binding site, but is efficient enough to quickly analyze trade-offs between affinity and selectivity, enabling the identification of optimal peptides.

Key words Interaction specificity, Computational protein design, PDZ–peptide interactions, Cluster expansion, Flexible peptide docking

1 Introduction

The loss of precise control over cellular protein interactions often results in disease [1]. Therefore, reagents that target protein interactions to rewire cellular signaling pathways in desired ways are of great relevance in both therapeutic development and mechanistic investigation [2]. A considerable fraction of the known cellular interactome is believed to be mediated by peptide-recognition domains (PRDs)—interaction-encoding modules that bind to short amino-acid stretches on their partner proteins [3–5]. Many PRD families are large, with members closely related in structure and sequence, but often having entirely divergent functions. Peptides are a natural choice for functional modulation of PRD-encoded interactions, because they are well suited to occupy the PRD binding site and are amenable to computational design. Further, recognition sequence preferences of several PRDs have

Barry L. Stoddard (ed.), *Computational Design of Ligand Binding Proteins*, Methods in Molecular Biology, vol. 1414,
DOI 10.1007/978-1-4939-3569-7_15, © Springer Science+Business Media New York 2016

been characterized experimentally [6–9], enabling the development of computational models for binding prediction either by direct training on high-throughput experimental data [10, 11], structure-based energy calculations [12–15], or combinations of the two [16, 17]. However, to effectively target a given interaction encoded by a PRD, the targeting peptide should in general be selective—i.e., it should avoid interactions with other proteins, including those within the same PRD family. Given the close similarity among family members, achieving such selectivity by design is not trivial. Peptides chosen purely for binding to the target are likely to also bind other family members, with unpredictable functional consequences.

Structure-based methods for modeling PRD–peptide binding have the potential to generalize across different PRDs [18]. However, the use of such techniques in designing selective recognition is complicated by the inherent flexibility of peptides, which places high computational demands on modeling. To mitigate this problem, we have developed a general computational framework that decouples the complexity of the structure-based simulation used to model PRD–peptide binding from the computational efficiency requirements imposed by the design of selectivity [19]. The framework uses the previously described method of cluster expansion (CE) [20, 21] to produce simple sequence-based expressions that rapidly estimate the results of detailed structure modeling techniques. The efficiency gained by CE enables the fast identification of optimal trade-offs between affinity for the targeted domain and selectivity against any number of undesired partners. The framework is detailed below.

2 Materials

The following resources or materials are needed to apply our framework:

1. A Unix-/Linux-based computing platform with:
   (a) A linear algebra engine (e.g., the proprietary MathWorks MATLAB or the open-source GNU Octave).
   (b) Macromolecular modeling suite Rosetta, version 3.4 or higher [22].
   (c) PyRosetta, a Python-based interface to Rosetta [23].
   (d) Highly desirable: access to a high-performance computing cluster with the ability to perform at least hundreds of jobs independently in parallel.
2. A basic understanding of and the capability to work with the computation resources in 1.

3. Optional, but highly desirable: experimentally validated examples of peptides that bind strongly and those that bind weakly (or undetectably) to members of the PRD family of interest.

3 Methods

In this section, we outline our framework for designing PRD-binding peptides. We will refer to our experience with using it to design PDZ-targeting peptides [19], but we believe that the framework should generalize to other systems. The procedure differs depending on whether the goal is to design high-affinity peptides for a single PRD or design selective peptides that bind one PRD (target) but not the others (competitors). In the latter case, binding to multiple PRDs has to be modeled. If not stated otherwise, it should be assumed that each discussed step is carried out for all PRDs being considered.

1. Download experimental structures of target and competitor PRDs from the Protein Data Bank (PDB), if they are available. The following preferences apply if multiple structures are available for a given PRD (in the order of priority): (a) an X-ray structure is preferred over an NMR structure, (b) a peptide-bound structure is preferred over an *apo* structure, and (c) a higher-resolution X-ray structure is preferred over a lower resolution one. If a given PRD has no experimental structures, use homology modeling (e.g., via the SWISS-MODEL server [24] or MODELLER [25]) to create a predicted structure. The template used in homology modeling should be a peptide-bound structure and otherwise as close in sequence to the relevant PRD as possible (*see* **Note 1**). If either an NMR structure or a homology model is used for a PRD, particular attention should be paid to the results in **step 5**.
2. Subject any homology models to continuous minimization in the presence of a known binding peptide. Because the backbone will be held fixed when sampling the bound state (*see* below), this step is recommended to make the PRD model resemble a peptide-bound state as much as possible. To this end, first align the homology model to the template by optimally superimposing the backbone of binding-site residues, and then copy the peptide backbone from the template to the PRD model. In PyRosetta [23], assign peptide side-chain identities according to a known ligand peptide (a ligand of a closely homologous PRD may be used if no ligand for the target is known) and repack all side chains in the model. Follow by applying full-atom minimization via the "dfpmin" algorithm in PyRosetta, with a tolerance of 0.01, allowing both

backbone torsion angles and side chain χ-angles to move. Note this assumes that the template used in homology modeling is close enough to the PRD of interest to have similar binding geometry and sequence preferences.

3. Collect a set of experimental PRD–peptide complex structures for use in seeding multiple simulation trajectories when modeling new PRD–peptide pairs. For example, for PDZ domains, we collected 51 unique complexes with peptides of at least six residues (Table 1). For each available complex, align its binding site onto that of the PRD of interest, and copy the peptide backbone from the complex onto the PRD (as in **step 2**). To automate the procedure of identifying binding sites in all experimental complexes, we recommend manually defining binding-site residues only in the PRD of interest and then using our substructure search engine MASTER [26] to automatically find corresponding residues in all complexes. We found the generation of diverse starting conformations to seed multiple sampling trajectories to be critical in modeling PDZ–peptide binding, presumably due to the considerable flexibility of the peptide in the binding site [19].
4. Given a peptide/PRD combination to be evaluated, run the Rosetta FlexPepDock ab initio protocol [27] for each of the starting conformations generated in **step 3**. We recommend asking each simulation to generate at least 500 structural models (from 500 independent Monte Carlo simulations). Therefore, in the PDZ example, for each peptide/PRD pair, $500 \times 51 = 25{,}500$ structural models would be generated. Rosetta FlexPepDock documentation is available at https://www.rosettacommons.org/docs/latest/application_documentation/docking/flex-pep-dock. Evaluate each model using the *talaris2013* Rosetta scoring function; in our experience, omitting backbone statistical energy terms "rama" and "omega" increases performance (*see* **Note 2**). The lowest score among all generated models should be used as the final predicted binding score for the given peptide/domain combination.
5. Use an experimental dataset as a benchmark to assess the accuracy of the structure-based simulation and the appropriateness of structural models used. Ideally, experimental data for the relevant PRDs should be used, but if such data are unavailable, results for highly homologous domains in the PRD family (those believed to share close binding preferences) may be used. Use the experimental data to build the benchmark dataset: sets of high-confidence binding peptides and weak/non-binding peptides for each PRD (*see* **Note 3**). Run the procedure in **step 4** to score each peptide/domain combination in the benchmark dataset. Use the Receiver Operating Characteristic (ROC) analysis to measure the ability of the simulation to sep-

Table 1
A set of experimental PDZ–peptide complex structures used to generate starting conformations for multiple simulation trajectories

PDB-ID	Chain-ID (domain)	Domain residue number range	Chain-ID (peptide)
1B8Q	A	11–90	B
1D5G	A	8–90	B
1KWA	A	3–82	B
1L6O	A	3–88	D
1N7F	A	5–84	C
1N7T	A	12–98	B
1OBY	B	2–74	Q
1Q3P	A	8–95	C
1RGR	A	4 88	B
1RZX	A	5–95	B
1TP3	A	13–91	B
1TP5	A	13–91	B
1U3B	A	4–88	A
1VJ6	A	8–90	B
1X8S	A	5–95	B
1YBO	A	88–160	C
1ZUB	A	23–107	B
2AIN	A	7–89	B
2EJY	A	3–81	B
2FNE	B	11–93	A
2HE2	A	7–85	B
2I04	B	3–83	D
2I0I	A	4–81	D
2I0L	A	2–83	C
2I1N	A	6–90	B
2IWP	A	3–83	B
2JIL	A	7–89	B
2JOA	A	5–88	B
2 K20	A	9–99	B
2KA9	A	5–89	B

(continued)

Table 1
(continued)

PDB-ID	Chain-ID (domain)	Domain residue number range	Chain-ID (peptide)
2KBS	A	4–83	B
2KPL	A	17–97	B
2KQF	A	8–91	B
2KYL	A	8–91	B
2L4T	A	17–110	B
2OPG	B	5–87	A
2OQS	A	2–86	B
2OS6	A	11–83	B
2PZD	A	1–85	B
2QBW	A	2–97	B
2UZC	B	3–81	A
2 V90	E	6–85	C
2VRF	B	7–87	A
3B76	B	11–94	A
3CBX	B	7–88	A
3CBY	B	4–86	A
3CC0	C	4–88	A
3CH8	A	2–95	P
3DIW	B	7–100	D
3GGE	A	9–88	B
3LNY	A	8–90	B

This table was created by filtering search results from extended PDZ domain database (http://bcz102.ust.hk/pdzex/) [31]

arate true binders from weak/non-binders, using Area Under the Curve (AUC) for quantification [28]. AUC values above ~0.7 would indicate a reasonable structural model and simulation approach.

6. Define amino acids allowed at each position of the peptide—i.e., the design alphabet. We strongly recommend constraining the alphabet based on any known information about the PRD family in general and the specific targeted domain(s). This keeps the sequence space from being unnecessarily large, limiting computational complexity. Further, patterning of allowed amino acids based on strong experimentally observed preferences limits the effect of error present in any modeling

approach. In our PDZ-targeting study, we were able to design highly selective binders by computationally considering a sequence space of only 8400 peptides [19].

7. Given the computationally complex modeling procedure described in **step 4**, it will likely be prohibitively expensive to enumerate even moderately large peptide sequence spaces (e.g., the procedure takes over 400 CPU hours per peptide in our PDZ example). On the other hand, given a specific PRD, the final score of the simulation depends only on the peptide sequence. Thus, the next step is to derive an analytical mapping from peptide sequence to predicted binding score, for each PRD of interest. We previously described a method for finding such a mapping, called cluster expansion (CE) [20]. In short, CE expresses the result of a structure-based computational procedure as a series expansion in contributions from amino-acid clusters of increasing size—we call these cluster functions or CFs. For example, if $E(\vec{\sigma})$ represents the binding score from the procedure in **step 4**, for a peptide sequence $\vec{\sigma}$ and a given domain, the CE expression states

$$E(\vec{\sigma}) = C + \sum_{\substack{i=1 \\ \sigma_i \neq \rho_i}}^{L} f_i(\sigma_i) + \sum_{\substack{i=1 \\ \sigma_i \neq \rho_i}}^{L-1} \sum_{\substack{j=i+1 \\ \sigma_j \neq \rho_j}}^{L} f_{ij}(\sigma_i, \sigma_j) + \ldots$$

where L is peptide length, $\vec{\rho}$ is a reference sequence, and σi and ρi are the amino acids in the i-th position of $\vec{\sigma}$ and $\vec{\rho}$, respectively. The significance of the reference sequence is that the summations in the expression extend only over combinations of positions (clusters) occupied by amino acids differing from the corresponding ones in $\vec{\rho}$. Thus, C represents the binding score for $\vec{\rho}$ (i.e., the reference CF), whereas the remaining terms capture the additional contributions of amino acids in $\vec{\sigma}$ that differ from $\vec{\rho}$ (i.e., higher-order CFs). The first summation considers point CFs, with $fi(\sigma i)$ representing the effective contribution of amino acid σi at position i. Similarly, the second summation considers pair CFs, with $fij(\sigma i, \sigma j)$ representing the additional pairwise contribution due to having σi at position i and σj at position j simultaneously. To be exact, the expansion must consider all higher-order contributions, up to L-tuples, but in most cases this is impractical. Instead, one can choose to preserve only lower-order CFs (e.g., including only up to pairwise contributions), and use a training set of sequences with pre-computed scores to deduce CF values that optimize the accuracy of the truncated expansion [20].

Based on our PDZ study, a CE with up to pair CFs should represent peptide–PRD interactions reasonably well [19], though higher-order terms can still be added if needed [20].

Point CFs at all positions should be included. To reduce computational complexity, pair CFs can be restricted to position pairs likely to host side chains that interact either directly or through a common site on the PRD. For example, when building CEs for PDZ–peptide interactions, we omitted pair CFs between adjacent peptide positions, as these alternate in pointing either into or away from the binding interface, making coupling between them less likely [19]. Once a cluster is included in a CE (e.g., a pair cluster), every combination of non-reference amino acids at the corresponding positions produces a unique CF. Thus, the number of CFs to be considered is related to the size of design alphabet. For example, in our PDZ study, allowing 2–8 amino acids at six peptide positions resulted in 77 CFs (the reference CF, 24 point CFs, and 52 pair CFs) [19].

8. Generate sequences for CE training by randomly drawing from the design alphabet. The number of sequences should be at least twice the number of CFs to be considered (determined in **steps 6** and 7). These sequences will be subjected to structure-based simulations, so choosing a design alphabet to be only as large as necessary (**step 6**) helps keep training time manageable. Figure 1 uses the PDZ example to show how the complexity of CE training increases with increasing number of amino acids allowed at each position. The random sequence

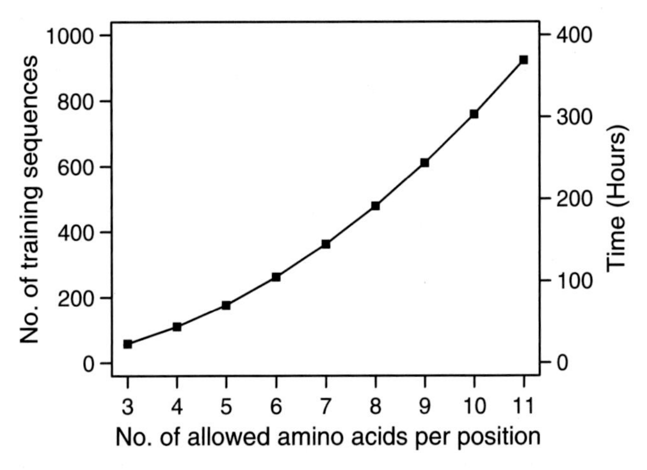

Fig. 1 The computational complexity of generating the CE training set increases with the number of amino acids allowed at each position. The clusters allowed in our PDZ study [19] are used in this estimation. The number of training sequences (*left-axis*) is estimated as twice the number of candidate cluster functions (CFs); time is estimated by assuming that a 1000-core compute cluster is available and that the simulations for one peptide take 400 wall-clock hours when run in serial (*see* **step 4**)

generation can be biased toward any known binding sequence preferences in order to concentrate the sampling (and ultimately CE accuracy) toward more relevant sequence spaces. No matter how the random set is generated, we recommend checking it for reasonable coverage of all CFs to be considered (e.g., at least three examples of each CF should be present). For any underrepresented CFs, sequences that contain them (but are otherwise random) should be added to balance the training set.

9. Run the simulation protocol in **step 4** for all sequences in the training set with all PRDs of interest, extracting the final binding score for each.
10. Train a CE model for each PRD by deriving optimal CF weights. In a linear algebra engine (e.g., MATLAB or Octave), create an $m \times n$ model matrix M, where m is the number of training sequences and n is the number of cluster functions considered ($m > n$). $M(i,j)$ should contain the number of times the j-th CF occurs in the i-th sequence. Typically, this will be either 1 or 0 (when the i-th sequence either does or does not involve the j-th CF, respectively), but can also be a larger integer in cases with structural symmetry, where a CF may occur multiple times within a sequence (e.g., with coiled coils; *see* Ref. [20]). Create also an $m \times 1$ vector E, whose i-th element is the structure-based binding score of the i-th sequence calculated in **step 9**. Optimal CF weights can then be obtained by finding the $m \times 1$ vector b that minimizes the mean squared difference between $E = Mb$ (CE-predicted scores) and E, with the j-th element of b representing the weight of the j-th CF. The least-square solution can be easily found using the method of pseudo-inverse as $\left(M^T M\right)^{-1} M^T E$. In MATLAB or Octave, this corresponds to the expression:

$$b = \left(M' * M\right)\hat{}\left(-1\right) * M' * E$$

Note that matrix M has to be rank n, meaning that CFs have to represent orthogonal information and may not be linear combinations of each other (if M is not rank n, it often means an error was made either in encoding the model matrix or in defining CFs). Rather than including all candidate CFs into M at once and obtaining the best-fitting b, we recommend using our previously described strategy to prevent overtraining. The quality of a CE model (with a specific subset of CFs included) can be conveniently estimated as the average error with which the score of each sequence is predicted when that sequence is left out of the training set—the cross-validation root-mean-square error (CV-RMS). This value can be computed in closed form as

$$\sqrt{\frac{1}{n}\sum_{n}^{i=1}\left(\frac{E_i - \breve{E}_i}{1 - M_{i\cdot}\left(M^T M\right)^{-1} M_{i\cdot}^T}\right)^2},$$

where $Mi.$ represents the i-th row of matrix M. In MATLAB or Octave, this can be computed via the expression:

```
sqrt(sum(((E-M*b)./(1-sum(M.*(M*((M'*M)^(-1))'),
2))).^2)/length(E))
```

Thus, first train a CE model including all CFs (constant, point, and pair)—the all-inclusive model. Next, train another CE model with only constant and point CFs—the current model. Then, consider pair CFs, in decreasing order of their weights in the all-inclusive model, for addition to the current model. Each time a pair CF is considered for addition, train a new CE model that includes all CFs in the current model and the candidate pair CF, and evaluate the resulting CV-RMS. If it is lower than that of the current model, update the current model to include the CF; otherwise, discard the pair CF. Repeat until all pair CFs are considered. We have found this simple procedure to work well in practice, as in our PDZ-targeting study, but we have also proposed a more principled and general-purpose statistical method for choosing CFs to maximize CE accuracy [29].

11. Randomly generate a test set containing sequences not included in the training set, following the same procedure as in **step 8**. The number of sequences in the test set need only be large enough to provide a reliable estimate of CE error. Run the protocol in **step 4** for these sequences, and compute the root-mean-square of the difference between the resulting binding scores and scores calculated by the CE model from above (test-set RMS). This metric is a better indicator of expected CE error and is generally marginally higher than CV-RMS. Evaluate the quality of the CE model in the context of the ROC analysis in **step 5**. CE error should be lower than the score differences that tend to differentiate known binders from non-binders. If this is not the case, then the CE model is not of sufficient accuracy for specificity design, with several possible root causes: (1) important clusters were missed in **step 7**; (2) training set for CE was too small, such that important CF contributions could not be discerned; or (3) the structure-based score being considered is not easily expandable in terms of low-order CFs and may require more context for higher accuracy (e.g., triplet CFs may be necessary; *see* Ref. [20]).

12. Identify optimal peptide sequences for experimental characterization. In an earlier study, we described CLASSY, a framework that feeds CE models into an integer linear programming (ILP)

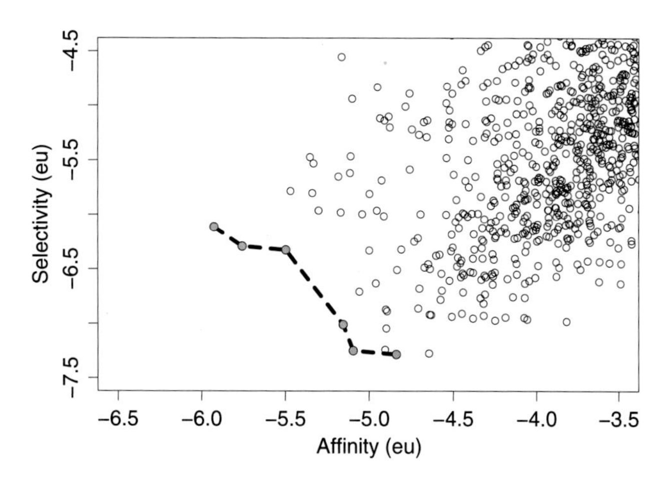

Fig. 2 An example predicted affinity/selectivity landscape, zoomed in around optimal sequences. Scores are shown in Rosetta energy units (eu). Each dot represents a peptide sequence; *X* and *Y* coordinates indicate affinity and selectivity scores, respectively (*see* Ref. [19]), with more negative numbers corresponding to higher affinity and selectivity. Sequences on the pareto-optimal front (i.e., those for which affinity and selectivity cannot be improved simultaneously; gray points) are connected with dashed lines. Adapted from Fig. 4a in Ref. [19]

framework to select sequences that make optimal trade-offs between affinity and selectivity [21]. Alternatively, in circumstances where the peptide sequence space is sufficiently small (i.e., $\leq 10^{10}$ sequences), given that the CE model typically takes less than 1 μs per peptide to evaluate, the entire sequence space can be simply enumerated. Either way, the goal is to find all peptide sequences that cannot be simultaneously improved in both predicted binding score and selectivity (i.e., the difference in binding scores between the target complex and the best-scoring off-target complex) [21]. These sequences lie at the edge of affinity/selectivity space (the so-called pareto-optimal front [30]) and are the only candidates worth considering, due to the simple fact that all other sequences can be simultaneously improved in both parameters. The pareto-optimal front is easy to visualize on a plot of affinity versus selectivity, where each point represents a sequence (Figure 2 shows a plot corresponding to one of the designs from our PDZ study [19]).

13. The number of sequences on the pareto-optimal front is often small enough to allow for the manual inspection of each [19, 21]. We recommend re-scoring each of these sequences by the structure-based framework in **step 4** to check for the possibility of anomalous CE error (discard any candidates scoring significantly less favorably in either affinity or selectivity by the structure-based framework than the CE model), manually analyzing the corresponding structural models for biophysical

plausibility (discard candidates with potential structural problems not properly recognized by the structural modeling framework), and finally choosing among remaining candidates based on the predicted scores. Depending on the availability of time and computational resources, one may also perform explicit-solvent molecular dynamics simulations of chosen candidates to build further support of at least local stability of the peptide in the binding site. Although relevant timescales will differ between systems, at least 10–100 ns of sampling is likely required in most situations to make any relevant observations. Additional issues in selecting candidate sequences are discussed in **Note 4**.

4 Notes

1. Our analysis showed that when a homologous template for a PDZ domain has around 35–45 % sequence identity to the target sequence, the Cα RMSD between the binding pockets of the true structure and the homology model has a median of 1.4 Å [19]. Also, when comparing *apo* and peptide-bound structures of PDZ domains, we noticed that PDZ binding sites tend to widen upon peptide binding [19]. Backbone rearrangements are not modeled in the Rosetta FlexPepDock, but it was shown that although these rearrangements are small, they are enough to affect the outcomes of the structural simulation significantly [27]. Therefore, peptide-bound structures are strongly preferred as homology-modeling templates. For example, in our previous work, we found that a PDZ domain homology model based on a peptide-bound structure with 40 % sequence identity performed much better in binding prediction than one based on an *apo* structure with 50 % sequence identity (unpublished data).
2. In our PDZ study, we conducted benchmark tests for two PDZ domains, NHERF-2 PDZ2 (N2P2) and MAGI-3 PDZ6 (M3P6), with Rosetta 3.4 [22] using the scoring function *score12*. We observed that dropping the backbone statistical terms "rama" and "omega" significantly improved performance [19]. The AUCs before and after omitting these terms were 0.57 and 0.77 for M3P6 (25 binders and 16 non-binders in the benchmark set; Fig. 3). In preparation of this manuscript, we also tested the performance of the new scoring function *talaris2013* used in a newer version of Rosetta (Rosetta_2014.35.57232_bundle), and the AUCs before and after dropping "rama" and "omega" were 0.71 and 0.76 for M3P6. This omission also marginally improves the performance on N2P2 (AUCs 0.86 and 0.91 before and after dropping),

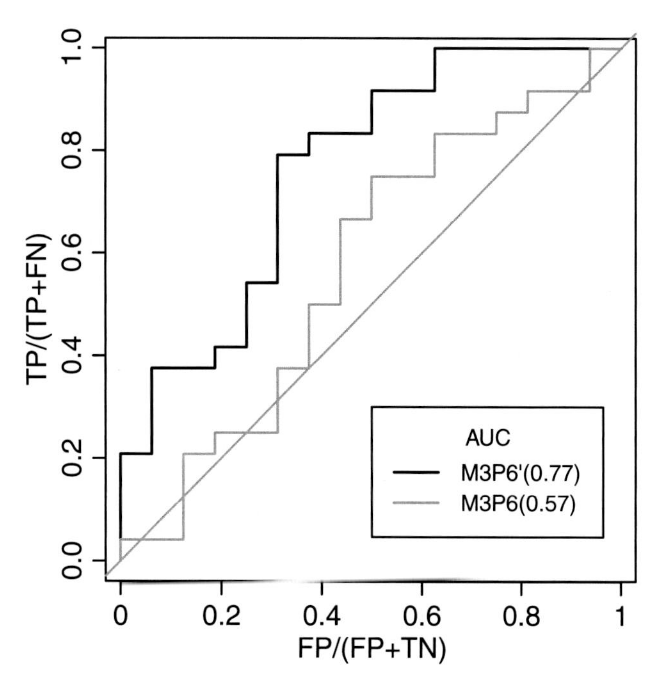

Fig. 3 An example ROC analysis, assessing the performance on differentiating binders from non-binders for M3P6. Default Rosetta scoring function *score12* (*gray line*, labeled as M3P6) and a modified version that omits "rama" and "omega" (*black line*, labeled as M3P6′) are compared. *Numbers in parenthesis* indicate the area under curve (AUC) for each case. *TP:* number of true positives, *FP:* number of false positives, *TN:* number of true negatives, *FN:* number of false negatives. Adapted from Fig. 2 in Ref. [19]

although this domain has fewer data points in our benchmark set (7 binders and 8 non-binders). Importantly, as no experimental structures of M3P6 were available, we used a homology model for simulating M3P6–peptide interactions in our study. Given that the improvement due to omitting "rama" and "omega" is larger for M3P6, it may be that the terms present more of an issue for homology models than crystal structures. Still, omitting the terms appears to improve the performance in general (including additional PDZ domains we have tested since our study; data not shown), and this may be due to the fact that Rosetta scoring functions are generally optimized to recognize/reproduce ground state-like conformations.

3. The benchmark dataset in our PDZ domain study came from the work of MacBeath and coworkers, which characterized binding affinities for a large number of PDZ–peptide pairs [7]. The authors reported dissociation constants if they were below 100 μM, or simply labeled interactions as "weak" in the opposite case. Thus, we naturally chose 100 μM as the cutoff for separating "binders" from "non-binders" for ROC analysis

[19]. If quantitative affinity measurements are not available, SPOT-array or phage-display data can also be used to classify sequences into two categories. However, one should use caution with such data, as they are in general more error prone, especially with respect to false negatives (i.e., true binders that are not detected in the assay).

4. It may be unnecessary to experimentally test all candidate sequences selected in **steps 12** and **13**. It is generally advantageous to characterize sequences spanning different levels of selectivity, to determine whether predicted affinity/selectivity trade-offs are correct. When possible and applicable, choose sequence subsets with diverse structural strategies for reaching either affinity or selectivity.

References

1. Ryan DP, Matthews JM (2005) Protein-protein interactions in human disease. Curr Opin Struct Biol 15(4):441–446
2. Bashor CJ, Horwitz AA, Peisajovich SG, Lim WA (2010) Rewiring cells: synthetic biology as a tool to interrogate the organizational principles of living systems. Annu Rev Biophys 39:515–537
3. Pawson T, Nash P (2003) Assembly of cell regulatory systems through protein interaction domains. Science 300(5618):445–452
4. Kuriyan J, Cowburn D (1997) Modular peptide recognition domains in eukaryotic signaling. Annu Rev Biophys Biomol Struct 26:259–288
5. Neduva V, Linding R, Su-Angrand I, Stark A, de Masi F, Gibson TJ, Lewis J, Serrano L, Russell RB (2005) Systematic discovery of new recognition peptides mediating protein interaction networks. PLoS Biol 3(12):2090–2099
6. Tonikian R, Zhang YN, Sazinsky SL, Currell B, Yeh JH, Reva B, Held HA, Appleton BA, Evangelista M, Wu Y, Xin XF, Chan AC, Seshagiri S, Lasky LA, Sander C, Boone C, Bader GD, Sidhu SS (2008) A specificity map for the PDZ domain family. PloS Biol 6(9):2043–2059
7. Stiffler MA, Chen JR, Grantcharova VP, Lei Y, Fuchs D, Allen JE, Zaslavskaia LA, MacBeath G (2007) PDZ domain binding selectivity is optimized across the mouse proteome. Science 317(5836):364–369
8. Jones RB, Gordus A, Krall JA, MacBeath G (2006) A quantitative protein interaction network for the ErbB receptors using protein microarrays. Nature 439(7073):168–174
9. Birnbaum ME, Mendoza JL, Sethi DK, Dong S, Glanville J, Dobbins J, Ozkan E, Davis MM, Wucherpfennig KW, Garcia KC (2014) Deconstructing the peptide-MHC specificity of T cell recognition. Cell 157(5):1073–1087
10. Chen JR, Chang BH, Allen JE, Stiffler MA, MacBeath G (2008) Predicting PDZ domain-peptide interactions from primary sequences. Nat Biotechnol 26(9):1041–1045
11. Kamisetty H, Ghosh B, Langmead CJ, Bailey-Kellogg C (2014) Learning sequence determinants of protein: protein interaction specificity with sparse graphical models, Research in computational molecular biology. Springer, New York, pp 129–143
12. London N, Lamphear CL, Hougland JL, Fierke CA, Schueler-Furman O (2011) Identification of a Novel Class of Farnesylation Targets by Structure-Based Modeling of Binding Specificity. PloS Comput Biol 7(10):e1002170
13. London N, Gulla S, Keating AE, Schueler-Furman O (2012) In silico and in vitro elucidation of BH3 binding specificity toward Bcl-2. Biochemistry 51(29):5841–5850
14. Roberts KE, Cushing PR, Boisguerin P, Madden DR, Donald BR (2012) Computational design of a PDZ domain peptide inhibitor that rescues CFTR activity. PloS Comput Biol 8(4)
15. Yanover C, Bradley P (2011) Large-scale characterization of peptide-MHC binding landscapes with structural simulations. Proc Natl Acad Sci U S A 108(17):6981–6986
16. DeBartolo J, Dutta S, Reich L, Keating AE (2012) Predictive Bcl-2 family binding models rooted in experiment or structure. J Mol Biol 422(1):124–144
17. DeBartolo J, Taipale M, Keating AE (2014) Genome-wide prediction and validation of

peptides that bind human prosurvival Bcl-2 proteins. PloS Comput Biol 10(6)

18. King CA, Bradley P (2010) Structure-based prediction of protein-peptide specificity in Rosetta. Proteins 78(16):3437–3449
19. Zheng F, Jewell H, Fitzpatrick J, Zhang J, Mierke DF, Grigoryan G (2015) Computational design of selective peptides to discriminate between similar PDZ domains in an oncogenic pathway. J Mol Biol 427(2):491–510
20. Grigoryan G, Zhou F, Lustig SR, Ceder G, Morgan D, Keating AE (2006) Ultra-fast evaluation of protein energies directly from sequence. PloS Comput Biol 2(6):551–563
21. Grigoryan G, Reinke AW, Keating AE (2009) Design of protein-interaction specificity gives selective bZIP-binding peptides. Nature 458(7240):859–U852
22. Leaver-Fay A, Tyka M, Lewis SM, Lange OF, Thompson J, Jacak R, Kaufman K, Renfrew PD, Smith CA, Sheffler W, Davis IW, Cooper S, Treuille A, Mandell DJ, Richter F, Ban YEA, Fleishman SJ, Corn JE, Kim DE, Lyskov S, Berrondo M, Mentzer S, Popovic Z, Havranek JJ, Karanicolas J, Das R, Meiler J, Kortemme T, Gray JJ, Kuhlman B, Baker D, Bradley P (2011) Rosetta3: an object-oriented software suite for the simulation and design of macromolecules. Methods Enzymol 487:545–574
23. Chaudhury S, Lyskov S, Gray JJ (2010) PyRosetta: a script-based interface for implementing molecular modeling algorithms using Rosetta. Bioinformatics 26(5):689–691
24. Arnold K, Bordoli L, Kopp J, Schwede T (2006) The SWISS-MODEL workspace: a web-based environment for protein structure homology modelling. Bioinformatics 22(2):195–201
25. Eswar N, Webb B, Marti-Renom MA, Madhusudhan MS, Eramian D, Shen MY, Pieper U, Sali A (2006) Comparative protein structure modeling using Modeller. Current protocols in bioinformatics Chapter 5: Unit 5 6
26. Zhou J, Grigoryan G (2015) Rapid search for tertiary fragments reveals protein sequence-structure relationships. Protein Sci 24(4):508–524
27. Raveh B, London N, Zimmerman L, Schueler-Furman O (2011) Rosetta FlexPepDock ab-initio: simultaneous folding, docking and refinement of peptides onto their receptors. PloS One 6(4)
28. Fawcett T (2006) An introduction to ROC analysis. Pattern Recogn Lett 27(8):861–874
29. Hahn S, Ashenberg O, Grigoryan G, Keating AE (2010) Identifying and reducing error in cluster-expansion approximations of protein energies. J Comput Chem 31(16):2900–2914
30. He L, Friedman AM, Bailey-Kellogg C (2012) A divide-and-conquer approach to determine the Pareto frontier for optimization of protein engineering experiments. Proteins 80(3):790–806
31. Wang CK, Pan LF, Chen J, Zhang MJ (2010) Extensions of PDZ domains as important structural and functional elements. Protein Cell 1(8):737–751

Chapter 16

Computational Design of DNA-Binding Proteins

Summer Thyme and Yifan Song

Abstract

Predicting the outcome of engineered and naturally occurring sequence perturbations to protein–DNA interfaces requires accurate computational modeling technologies. It has been well established that computational design to accommodate small numbers of DNA target site substitutions is possible. This chapter details the basic method of design used in the Rosetta macromolecular modeling program that has been successfully used to modulate the specificity of DNA-binding proteins. More recently, combining computational design and directed evolution has become a common approach for increasing the success rate of protein engineering projects. The power of such high-throughput screening depends on computational methods producing multiple potential solutions. Therefore, this chapter describes several protocols for increasing the diversity of designed output. Lastly, we describe an approach for building comparative models of protein–DNA complexes in order to utilize information from homologous sequences. These models can be used to explore how nature modulates specificity of protein–DNA interfaces and potentially can even be used as starting templates for further engineering.

Key words Protein–DNA interactions, Computational design, Rosetta, Specificity, In silico prediction, Direct readout, Homology model

1 Introduction

Sequence-specific protein–DNA interactions play a key role in fundamental cellular processes. Alterations to gene regulatory networks, via changes to transcription factor binding site affinity, drive disease progression [1–5] and potentially species evolution [6–10]. Being able to accurately model these interactions can enhance understanding of the biophysical basis behind such changes [1], enabling the development of tools to test predictions and modulate the interactions. The Rosetta program for macromolecular modeling and design [11], the focus of this chapter, has been used to redesign protein–DNA interfaces. The design algorithm in Rosetta searches protein sequence and rotameric [12] (*see* **Note 1**) space, finding amino acid combinations that are energetically compatible with the DNA sequence being targeted. Evaluation of each

Barry L. Stoddard (ed.), *Computational Design of Ligand Binding Proteins*, Methods in Molecular Biology, vol. 1414, DOI 10.1007/978-1-4939-3569-7_16, © Springer Science+Business Media New York 2016

amino acid combination with a physically based energy function identifies the lowest-energy designed sequence [11, 13].

Proteins recognize DNA partners through direct interactions between side chains and bases, water-mediated contacts, and indirect readout, the sequence-dependent shape, and conformation of the DNA [14, 15]. High specificity positions in the binding sites of many DNA-interacting proteins, where one nucleotide is much preferred over others, are often characterized by strong direct contacts that are disrupted when the favored base is replaced [14, 16, 17]. Computational protein–DNA interface design has mainly been successful at altering these direct interactions to shift binding specificity for small numbers of nucleotide substitutions [18–22]. The main drivers of direct readout are hydrogen bonding and hydrophobic packing, both crucial components of computational design algorithms that are actively being improved upon [23–25]. Water-mediated interactions are generally captured through implicit solvent models [26, 27], although explicit water molecules have recently been incorporated into computational design algorithms [28]. Modeling indirect readout is arguably the current biggest challenge for computational protein–DNA design. All previous redesign successes maintained the DNA backbone conformation from the starting crystal structure, although it is clear from crystal structures of computational designs [22] and evolved interfaces that extensive movements of the DNA can occur [29–31]. There is some knowledge of how DNA bending preferences influence target site specificity [14, 15], but these energetic components are just beginning to be incorporated into the Rosetta program [27] (*see* **Note 2**).

One way to go beyond limits in state-of-the-art computational models, while simultaneously gathering experimental data to improve them, is to combine design with directed evolution. Computational design results can be used for low-activity starting points for directed evolution [32–35] or can guide initial library design. Directed evolution is itself limited in how many amino acids can be simultaneously randomized, and computational design can enable many more positions to be concurrently explored by suggesting the inclusion of only certain amino acid types at each position in a protein library [36–39]. There are a number of approaches for directed evolution of protein–DNA interactions [29, 36, 40–43] that can be used in conjunction with computational design to increase the likelihood of engineering success and potentially feedback to the models to improve future outcomes.

Utilizing all available information about a particular protein sequence is important for success in protein engineering, particularly if the information can be merged with a high-throughput screening method. In this chapter we describe several protocols to diversify computational design results over the standard fixed-backbone approach: using libraries of native-like interactions (called motifs) to guide rotamer sampling [13, 36, 44, 45], explicit design for specificity using a genetic algorithm [20, 22], and flexibility of the protein backbone [22]. In addition to these methods for increased design diversity, sequence information from protein homologues can increase our understanding of how the specificities of a protein of interest are modulated by natural evolution [46–48]. One way to incorporate this information into the design process is by building high-resolution homology models of protein–DNA complexes and predicting specificities of homologues from the models [27, 47]. Here we describe protein–DNA homology modeling and target site prediction with Rosetta. Homology models can be used in conjunction with directed evolution in engineering pipelines [49] and can potentially even be used as starting templates for computational design.

2 Materials

1. The Rosetta software suite. The release version of Rosetta (Rosetta 2015.19 as of May, 2015) is free of charge for academics and nonprofit users and is available from https://www.rosettacommons.org/software/license-and-download. While the majority of the protocols described in this chapter can be completed with this release version, some advanced design modes, such as using motifs [13, 36, 44, 45], require the developer's version of Rosetta. Access to the developer's repository can be obtained through a sponsor from or collaboration with a lab that is a member of RosettaCommons (*see* **Note 3**), and protocols that require these extended capabilities are noted throughout.
2. Python (version ≥ 2.4 and <3.0) to compile the Rosetta code. A local version of the compiling software SCons comes packaged with Rosetta and is run via the scons.py script that is also included with the Rosetta download (*see* **Note 4**).
3. A Unix or Linux server or cluster for running Rosetta jobs (*see* **Note 5**). The Rosetta software can run on multiple platforms (*see* **Note 6**), however they may not all be fully supported. The majority of experiments, any protocols other than the standard design method (Subheading 3.1), will require submitting many runs in parallel to a Unix or Linux cluster to achieve adequate results with reasonable calculation times.
4. A high-resolution (preferably <3.0 Å) structure of a protein–DNA complex. Alternatively, a homology model can be used if a protein–DNA complex of a related protein is available to use as a template.

5. The homology modeling protocol (Subheading 3.4) currently requires several in-house scripts (*see* **Note 7**) and the following databases: NCBI NR (*see* **Note 8**) and HH-suite (*see* **Note 9**).

3 Methods

3.1 Standard Protein–DNA Interface Design

1. Open a terminal window (*see* **Note 5**).
2. Enter the Rosetta source directory that contains the scons.py file. Compile a version of the code that can be ported to a different computer system and operating platform by typing "scons bin mode=release extras=static" (*see* **Notes 10** and **11**).
3. Make a directory where the code will be run and the output collected by typing "mkdir nameofdirectory" (*see* **Notes 12** and **13**).
4. Make a file that contains the arguments read by Rosetta (Fig. 1) with your favorite text editor. The text editor Vi (*see* **Note 14**) is likely present in your Linux/Unix system. To make the arguments file using Vi by typing "vi nameofargsfile", entering insertion mode by typing "i", and then typing the desired flags using Fig. 1 as a guide.
5. Make an XML script file (*see* **Note 15**) that contains protocol instructions given to the program through RosettaScripts [51], using Fig. 2 as a guide for the content.
6. The amino acid positions in the protein–DNA interfaces that will be designed are automatically calculated based on the "dna_defs" and "z_cutoff" flags that are part of the operations (TASKOPERATIONS) included in the XML file (Fig. 2). If the user would instead prefer to only allow a subset of amino acid types and designed positions, a resfile (Fig. 3) can be used. The resfile will override automatic detection of interface residues by the addition of the line "-resfile nameoffile" to the args file (Fig. 1). The XML script should also be modified to add the task operation "<ReadResfile name=RRF/>" and replace the use of AUTOprot with RRF in the mover. The "dna_def" option in the DnaInt operation is no longer necessary because the target base is specified in the resfile.
7. Choose an energy function that is optimized for protein–DNA interactions (*see* **Note 16**), and make a file containing the necessary weights for energy function components (Fig. 4). The name of the weights file is the input for the flag "-score::weights nameoffile" (Fig. 1).
8. If necessary, modify the Rosetta database to go with the energy function shown in Fig. 4. Previously used optimized energy functions [13] have required database changes (*see* **Note 17**).

```
-in:ignore_unrecognized_res # ignore anything in the pdb structure that is
not recognizable
-file:s 2QOJ.pdb # input structure
-mute all # no output into an output file, skip this flag off when debugging
and include for large-scale runs
-unmute protocols.dna # unmute a subset of the output if desired
-score::weights rosetta_database/scoring/weights/optimizedenergyfxn.wts
# energy function for evaluating structures (see Fig. 5)
-score:output_residue_energies # include information in the pdb about the
interaction energies of residues in the design
-run:output_hbond_info # include information in the pdb about the
hydrogen bonding of residues in the design
-database rosetta_database # required Rosetta database, see Note 17 for
useful changes to the database
-ex1 # extra rotamer sampling around chi angle 1
-ex2 # extra rotamer sampling around chi angle 2
-ex1aro::level 6 # even more extra rotamer sampling for aromatic residues
around chi angle 1. This flag is recommended because aromatic residues
can have large repulsion scores if the rotamer is not in the optimal position.
-ex2aro::level 6 # even more extra rotamer sampling for aromatic residues
around chi angle 2
-exdna::level 4 # use DNA rotamers and include extra sampling (inclusion
of this flag is highly advised for protein-DNA design)
-jd2:dd_parser # use the parser protocols
-parser:protocol XML.scriptfile # XML script (see Fig. 3)
-overwrite # if a pdb with the same name already exists in the directory
where the design occurring, then overwrite the old pdb
-out:prefix design_ # an optional prefix to add to the name of designs
```

Fig. 1 Example arguments file. This file controls the parameters of the design run or specificity calculation. All writing after the # mark is a comment that is not read in by the Rosetta program. This figure is reproduced with publisher's permission from Ref. [50]

9. Run code by submitting to whatever computer cluster you are using or by typing "rosettaDNA.static.linuxgccrelease @ nameofargsfile" (*see* **Notes 18** and **19**).

3.2 Assessment of Designs Using Specificity and Binding Energy Calculations

Calculations of specificity and binding energy are used to identify designed sequences with properties of interest (*see* **Note 20**). These methods can also be used to predict the binding sites for proteins with unknown target preferences.

3.2.1 Automatic Specificity and Binding Energy Prediction Following Fixed-Backbone Design

The simplest method of specificity prediction [1, 16] is the addition of the two lines to the XML file. This method is not suitable for protocols that involve any backbone movement because the backbone is optimized for the base-pair originally designed for and the energy would be biased for this base-pair.

```
<dock_design>
  <TASKOPERATIONS>
   <InitializeFromCommandline name=IFC/> # use the information in the args file to supplement this XML
   <IncludeCurrent name=IC/> # includes the rotamers in the input structure (may not want to use)
   <RestrictDesignToProteinDNAInterface name=DnaInt base_only=1 z_cutoff=6.0 dna_defs=Z.409.GUA/> #
make the target site substitution of interest (chainID.crystalposition.type) and designate the sphere of residues
surrounding it that are designable and packable
   <OperateOnCertainResidues name=AUTOprot> # works with the DnaInt operation to enable residues to be
chosen for design and packing if they are marked as AUTO
     <AddBehaviorRLT behavior=AUTO/>
     <ResidueHasProperty property=PROTEIN/>
   </OperateOnCertainResidues>
  </TASKOPERATIONS>
  <SCOREFXNS>
   <DNA weights=optimizedenergyfxn/> # energy function for design evaluation, this file must be put in the
directory (ie, rosetta_database/scoring/weights/optimizedenergyfxn.wts)
  </SCOREFXNS>
  <FILTERS>
   <FalseFilter name=falsefilter/> # RosettaScripts has the ability to only output designs that pass a designated
filter. This functionality is not being used here.
  </FILTERS>
  <MOVERS>
   <DnaInterfacePacker name=DnaPack scorefxn=DNA task_operations=IFC,IC,AUTOprot,DnaInt/>
  </MOVERS>
  <PROTOCOLS>
   <Add mover_name=DnaPack/>
  </PROTOCOLS>
</dock_design>
```

Fig. 2 Example RosettaScripts XML file. This file can be used to set up and modify Rosetta protocols with. All writing after the # mark is a comment that is not read in by the Rosetta program. This figure is reproduced with publisher's permission from Ref. [50]

1. Follow instructions in Subheading 3.1 with the following described variations to the XML script (Fig. 2) and arguments files (Fig. 1).
2. Replace *line a* with *line b* in the XML file (Fig. 2) and run the protocol exactly as described in Subheading 3.1, but with this new XML file instead of the original.

```
line a: <DnaInterfacePacker name=DnaPack scorefxn=DNA
task_operations=IFC,IC,AUTOprot,DnaInt/>
```

```
line b: <DnaInterfacePacker name=DnaPack scorefxn=DNA
task_operations=IFC,IC,AUTOprot,DnaInt binding=1 probe_
specificity=3/>
```

3. The number following the added options refers to the number of "repacks," the lowest energy of which is used in the calculations. A repack is a search similar to the design procedure except that only rotameric state is varied while amino acid types are fixed. The recommended number of repacks is at least three, to reduce noise in the resulting energies.
4. The calculation results are located inside the output pdb file for each design. Open the file with a text-editing program to view the data. If multiple files need to be analyzed, a script may be necessary to parse the information.

```
AUTO # all protein positions not
explicitly noted are to be marked as
AUTO, the same as using the
AUTOprot operation
start
28 A PIKAA L # forces amino acid L
at position 28 on chain A
83 A PIKAA R
-12 C NATRO # g, fixes the native
rotamer
-11 C NATRO #   c
-10 C NATRO #   a
-9 C NATRO # g
-8 C NATRO #   a
-7 C NATAA #   a, fixes the native
residue type, but allows different
rotamers
-6 C TARGET GUA #   c, target
base, same as using the dna_def
option, but DNA is required to be
explicit in the resfile
-5 C NATAA # g
-4 C NATRO # t
-3 C NATRO #   c
-2 C NATRO # g
-1 C NATRO # t
1 D NATRO #   a
2 D NATRO #   c
3 D NATRO # g
4 D NATRO #   a
5 D NATAA #   c
6 D TARGET CYT # g
7 D NATAA # t
8 D NATRO # t
9 D NATRO #   c
10 D NATRO # t
11 D NATRO # g
12 D NATRO #   c
```

Fig. 3 Example resfile. This file is used if specific protein positions or amino acid types need to be forced in the design run. It is an alternative to allowing the location of the target substitution to control the designable protein positions. All writing after the # mark is a comment that is not read in by the Rosetta program. This figure is reproduced with publisher's permission from Ref. [50]

3.2.2 Protocol for Specificity Calculation that Is Suitable Following Any Design Procedure

The main goal of a specificity calculation is to find and compare the energy of a set of given sequences by exploring rotameric and potentially backbone space. The computational program used to generate a design is not always the best choice for the specificity prediction. For example, a crystal structure backbone may have an energetic bias for the native base-pair and a flexible backbone specificity calculation can overcome this bias by enabling the protein backbone to be optimized for each base.

```
METHOD_WEIGHTS ref -0.3 -0.7 -0.75 -0.51 0.95 -0.2 0.8
-0.7 -1.1 -0.65 -0.9 -0.8 -0.5 -0.6 -0.45 -0.9 -1.0 -0.7 2.3 1.1 #
reference weights that are for each amino acid type

fa_atr 0.95 # attractive forces between residues
fa_rep 0.44 # repulsive forces between residues
fa_intra_rep 0.004 # repulsion within a sidechain
fa_sol 0.65 # one component of desolvation
lk_ball 0.325 # newer orientation-dependent desolvation
lk_ball_iso -0.325 # newer orientation-dependent desolvation
hack_elec 0.5 # coulombic electrostatics
fa_dun 0.56 # probability for each approximated rotamer
ref 1 # weight for the reference energies
hbond_lr_bb 1.17 # hydrogen bonding
hbond_sr_bb 1.17 # hydrogen bonding
hbond_bb_sc 1.17 # hydrogen bonding
hbond_sc 1.17 # hydrogen bonding
p_aa_pp 0.64 # probability of amino acid type given
backbone
dslf_ss_dst 0.5 # disulphides
dslf_cs_ang 2 # disulphides
dslf_ss_dih 5 # disulphides
dslf_ca_dih 5 # disulphides
pro_close 1.0 # proline ring closure
```

Fig. 4 Example energy function file. This energy function was optimized to produce high sequence recovery of protein–DNA interactions over a benchmark set of proteins [13]. All writing after the # mark is a comment that is not read in by the Rosetta program. This figure is reproduced with publisher's permission from Ref. [50]

1. Modify the XML script to fix the protein sequence of the structure being analyzed. In the TASKOPERATIONS section of the XML file, the operation to fix the protein sequence is added with the following four lines:

```
<OperateOnCertainResidues name=ProtNoDes>
<RestrictToRepackingRLT/>
<ResidueHasProperty property=PROTEIN/>
</OperateOnCertainResidues>
```

   To use this operation, the DnaInterfacePacker mover must be changed to the following:

```
<DnaInterfacePacker name=DnaPack scorefxn=DNA task_
operations=IFC,IC,AUTOprot,ProtNoDes,DnaInt/>
```

2. If desired, modify the arguments file to increase the number of rotamers (*see* **Note 21**). The addition of the flags "-ex3" and "-ex4" is a reasonable increase. Further increases can be enabled by using the "::level #" addition to any of the -ex flags. The available levels are 1–7 (*see* **Note 22**).

3. Set up four runs, one for each base type (or more if the target has multiple base-pair substitutions, do runs for whichever competing states are to be compared).
4. Complete a minimum of 10 runs per base type for a fixed-backbone approach and at least 50 (×4 or more) for any approach involving flexible backbone (*see* Subheading 3.3.3).
5. Collect the total_score value from inside of each pdb (*see* **Note 23**). The specificity can be calculated from the lowest-energy structure or from the mean or median of the energies of all structures. A comparison of all these three specificity calculations is most informative (*see* **Notes 24** and **25**).

3.3 Rosetta Modes for Increasing Diversity of Designed Sequences

Using computation to guide directed evolution libraries depends on having multiple designs to combine in the selection process. The standard fixed-backbone approach yields a single or very limited number of design solutions with a given energy function. This most energetically favorable computational model is not always the optimal experimental solution, yet it can contain individual high-quality interactions. Incorporating information from multiple low-energy solutions by using directed evolution is one way to more fully take advantage of the information available from modeling.

3.3.1 Motifs

Design procedures are computationally limited in how many rotamers can be included in the design search. This reliance of design on the rotamer approximation means that sometimes energetically favorable interactions will be missed. One way to get around this limit is to increase rotamer sampling in regions likely to form favorable interactions by using motifs [13, 36, 44, 45], libraries of interactions seen in crystal structures. In one type of motif-based protocol a vastly expanded rotamer set is compared to the motif library, and those rotamers that can form one of these native-like interactions with a target base-pair are identified. The design procedure can then be biased with these favorable rotamers by adding them to the standard rotamer set and giving them an energetic bonus, overcoming rotamer sampling limitations and also potential inaccuracies in the energy function. Expanded instructions for running this protocol are available in Ref. [13].

1. Acquire access to the developer's version of the code (*see* Subheading 2).
2. Compile the dna_motif_collector application in order to build a library of protein–DNA motifs.
3. Download all crystallized protein–DNA complexes under some resolution cutoff (<2.8 is reasonable).
4. Run the following command (or a slight variation of it): /rosetta/bin/dna_motifs_collector.linuxgccrelease *-motif_output_directory* <directory name> *-ignore_unrecognized_res -adducts* dna_

major_groove_water -*database* <rosetta database> -*l* <name of output motif list>

5. Compile the motif_dna_packer_design app.
6. Add the line "special_rot 1.0" to the energy function (Fig. 4).
7. Add the line "-patch_selectors SPECIAL_ROT" to the args file (Fig. 1).
8. Add flags to the args file (Fig. 1) to load in the motif library, set up cutoffs for acceptance of a motif rotamer, pick a rotamer level for the expanded motif rotamer library, and pick the energetic bonuses to try for these added rotamers. An example command line is shown in the following step. All command line options are explored in extensive detail in the supplemental methods of Ref. [13].
9. /rosetta/bin/motif_dna_packer_design.linuxgccrelease -*run_motifs* -*dtest* 2.0 -*z1* 0.97 -*z2* 0.97 -*r1* 1.0 -*r2* 1.0 -*dna::design::z_cutoff* 6.0 -*motifs::rotlevel* 8 -*motifs::list_motifs* <name of output motif list> -*motifs::output_file* <output file for motifs> -*s* <PDB file being designed> -*score::weights* <energy function file> -*ignore_unrecognized_res* -*database* <rosetta database> -*ex1* -*ex2* -*ex1aro::level* 6 -*ex2aro::level* 6 -*extrachi_cutoff* 0 -*dna::design::dna_defs* <position being designed with motifs, e.g. X.409.CYT> -*special_rotweight* <weight for motif rotamers, e.g. -1.25> -*num_repacks* 4

3.3.2 Multistate Design

Multistate design relies on a genetic algorithm method to explicitly design for one state and against others [52–54]. In protein–DNA design, those states are the targeted bases and the alternative possible bases [20, 22].

1. Follow instructions in Subheading 3.1 with the following variations to the XML script (Fig. 2).
2. Modify the XML file by replacing the standard DNA design mover with the following mover for doing multistate:

```
<DnaInterfaceMultiStateDesign name=msd scorefxn=
DNA task_operations=IFC,IC,AUTOprot,DnaInt pop_
size=20 num_packs=1 numresults=0 boltz_temp=2
anchor_offset=15 mutate_rate=0.8 generations=5/>
```

3. Additionally, the line "<Add mover_name=DnaPack/>" must be replaced with the line:

```
<Add mover_name=msd/>
```

4. All of the parameters of the genetic algorithm can be varied, and the ones in the above line are parameters to test the procedure, rather than do a complete run. Refer to cited literature [20, 22, 52] to identify reasonable starting parameters.

3.3.3 Protein Flexibility

1. Follow instructions in Subheading 3.1 with the following described variations to the XML script (Fig. 2) and arguments files (Fig. 1).
2. Modify the XML file to include a second mover before the standard design mover (DnaInterfacePacker). The line to add is:

```
<DesignProteinBackboneAroundDNA name=bb scorefxn=DNA task_operations=IFC,IC,AUTOprot,DnaInt type=ccd gapspan=4 spread=3 cycles_outer=3 cycles_inner=1 temp_initial=2 temp_final=0.6/>
```

3. Additionally, the following line must be added after the line "<Add mover_name=DnaPack/>":

```
<Add mover_name=bb/>
```

4. The DesignProteinBackboneAroundDNA enables the ccd backbone movement [55, 56]. An advanced user of Rosetta and RosettaScripts format could explore protein backbone space with alternative protocols [57–61] and then use those structures as input for standard design (*see* **Note 26**).
5. Many more design runs, at least 50 for a single base-pair substitutions, are required to explore the range of design possibilities when using flexible backbone simulation, as the diversity of results will be significantly increased.

3.3.4 High-Temp Packer

The High-Temp packer approach increases the temperature that the simulated annealing algorithm driving the design process converges to. Using this approach increases the chance of producing a design that is low-energy, but not the absolute lowest energy. The supplemental methods of Ref. [13] describes the two code changes required to use this method. These changes can be made to any version of Rosetta and then the code must be recompiled.

3.4 Design Starting from Homology Models of Protein–DNA Complexes

The current protocol of protein–DNA complex homology modeling is based on a modified version of RosettaCM [62]. It models proteins structures in the same way as RosettaCM and treats the DNA as a rigid body. The interactions between protein and DNA are optimized during the RosettaCM protein structure modeling. This procedure requires that there is a homologue of the protein of interest that has been crystallized bound to DNA. In this example, we use the structure of I-OnuI (PDB code: 3QQY).

1. Set up standard homology modeling input files by running "setup_cm.pl sequence_file" (*see* **Note 7**).
2. Choose a template structure of which the DNA structure (example, 3qqy) will be used for modeling.

3. Superimpose all input templates for the protein homology modeling onto the template structure with the DNA, using pymol or superposition scripts.
4. Thread the new DNA sequence onto the structure with DNA "change_base.py --inpdb 3QQY.pdb --dna_seq ACGT --outpdb out.pdb --chain B" (chain B is the DNA chain grafted from 3QQY, dna_seq gives the DNA sequence input being threaded to chain B of 3QQY.pdb).
5. Copy and paste DNA coordinates from 3QQY to all the superimposed template structures in a text editor.
6. If any protein segments in template structures clashes with DNA (this can be visually identified in pymol), remove the coordinates of the clashed segments in a text editor.
7. A rosetta_scripts xml input is created in **step 1**. Edit the xml input and add "add_hetatm=1" to the <Hybridize …> line, so that DNA structure from the templates is added for modeling.
8. Run RosettaCM using the flags set up by **step 1**. (rosetta_scripts.xxx @flags_common @flags0_C1 -nstruct 100 (generating 100 models)
9. Using the energy output (score.sc) to identify 20 lowest-energy models.
10. Using the ΔΔG protocol (*see* rosetta scripts documentation) to calculate the interaction between protein and DNA of the 20 low-energy models (Fig. 5), select the model with the strongest interaction.

4 Notes

1. A rotamer is a low-energy conformation of an amino acid [12]. The protocol to identify the lowest-energy design is based in a simulated annealing algorithm [11].
2. The DNA movement protocols in Rosetta are currently experimental and undergoing development. Contact Philip Bradley at pbradley@fhcrc.org for information on the most up-to-date methods for designing with DNA flexibility.
3. See https://www.rosettacommons.org/about for a comprehensive list of all members of RosettaCommons available for collaborations.
4. Other compilation software, such as CMake, can be used but are not as well supported. Further details on building Rosetta can be found here: https://www.rosettacommons.org/docs/latest/Build-Documentation.html.

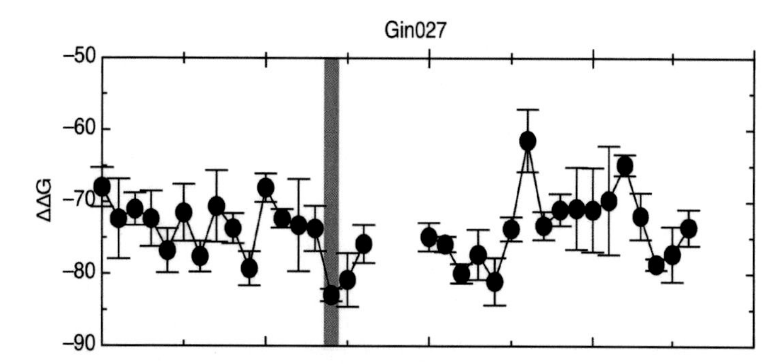

Fig. 5 Example prediction of target site preferences for a homology model. A protein–DNA complex was modeled for the homing endonuclease homologue Gin027, and the interface binding energy (ΔΔG) was calculated for the model with 34 possible target site orientations [47]. The predicted target site for this endonuclease is highlighted with a *magenta bar* and corroborated by experimental characterization [47]. The reverse complement of this target, a binding mode that cannot be ruled out without a crystal structure, is shown with a *gray dashed line*

5. Running Rosetta requires a basic understanding of Linux/Unix commands. There are many available resources, and one tutorial for a beginner user is located at the following web address: http://www.ee.surrey.ac.uk/Teaching/Unix/.
6. A partial list of the supported platforms is available at the following web address: https://www.rosettacommons.org/docs/latest/platforms.html.
7. Contact Yifan Song at Cyrus Biotechnology, Inc. (yifan@cyrus-bio.com).
8. This database can be downloaded at the following web address: ftp://ftp.ncbi.nih.gov/blast/db/FASTA/nr.gz.
9. More information on the program and a download of the database is available at the following web address: http://toolkit.tuebingen.mpg.de/.
10. The mode=release command builds the release version, and it is at least 10 times faster of an executable than the default debug version. The only reason to leave out the "mode=release" command is if you are developing code that will need to be debugged. The "extras=static" command enables porting of the complied code to other platforms because static linking of shared libraries is completed. The only downside to the static complication is that the executable size is large. The command "-j #" can be used to parallelize the build into multiple threads if you are compiling on a multiprocessor machine (ie, -j 20 for splitting compilation over 20 machines).
11. If the code is going to be run on a different computer system than it was compiled, the rosettaDNA executable and entire rosetta_

database folder be moved to that system by typing "scp ./bin/rosettaDNA.static.linuxgccrelease computerwhereitwillberun" and "scp -r ../rosetta_database/ computerwhereitwillberun".

12. If the user plans on running parallel multiple trajectories of the same code, the output of these trajectories needs to go into different directories so that the output files do not overwrite each other. One strategy is to create directories labeled job0-job55 (if 55 trajectories are being completed) by using the Unix command "mkdir job {0..55}". A second, but much less time-effective, option is to run jobs sequentially by using commands in the arguments file or by using capabilities within RosettaScripts [51]. This approach is not recommended if the job is long, such as for multiple base-pair designs in which many interface positions must be designed simultaneously.

13. If you are running on a multiprocessor system that does not have a job submission system, the program GNU parallel [63] is a highly recommended way to run parallelized jobs. The website explaining the program is http://www.gnu.org/software/parallel/. The following example command will use GNU parallel to submit jobs 5 and have the results go into separate job# directories: nice -19 ./bin/parallel -j 5 'cd {.}*f*; ./bin/rosettaDNA.static.linuxgccrelease @../args > log;cd ../' ::: job* &.

14. Many beginner Vi tutorials are available online (i.e., http://www.infobound.com/vi.html).

15. The XML files are a part of RosettaScripts [51]. This system for protocol development provides a flexible environment in which movers and operations can be recombined into different algorithms without having to recompile Rosetta.

16. The energy function used in modeling makes a substantial difference in the design outcome. Energy functions optimized specifically for protein–DNA interactions should be used in protein–DNA design calculations for best results [13]. There have been recent advancements in the Rosetta program, such as the development of a new way to capture hydrogen-bonding interactions [23], however the energy function must go through an optimization process [13, 64] for the problem of interest before using new functionality.

17. Change the 5th and 7th columns of the following five lines in the atom_properties.txt file (./rosetta_database/chemical/atom_type_sets/fa_standard/atom_properties.txt) to the values shown here:

    Phos P 2.1500 0.5850 -4.1000 3.5000 14.7000

    Narg N 1.7500 0.2384 -10.0000 6.0000 11.2000 DONOR ORBITALS

NH2O N 1.7500 0.2384 -7.8000 3.5000 11.2000 DONOR ORBITALS

Nlys N 1.7500 0.2384 -16.0000 6.0000 11.2000 DONOR

ONH2 O 1.5500 0.1591 -5.8500 3.5000 10.8000 ACCEPTOR SP2_HYBRID ORBITALS

Also change the fifth column of the three HC atoms in the LYS.params file to the value 0.48 from 0.33 to increase the positive charge of lysine. The LYS.params file is found here:

"./rosetta_database/chemical/residue_type_sets/fa_standard/residue_types/l-caa/LYS.params".

18. If running many jobs on a multiprocessor system, always submit a single test run to confirm that all paths are correct and that all necessary files are included.
19. The number of runs that should be completed depends on how many base pairs are being mutated in the target site. The number of base pairs controls the number of interface positions that are designed (unless a resfile is used, *see* Fig. 4). As a starting point, a minimum of 10 runs should be completed for a fixed-backbone standard design for a one base-pair substitution. At least 50 runs should be completed for a single base-pair pocket with flexibility (either protein or DNA). A triple base-pair pocket with backbone flexibility needs several hundred runs (300–500) to assess the full range of low-energy solutions.
20. DNA-interacting proteins can have either or both high activity and specificity [16]. Even without explicit design for specificity, computational design procedures are biased to generate high specificity designs if it is possible because they optimize for direct base-pair interactions. Nonspecific proteins, such as DNA polymerase, often use DNA backbone contacts to gain binding energy.
21. The discreteness of rotamers is an approximation that is necessary because of computational limits when all amino acids are being considered. Increasing the number of rotamers can improve design results [13]. When the amino acid sequence is fixed the number of rotamers included in the calculation can be greatly increased and any negative effect of the approximation is lessened.
22. An advanced Rosetta XML user can add the extra rotamers through the ExtraRotamersGeneric operation and complete this specificity calculation directly after design in one run.
23. The simplest way to access these values without writing a script is to execute the command "grep total_score *pdb" in the directory that contains the pdbs you are interested in analyzing.

24. For specificity predictions, it is recommended that either the mean or median value of the total_energy over many structures be used, rather than the score of the lowest-energy structure. This recommendation is especially true for protocols involving any amount of backbone flexibility, as design protocols can generate outlier structures with energies much lower than the majority and these outliers are not likely to represent that actual energetic and structural state of the complex.
25. The calculation of specificity is based on the Boltzmann distribution. The value of k_BT can be changed, but a value of 1 is reasonable. The equation for calculating specificity for a guanine base-pair is (2.718^0) / (2.718^0 + 2.178^(-$\Delta E_{G\text{-}A}$) + 2.178^(-$\Delta E_{G\text{-}C}$) + 2.178^(-$\Delta E_{G\text{-}T}$)).
26. Only the DesignProteinBackboneAroundDNA mover will limit protein backbone movement to around the target base-pair. Other methods of protein backbone movement will require another way of designating the regions that should be flexible.

Acknowledgements

The authors would like to thank Justin Ashworth, Phil Bradley, and Jim Havranek for their vast contributions to improving protein–DNA interface design, as well as the entire ROSETTA Commons community for contributions to the Rosetta code base. This work was supported by the US National Institutes of Health (#GM084433 and #RL1CA133832 to D.B.), the Foundation for the National Institutes of Health through the Gates Foundation Grand Challenges in Global Health Initiative, and the Howard Hughes Medical Institute.

References

1. Alibes A, Nadra AD, De Masi F, Bulyk ML, Serrano L, Stricher F (2010) Using protein design algorithms to understand the molecular basis of disease caused by protein–DNA interactions: the Pax6 example. Nucleic Acids Res 38(21):7422–7431. doi:10.1093/nar/gkq683
2. Epstein DJ (2009) Cis-regulatory mutations in human disease. Brief Funct Genomics 8(4):310–316. doi:10.1093/bfgp/elp021
3. VanderMeer JE, Ahituv N (2011) cis-regulatory mutations are a genetic cause of human limb malformations. Dev Dyn 240(5):920–930. doi:10.1002/dvdy.22535
4. Muller PA, Vousden KH (2013) p53 mutations in cancer. Nat Cell Biol 15(1):2–8. doi:10.1038/ncb2641
5. D'Elia AV, Tell G, Paron I, Pellizzari L, Lonigro R, Damante G (2001) Missense mutations of human homeoboxes: a review. Hum Mutat 18(5):361–374. doi:10.1002/humu.1207
6. Wray GA (2007) The evolutionary significance of cis-regulatory mutations. Nat Rev Genet 8(3):206–216. doi:10.1038/nrg2063
7. Wittkopp PJ, Kalay G (2012) Cis-regulatory elements: molecular mechanisms and evolutionary processes underlying divergence. Nat Rev Genet 13(1):59–69. doi:10.1038/nrg3095

8. Borneman AR, Gianoulis TA, Zhang ZD, Yu H, Rozowsky J, Seringhaus MR, Wang LY, Gerstein M, Snyder M (2007) Divergence of transcription factor binding sites across related yeast species. Science 317(5839):815–819. doi:10.1126/science.1140748
9. Schmidt D, Wilson MD, Ballester B, Schwalie PC, Brown GD, Marshall A, Kutter C, Watt S, Martinez-Jimenez CP, Mackay S, Talianidis I, Flicek P, Odom DT (2010) Five-vertebrate ChIP-seq reveals the evolutionary dynamics of transcription factor binding. Science 328(5981):1036–1040. doi:10.1126/science.1186176
10. Prud'homme B, Gompel N, Carroll SB (2007) Emerging principles of regulatory evolution. Proc Natl Acad Sci U S A 104(Suppl 1):8605–8612. doi:10.1073/pnas.0700488104
11. Leaver-Fay A, Tyka M, Lewis SM, Lange OF, Thompson J, Jacak R, Kaufman K, Renfrew PD, Smith CA, Sheffler W, Davis IW, Cooper S, Treuille A, Mandell DJ, Richter F, Ban YE, Fleishman SJ, Corn JE, Kim DE, Lyskov S, Berrondo M, Mentzer S, Popovic Z, Havranek JJ, Karanicolas J, Das R, Meiler J, Kortemme T, Gray JJ, Kuhlman B, Baker D, Bradley P (2011) ROSETTA3: an object-oriented software suite for the simulation and design of macromolecules. Methods Enzymol 487:545–574. doi:10.1016/B978-0-12-381270-4.00019-6
12. Dunbrack RL Jr, Cohen FE (1997) Bayesian statistical analysis of protein side-chain rotamer preferences. Protein Sci 6(8):1661–1681. doi:10.1002/pro.5560060807
13. Thyme SB, Baker D, Bradley P (2012) Improved modeling of side-chain--base interactions and plasticity in protein–DNA interface design. J Mol Biol 419(3-4):255–274. doi:10.1016/j.jmb.2012.03.005
14. Rohs R, Jin X, West SM, Joshi R, Honig B, Mann RS (2010) Origins of specificity in protein–DNA recognition. Annu Rev Biochem 79:233–269. doi:10.1146/annurev-biochem-060408-091030
15. Harteis S, Schneider S (2014) Making the bend: DNA tertiary structure and protein–DNA interactions. Int J Mol Sci 15(7):12335–12363. doi:10.3390/ijms150712335
16. Ashworth J, Baker D (2009) Assessment of the optimization of affinity and specificity at protein–DNA interfaces. Nucleic Acids Res 37(10), e73. doi:10.1093/nar/gkp242
17. Morozov AV, Havranek JJ, Baker D, Siggia ED (2005) Protein-DNA binding specificity predictions with structural models. Nucleic Acids Res 33(18):5781–5798. doi:10.1093/nar/gki875
18. Ashworth J, Havranek JJ, Duarte CM, Sussman D, Monnat RJ Jr, Stoddard BL, Baker D (2006) Computational redesign of endonuclease DNA binding and cleavage specificity. Nature 441(7093):656–659. doi:10.1038/nature04818
19. Nadra AD, Serrano L, Alibes A (2011) DNA-binding specificity prediction with FoldX. Methods Enzymol 498:3–18. doi:10.1016/B978-0-12-385120-8.00001-2
20. Thyme SB, Jarjour J, Takeuchi R, Havranek JJ, Ashworth J, Scharenberg AM, Stoddard BL, Baker D (2009) Exploitation of binding energy for catalysis and design. Nature 461(7268):1300–1304. doi:10.1038/nature08508
21. Ulge UY, Baker DA, Monnat RJ Jr (2011) Comprehensive computational design of mCreI homing endonuclease cleavage specificity for genome engineering. Nucleic Acids Res 39(10):4330–4339. doi:10.1093/nar/gkr022
22. Ashworth J, Taylor GK, Havranek JJ, Quadri SA, Stoddard BL, Baker D (2010) Computational reprogramming of homing endonuclease specificity at multiple adjacent base pairs. Nucleic Acids Res 38(16):5601–5608. doi:10.1093/nar/gkq283
23. O'Meara MJ, Leaver-Fay A, Tyka M, Stein A, Houlihan K, DiMaio F, Bradley P, Kortemme T, Baker D, Snoeyink J, Kuhlman B (2015) A combined covalent-electrostatic model of hydrogen bonding improves structure prediction with Rosetta. J Chem Theory Comput 11(2):609–622. doi:10.1021/ct500864r
24. Sheffler W, Baker D (2010) RosettaHoles2: a volumetric packing measure for protein structure refinement and validation. Protein Sci 19(10):1991–1995. doi:10.1002/pro.458
25. Borgo B, Havranek JJ (2012) Automated selection of stabilizing mutations in designed and natural proteins. Proc Natl Acad Sci U S A 109(5):1494–1499. doi:10.1073/pnas.1115172109
26. Lazaridis T, Karplus M (1999) Effective energy function for proteins in solution. Proteins 35(2):133–152
27. Yanover C, Bradley P (2011) Extensive protein and DNA backbone sampling improves structure-based specificity prediction for C2H2 zinc fingers. Nucleic Acids Res 39(11):4564–4576. doi:10.1093/nar/gkr048
28. Li S, Bradley P (2013) Probing the role of interfacial waters in protein–DNA recognition using a hybrid implicit/explicit solvation model. Proteins 81(8):1318–1329. doi:10.1002/prot.24272
29. Redondo P, Prieto J, Munoz IG, Alibes A, Stricher F, Serrano L, Cabaniols JP, Daboussi

F, Arnould S, Perez C, Duchateau P, Paques F, Blanco FJ, Montoya G (2008) Molecular basis of xeroderma pigmentosum group C DNA recognition by engineered meganucleases. Nature 456(7218):107–111. doi:10.1038/nature07343

30. Takeuchi R, Lambert AR, Mak AN, Jacoby K, Dickson RJ, Gloor GB, Scharenberg AM, Edgell DR, Stoddard BL (2011) Tapping natural reservoirs of homing endonucleases for targeted gene modification. Proc Natl Acad Sci U S A 108(32):13077–13082. doi:10.1073/pnas.1107719108
31. Grizot S, Duclert A, Thomas S, Duchateau P, Paques F (2011) Context dependence between subdomains in the DNA binding interface of the I-CreI homing endonuclease. Nucleic Acids Res 39(14):6124–6136. doi:10.1093/nar/gkr186
32. Fleishman SJ, Whitehead TA, Ekiert DC, Dreyfus C, Corn JE, Strauch EM, Wilson IA, Baker D (2011) Computational design of proteins targeting the conserved stem region of influenza hemagglutinin. Science 332(6031):816–821. doi:10.1126/science.1202617
33. Strauch EM, Fleishman SJ, Baker D (2014) Computational design of a pH-sensitive IgG binding protein. Proc Natl Acad Sci U S A 111(2):675–680. doi:10.1073/pnas.1313605111
34. Azoitei ML, Correia BE, Ban YE, Carrico C, Kalyuzhniy O, Chen L, Schroeter A, Huang PS, McLellan JS, Kwong PD, Baker D, Strong RK, Schief WR (2011) Computation-guided backbone grafting of a discontinuous motif onto a protein scaffold. Science 334(6054):373–376. doi:10.1126/science.1209368
35. Rothlisberger D, Khersonsky O, Wollacott AM, Jiang L, DeChancie J, Betker J, Gallaher JL, Althoff EA, Zanghellini A, Dym O, Albeck S, Houk KN, Tawfik DS, Baker D (2008) Kemp elimination catalysts by computational enzyme design. Nature 453(7192):190–195. doi:10.1038/nature06879
36. Thyme SB, Boissel SJ, Arshiya Quadri S, Nolan T, Baker DA, Park RU, Kusak L, Ashworth J, Baker D (2014) Reprogramming homing endonuclease specificity through computational design and directed evolution. Nucleic Acids Res 42(4):2564–2576. doi:10.1093/nar/gkt1212
37. Voigt CA, Mayo SL, Arnold FH, Wang ZG (2001) Computational method to reduce the search space for directed protein evolution. Proc Natl Acad Sci U S A 98(7):3778–3783. doi:10.1073/pnas.051614498
38. Chen MM, Snow CD, Vizcarra CL, Mayo SL, Arnold FH (2012) Comparison of random mutagenesis and semi-rational designed libraries for improved cytochrome P450 BM3-catalyzed hydroxylation of small alkanes. Protein Eng Des Sel 25(4):171–178. doi:10.1093/protein/gzs004
39. Khersonsky O, Rothlisberger D, Wollacott AM, Murphy P, Dym O, Albeck S, Kiss G, Houk KN, Baker D, Tawfik DS (2011) Optimization of the in-silico-designed kemp eliminase KE70 by computational design and directed evolution. J Mol Biol 407(3):391–412. doi:10.1016/j.jmb.2011.01.041
40. Jarjour J, West-Foyle H, Certo MT, Hubert CG, Doyle L, Getz MM, Stoddard BL, Scharenberg AM (2009) High-resolution profiling of homing endonuclease binding and catalytic specificity using yeast surface display. Nucleic Acids Res 37(20):6871–6880. doi:10.1093/nar/gkp726
41. Takeuchi R, Certo M, Caprara MG, Scharenberg AM, Stoddard BL (2009) Optimization of in vivo activity of a bifunctional homing endonuclease and maturase reverses evolutionary degradation. Nucleic Acids Res 37(3):877–890. doi:10.1093/nar/gkn1007
42. Chames P, Epinat JC, Guillier S, Patin A, Lacroix E, Paques F (2005) In vivo selection of engineered homing endonucleases using double-strand break induced homologous recombination. Nucleic Acids Res 33(20), e178. doi:10.1093/nar/gni175
43. Doyon JB, Pattanayak V, Meyer CB, Liu DR (2006) Directed evolution and substrate specificity profile of homing endonuclease I-SceI. J Am Chem Soc 128(7):2477–2484. doi:10.1021/ja057519l
44. Havranek JJ, Baker D (2009) Motif-directed flexible backbone design of functional interactions. Protein Sci 18(6):1293–1305. doi:10.1002/pro.142
45. Borgo B, Havranek JJ (2014) Motif-directed redesign of enzyme specificity. Protein Sci 23(3):312–320. doi:10.1002/pro.2417
46. Szeto MD, Boissel SJ, Baker D, Thyme SB (2011) Mining endonuclease cleavage determinants in genomic sequence data. J Biol Chem 286(37):32617–32627. doi:10.1074/jbc.M111.259572
47. Thyme SB, Song Y, Brunette TJ, Szeto MD, Kusak L, Bradley P, Baker D (2014) Massively parallel determination and modeling of endonuclease substrate specificity. Nucleic Acids Res 42(22):13839–13852. doi:10.1093/nar/gku1096
48. Combs SA, Deluca SL, Deluca SH, Lemmon GH, Nannemann DP, Nguyen ED, Willis JR, Sheehan JH, Meiler J (2013) Small-molecule

ligand docking into comparative models with Rosetta. Nat Protoc 8(7):1277–1298. doi:10.1038/nprot.2013.074

49. Jha RK, Chakraborti S, Kern TL, Fox DT, Strauss CE (2015) Rosetta comparative modeling for library design: Engineering alternative inducer specificity in a transcription factor. Proteins. doi:10.1002/prot.24828
50. Thyme S, Baker D (2014) Redesigning the specificity of protein–DNA interactions with Rosetta. Methods Mol Biol 1123:265–282. doi:10.1007/978-1-62703-968-0_17
51. Fleishman SJ, Leaver-Fay A, Corn JE, Strauch EM, Khare SD, Koga N, Ashworth J, Murphy P, Richter F, Lemmon G, Meiler J, Baker D (2011) RosettaScripts: a scripting language interface to the Rosetta macromolecular modeling suite. PLoS One 6(6):e20161. doi:10.1371/journal.pone.0020161
52. Havranek JJ, Harbury PB (2003) Automated design of specificity in molecular recognition. Nat Struct Biol 10(1):45–52. doi:10.1038/nsb877
53. Mitchell M (1996) An introduction to genetic algorithms. Complex adaptive systems. MIT Press, Cambridge, MA
54. Coley DA (2010) An introduction to genetic algorithms for scientists and engineers. World Scientific, River Edge, NJ
55. Canutescu AA, Dunbrack RL Jr (2003) Cyclic coordinate descent: a robotics algorithm for protein loop closure. Protein Sci 12(5):963–972. doi:10.1110/ps.0242703
56. Wang C, Bradley P, Baker D (2007) Protein-protein docking with backbone flexibility. J Mol Biol 373(2):503–519. doi:10.1016/j.jmb.2007.07.050
57. Smith CA, Kortemme T (2008) Backrub-like backbone simulation recapitulates natural protein conformational variability and improves mutant side-chain prediction. J Mol Biol 380(4):742–756. doi:10.1016/j.jmb.2008.05.023
58. Mandell DJ, Coutsias EA, Kortemme T (2009) Sub-angstrom accuracy in protein loop reconstruction by robotics-inspired conformational sampling. Nat Methods 6(8):551–552. doi:10.1038/nmeth0809-551
59. Huang PS, Ban YE, Richter F, Andre I, Vernon R, Schief WR, Baker D (2011) RosettaRemodel: a generalized framework for flexible backbone protein design. PLoS One 6(8), e24109. doi:10.1371/journal.pone.0024109
60. Ollikainen N, Smith CA, Fraser JS, Kortemme T (2013) Flexible backbone sampling methods to model and design protein alternative conformations. Methods Enzymol 523:61–85. doi:10.1016/B978-0-12-394292-0.00004-7
61. Das R (2013) Atomic-accuracy prediction of protein loop structures through an RNA-inspired Ansatz. PLoS One 8(10):e74830. doi:10.1371/journal.pone.0074830
62. Song Y, DiMaio F, Wang RY, Kim D, Miles C, Brunette T, Thompson J, Baker D (2013) High-resolution comparative modeling with RosettaCM. Structure 21(10):1735–1742. doi:10.1016/j.str.2013.08.005
63. Tange O (2011) GNU Parallel - the command-line power tool. The USENIX Magazine: pp. 42–47
64. Leaver-Fay A, O'Meara MJ, Tyka M, Jacak R, Song Y, Kellogg EH, Thompson J, Davis IW, Pache RA, Lyskov S, Gray JJ, Kortemme T, Richardson JS, Havranek JJ, Snoeyink J, Baker D, Kuhlman B (2013) Scientific benchmarks for guiding macromolecular energy function improvement. Methods Enzymol 523:109–143. doi:10.1016/B978-0-12-394292-0.00006-0

Chapter 17

Motif-Driven Design of Protein–Protein Interfaces

Daniel-Adriano Silva, Bruno E. Correia, and Erik Procko

Abstract

Protein–protein interfaces regulate many critical processes for cellular function. The ability to accurately control and regulate these molecular interactions is of major interest for biomedical and synthetic biology applications, as well as to address fundamental biological questions. In recent years, computational protein design has emerged as a tool for designing novel protein–protein interactions with functional relevance. Although attractive, these computational tools carry a steep learning curve. In order to make some of these methods more accessible, we present detailed descriptions and examples of ROSETTA computational protocols for the design of functional protein binders using seeded protein interface design. In these protocols, a motif of known structure that interacts with the target site is grafted into a scaffold protein, followed by design of the surrounding interaction surface.

Key words Computational protein design, Protein–protein interaction, ROSETTA, Motif grafting, Interface design

1 Introduction

Computational design of protein–protein interactions has steadily progressed in recent years, including the creation of inhibitors that block enzymatic sites [1], small proteins that prevent viral entry [2], and antitumor agents that sequester oncogenic factors [3]. The ability to design in silico new functional binding proteins from minimal starting components opens tremendous possibilities for engineering innovative therapeutics and may eventually challenge antibody technology as the premiere method for generating protein-based drugs. However, designing a truly de novo protein–protein interface is a challenging problem that remains largely unsolved. This is due to several factors, most importantly the inaccuracies in energy functions used to evaluate protein designs and the intrinsic difficulties in efficiently sampling docked protein configurations that allow the design of side chains for favorable interactions. Therefore, to overcome these limitations, protein designers often use a "seeded interface design" approach, in which a small

Barry L. Stoddard (ed.), *Computational Design of Ligand Binding Proteins*, Methods in Molecular Biology, vol. 1414,
DOI 10.1007/978-1-4939-3569-7_17, © Springer Science+Business Media New York 2016

motif of known structure that binds to the target site is used to initiate the design process. This motif is then grafted (i.e., embedded) into a larger protein scaffold that in turn is designed to achieve optimal packing and interactions with the target protein. This approach solves two problems: (1) by beginning with a motif that is known to bind the target, the design immediately starts with some favorable interactions, and (2) the scaffold orientation against the target surface is guided by the motif itself. By using this information, the design is biased toward sampling only a small number of permissible docked configurations. Seeded protein–protein interface design strategies are indeed extremely powerful for creating novel protein binders, but the methods are also daunting for newcomers.

In this chapter, we describe a step-by-step workflow for the design of new protein binders based on motif grafting and "seeded" interface design. The majority of the protocols described can easily be run on a single personal computer, though large clusters and supercomputers will increase sampling and help find better solutions.

2 Materials (Required Software)

ROSETTA. The ROSETTA software suite includes algorithms for protein modeling and design [4]. ROSETTA is free for academic users and can be downloaded from: https://www.rosettacommons.org/software.

In the examples given here, ROSETTA was compiled and executed on a MacBook Pro with a 2.5 GHz quad-core Intel i7 processor. Basic knowledge of UNIX-style terminal commands is necessary.

For any design or structure prediction problem within ROSETTA, the potential energy is calculated using ROSETTA's energy function, which includes terms for attributes such as rotamer energies, van der Waals interactions, and hydrogen bonding, among others [5]; the process of applying the energy function to a given protein conformation is simply referred to as "scoring." As with free energy, a conformation or sequence with a lower energy in ROSETTA is more favorable. During protein structure prediction, the conformation of lowest energy is determined for a given amino acid sequence. During protein–protein interface design, the problem is reversed. Since the basic docked configuration of the binding partners is now known, the aim is to design the lowest energy sequence to stabilize the bound state of the two proteins.

ROSETTA and RosettaScripts. ROSETTA protocols are written in an XML-script format. The script is interpreted using the RosettaScripts parser, which is packaged within the ROSETTA suite [6]. Using a simple analogy, RosettaScripts protocols are like cooking recipes; they first define the ingredients (energy functions, task operations, filters, and movers) and then outline the protocol by which these are combined. RosettaScripts is easy to use, even for novices with minimal programming experience. Wiki-style documentation can be accessed at: https://www.rosettacommons.org/docs/latest/scripting_documentation/RosettaScripts/RosettaScripts.

This website provides an index of available operations and is an excellent resource when creating or modifying scripts.

Important: For the examples presented here, command lines contain the environment variable ${Rosetta}, which means the directory path in which ROSETTA is installed on the user's computer.

Molecular Visualization. A molecular graphics-viewing program is required. PyMol (Schrödinger, LLC) is recommended, as it has excellent and easy-to-use features for visualization, simple structural alignments, and even allows modifying proteins. A limited educational version (precompiled for several platforms) is available for free from: https://www.pymol.org/.

A full-featured open-source branch from SourceForge (Slashdot Media, requires compilation) is available at: http://sourceforge.net/projects/pymol/.

3 Methods

The workflow (Fig. 1) for computational interface design using motif grafting is comprised of the following steps:

1. Definition of the binding motif for seeded interface design.
2. Preparing a scaffold database.
3. Matching for putative scaffolds (i.e., motif grafting).
4. Sequence design.
5. Selection and improvement of designs.

3.1 Definition of the Binding Motif for Seeded Interface Design

To guide readers through each of these steps, we present the example of designing a protein binder for the estrogen receptor (ERα) based on a known peptide interaction. The crystal structure of ERα has been solved with a bound helical peptide from a transcriptional coactivator (PDB ID 1GWQ; Fig. 2) [7]. This natural protein–peptide complex provides an initial structural motif for seeded interface design. The bound peptide provides the core of the interface, and the design process involves transplanting/grafting the motif into alternative protein scaffolds, followed by design of neighboring residues close to the target protein surface, creating an extended interface for improved affinity and specificity.

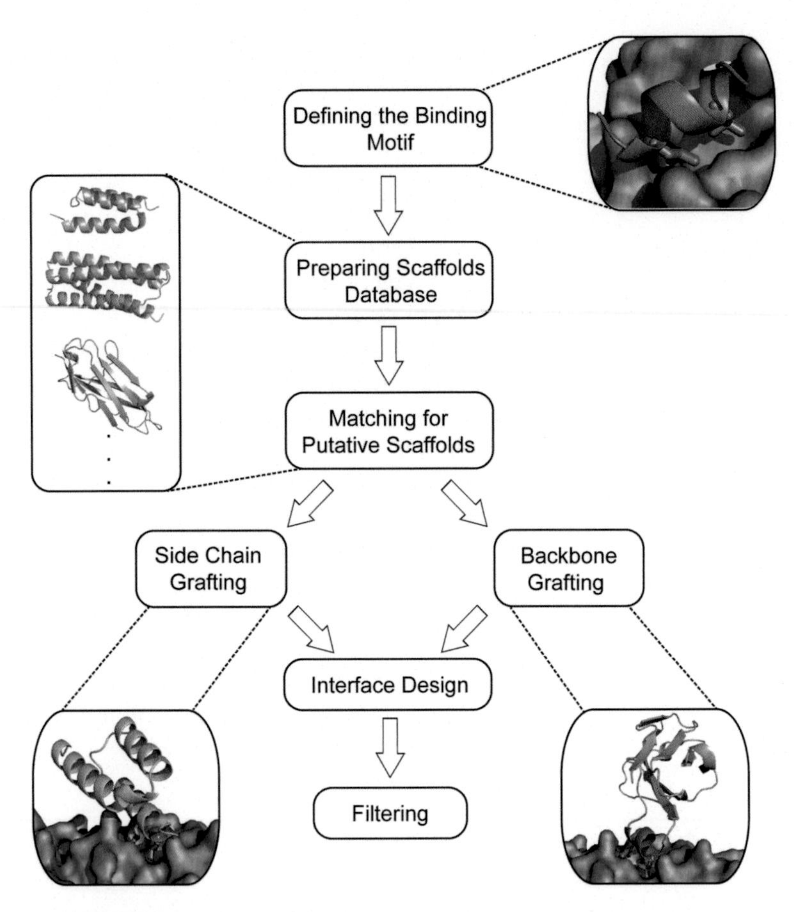

Fig. 1 Workflow for seeded interface design. In the *inset panels*, the target protein surface is colored in *green*, the motif to be grafted in *orange*, and scaffolds are shown in *grey*

ERα is a steroid hormone-activated transcription factor that recruits coactivators to a target gene [8]. The ERα-coactivator interaction is established through a helical motif that bears the signature sequence LXXLL (where L is leucine and X is any amino acid), with the leucine residues (hot spots) binding a hydrophobic cleft on the ERα surface (Fig. 2b) [7]. In the following sections, we show how to graft the helical motif into a new protein scaffold. The assumptions guiding this design strategy are: (1) stabilization of the bound conformation of the LXXLL motif by embedding it within a stable scaffold reduces the entropic penalty of binding a flexible peptide, and (2) expanding the interfacial contact area can create new favorable interactions with the target. If successful, a design that combines these two factors can achieve an interaction with enhanced affinity and specificity.

First, the PDB of the protein–peptide complex is formatted for compatibility with ROSETTA and the structure is minimized (*see*

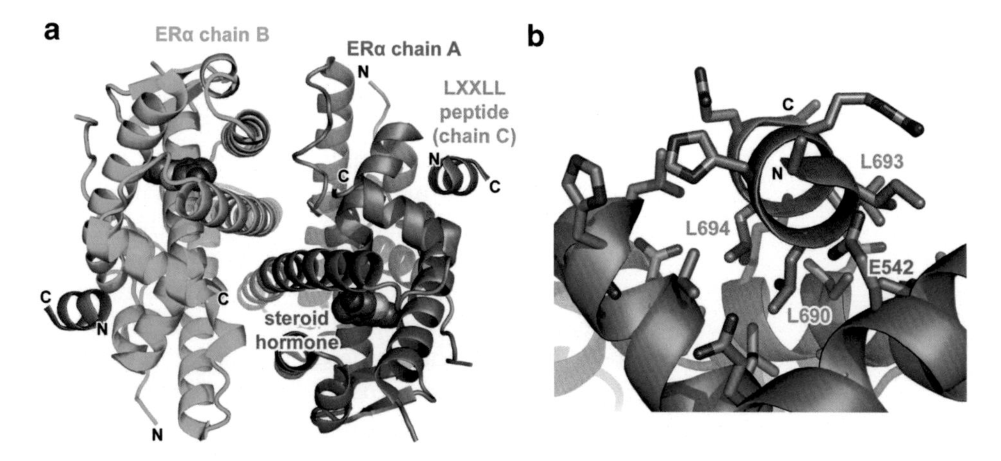

Fig. 2 The ERα-LXXLL peptide complex. (**a**) The crystal structure of the ligand-binding domain of ERα (a dimer; two chains are shown in *light* and *dark green*) bound to the aroylbenzothiophene core of raloxifene (*grey spheres*) and a peptide (*orange*) spanning the helical LXXLL motif from the transcriptional coactivator TIF2 (PDB 1GWQ). PDB files of the motif (chain C) and target (chain A) are prepared. (**b**) The three conserved leucines of the LXXLL motif interact with a hydrophobic cavity on the ERα surface, while glu-542 of ERα caps the peptide's N-terminus

Note 1 at the end for a detailed description on preparing input PDB files). Next, the structure is divided into two new PDB files, referred to as the "context" and "motif." The "context" file contains the target structure (i.e., ERα; only chain A of PDB ID 1GWQ), while the "motif" file contains the LXXLL peptide (chain C of PDB ID 1GWQ). In different scenarios, the motif could also be a small segment of a much larger protein, for example, an interacting loop extracted from an antibody–antigen structure.

3.2 Preparing a Scaffold Database

To prepare an inclusive scaffold database that can be searched for a variety of structural motifs, we downloaded 1519 structures from the PDB (www.rcsb.org) based on the following four criteria: (1) crystal structures with high-resolution x-ray diffraction data (<2.5 Å), (2) the proteins had been reported to be expressable in *E. coli* (this simplifies later experimental characterization), (3) a single protein chain in the asymmetric unit (MotifGraft only works with monomeric scaffolds as grafting targets), and (4) no bound ligands or modified residues. The scaffold PDB files were formatted for ROSETTA and subjected to an energy minimization step (*see* **Note 2**).

In some circumstances, a focused scaffold library may produce more useful matches. For our particular example, the peptide that seeds interface design has an α-helical conformation. Therefore, we also prepared a small focused scaffold library of 28 helical proteins.

3.3 Matching for Putative Scaffolds

The scaffold library is computationally scanned for possible graft sites. If the motif and scaffold backbones superimpose with very low root mean squared deviation (RMSD < 0.5 A), then only hot spot side chains need be transplanted from the motif to the corresponding positions in the matching site of the scaffold [9, 10]. This is known as "side chain grafting." Subsequently, surrounding residues on the scaffold surface that are in contact with the target are designed for favorable interactions [3]. We suggest that side chain grafting should be attempted first, as it makes the minimal number of changes to the scaffold, increasing the chances of obtaining correctly folded designs during experimental validation. However, often side chain grafting is not possible because the motif and scaffold structures are too dissimilar. In these cases, even though the motif and scaffold may have very different structures, it is still possible to use an alternative method known as "backbone grafting" [11, 12].

During backbone grafting, the algorithm looks for segments of the scaffold backbone that align closely to the termini of the motif (both N- and C-terminal sides), and then the scaffold segment between these alignment points is replaced by the motif. This technique is extremely versatile, for example, a loop in the scaffold might be replaced by a peptide motif with different secondary structure, or even with a different amino acid length. Since the changes to the scaffold structure following backbone grafting can disrupt the overall fold, it is important to design the hydrophobic core to support the new backbone structure of the scaffold, followed by design of the protein–protein interface. The backbone grafting procedure often introduces many mutations to the scaffold, requiring careful filtering of designs to select those that present quality interfaces and high stability of the new scaffold.

The flow chart in Fig. 1 details the steps involved for both design strategies. We begin by describing side chain grafting, followed by backbone grafting.

3.4 Sequence Design

3.4.1 Side Chain Grafting with RosettaScripts

Motif matching and interface design are distinct conceptual steps, but due to the flexibility of the RosettaScripts framework, both can be included in a single computational step. First, a list is generated containing all PDB files within the scaffold database:

```
#> ls -1 scaffolds_directory/*.pdb > scaffolds.list
```

Then RosettaScripts is executed using the following command:

```
#> ${Rosetta}/main/src/bin/rosetta_scripts -database ${Rosetta}/main/database/ -l scaffolds.list -use_input_sc -ex1 -ex2 -nstruct 1 -parser:protocol MotifGraft_sc.xml
```

The command line includes several important options. First, the location of the ROSETTA database must be specified using `-database`. Option `-l scaffolds.list` specifies the input list of scaffold PDB files. (Option `-s scaffold.pdb` would specify a single PDB file.) The options `-ex1` and `-ex2` allow ROSETTA to explore additional side chain rotamers, and `-use_input_sc` means that rotamers in the input structure are included in the rotamer library. Finally, option `-nstruct 1` means that the design script will be launched once per input scaffold. This can be increased if the user wishes to filter through more designs, but requires usage of the MultiplePoseMover in the XML script (for further information see RosettaScripts documentation).

In the case of grafting by side chain replacement, it took less than an hour to scan through the focused scaffold library of 28 helical proteins on a laptop computer and generate 23 designs. (Since several steps in the design process are stochastic, the number of results that pass the filters might vary if the protocol is re executed.). The XML file MotifGraft_sc.xml reads as follows:

```
<ROSETTASCRIPTS>
<TASKOPERATIONS>
  <ProteinInterfaceDesign   name=pido   repack_chain1=1
  repack_chain2=1    design_chain1=0    design_chain2=1
  interface_distance_cutoff=8.0/>
  <OperateOnCertainResidues name="hotspot_repack">
    <RestrictToRepackingRLT/>
    <ResiduePDBInfoHasLabel property="HOTSPOT"/>
  </OperateOnCertainResidues>
</TASKOPERATIONS>
<SCOREFXNS>
</SCOREFXNS>
<FILTERS>
  <Ddg name=ddg confidence=0/>
  <BuriedUnsatHbonds name=unsat confidence=0/>
  <ShapeComplementarity name=Sc confidence=0/>
</FILTERS>
<MOVERS>
  <MotifGraft  name="motif_grafting"  context_structure=
  "context.pdb" motif_structure="motif.pdb" RMSD_toler-
  ance="0.3" NC_points_RMSD_tolerance="0.5" clash_score_
  cutoff="5"   clash_test_residue="GLY"   hotspots="3:7"
  combinatory_fragment_size_delta="2:2"   full_motif_bb_
  alignment="1"graft_only_hotspots_by_replacement="1"
  revert_graft_to_native_sequence="1"/>
  <build_Ala_pose   name=ala_pose   partner1=0   partner2=1
  i n t e r f a c e _ c u t o f f _ d i s t a n c e = 8 . 0
  task_operations=hotspot_repack/>
  <Prepack name=ppk jump_number=0/>
  <PackRotamersMover    name=design    task_operations=
  hotspot_repack,pido/>
  <MinMover name=rb_min bb=0 chi=1 jump=1/>
</MOVERS>
```

```
<PROTOCOLS>
  <Add mover_name=motif_grafting/>
  <Add mover_name=ala_pose/>
  <Add mover_name=ppk/>
  <Add mover_name=design/>
  <Add mover_name=rb_min/>
  <Add mover_name=design/>
  <Add filter_name=unsat/>
  <Add filter_name=ddg/>
  <Add filter_name=Sc/>
</PROTOCOLS>
</ROSETTASCRIPTS>
```

Within the XML file, the user may first specify which score/ energy function to use from the ROSETTA database or reweight specific score terms; if no score function is defined, the default is used (currently "talaris2013," but this will likely change in future ROSETTA releases). Next, task operations define which residues can be altered. The ProteinInterfaceDesign task operation restricts design to residues of chain 2 (the scaffold) within 8 Å of the interface, while target residues within 8 Å of the interface may repack to alternative low-energy rotamers. By default, the design of nonnative prolines, glycines, and cysteines (which can have important structural consequences) is forbidden. The second task operation, RestrictToRepackingRLT, prevents the two grafted hot spot leucines from being mutated in later design steps, though they can repack to alternative rotamers. (For polar hot spot residues, alternative rotamers would disrupt hydrogen-bonding networks, and we would advise using the more restrictive task operation PreventRepackingRLT, which prevents both design and repacking.) The MotifGraft mover (described below) keeps track of which residues correspond to the target, scaffold, or motif and which critical side chains are grafted. These are labeled CONTEXT, SCAFFOLD, MOTIF, and HOTSPOT, respectively. These residue classes are then available for task operations, as used here. The details for these task operations are given on the wiki website: https://www.rosettacommons.org/docs/latest/scripting_documentation/RosettaScripts/TaskOperations/taskoperations_pages/OperateOnCertainResiduesOperation.

Movers dictate how the protein complex is manipulated, such as sequence design, side chain and backbone minimization, or rigid-body docking. The protocol begins with the MotifGraft mover, which searches for alignments between the scaffold and motif that do not produce steric clashes with the target structure. The MotifGraft mover has many options. First, the names of the PDB files for the target (`context_structure`) and motif (`motif_structure`) must be specified. The option `RMSD_tolerance` sets the maximum RMSD allowed between the motif and scaffold alignment. For side chain grafting, the motif should closely match the scaffold segment it is aligned with, so that the backbones are virtually superimposable. In this XML script, the RMSD tolerance was set to 0.3 Å (maximum recommended is ~0.5 Å). The option NC_points_RMSD_tolerance sets the maximum RMSD allowed between the N-/C-termini of the motif and scaffold graft site (recommended 0.5 Å). Once the scaffold has been aligned, the configuration of the system must be checked for clashes. After it is grafted, the motif cannot clash with other parts of the scaffold (this is not an issue for side chain grafting when the motif closely matches a native structural region within the scaffold, but is of serious concern when performing backbone grafting).

In addition, the orientation of the scaffold when aligned with the motif cannot clash with the target surface. Since residues can be designed to smaller amino acids in later steps, clashes are checked after first mutating the motif to small amino acids, such as alanine or glycine (using option `clash_test_residue="GLY"` in this XML script). All the atomic clashes are computed, and if the score is above the clash_score cutoff, the graft fails and an alternative alignment in the scaffold is attempted (it is recommended to set the clash_score_cutoff at ≤ 5). The options `full_motif_bb_alignment="1"` and `graft_only_hotspots_by_replacement="1"` indicate that side chain grafting is being performed. Option `hotspots="3:7"` defines which positions in the motif PDB correspond to the two leucine hot spots of the LXXLL peptide. Additional hot spots are each separated by colons. Option `combinatory_fragment_size_delta="2:2"` indicates by how many amino acids the motif may be shortened at each terminus (N-terminus:C-terminus), i.e., whether the full motif must align ("0:0") or only a partial fragment. Here, the algorithm will attempt to match the full-length motif, as well as each motif fragment shorter by up to two residues at one or both termini. The final option, `revert_graft_to_native_sequence="1"`, means that after the motif has been placed into the scaffold, all residues except for the hot spots are reverted back to their native identities. Therefore, only the two hot spot amino acids are effectively transferred as changes to the scaffold sequence.

After side chain grafting, the protocol continues by replacing scaffold side chains within 8 Å of the target with alanine using the build_Ala_pose mover. Task operations prevent the hot spots from changing. Side chains are now repacked with the Prepack mover. During this step, target protein residues that sterically clash with the scaffold have the opportunity to find alternative, non-clashing rotamers. Next, the interface surrounding the grafted hot spots is designed using the PackRotamersMover. Task operations ensure that hot spot and target residues can only change rotamer conformations, whereas scaffold residues within 8 Å of the target surface are available for design. Side chains and rigid-body orientations of the designed complex are then minimized with MinMover, followed by a second round of design.

Multiple rounds of minimization and design are recommended as they may improve results. Further details about movers can be found at: https://www.rosettacommons.org/docs/latest/scripting_documentation/RosettaScripts/Movers/Movers-RosettaScripts.

Finally, three filters are used to assess the designs' structural features: binding energy (ΔΔG), interface shape complementarity, and buried unsatisfied hydrogen-bonding atoms at the interface. In this example XML script, each filter is assigned a confidence of 0, such that all designs will pass. Rather than acting to terminate design calculations, these filters are instead being used to report interface quality. Based on these reported values, the user can determine which are the best designs of the pool. A full list of available filters can be found at: https://www.rosettacommons.org/docs/latest/scripting_documentation/RosettaScripts/Filters/Filters-RosettaScripts.

Some examples of the designs generated by the aforementioned script are shown in Fig. 3. XML scripting is amenable to rapid protocol modifications, and users are encouraged to attempt their own variations of the protocols. The RosettaScripts online documentation is an excellent resource to understand the functionality that different options provide.

3.4.2 Backbone Grafting with RosettaScripts

Using the same motif and target PDB files described above, we present an example XML script that scans scaffolds for potential backbone graft sites and subsequent design. The script can be executed as follows:

```
#> ${Rosetta}/main/source/bin/rosetta_scripts.macosclang-release -database ${Rosetta}/main/database/ -l scaffolds.list -use_input_sc -nstruct 1 -parser:protocol MotifGraft_bb.xml
```

The XML script reads:

```
<ROSETTASCRIPTS>
<TASKOPERATIONS>
  <ProteinInterfaceDesign name=pido_far interface_distance
  _cutoff=15.0/>
  <ProteinInterfaceDesign name=pido_med interface_distance_
  cutoff=12.0/>
  <ProteinInterfaceDesign name=pido_near interface_distance_
  cutoff=8.0/>
  <OperateOnCertainResidues name="hotspot_repack">
    <RestrictToRepackingRLT/>
    <ResiduePDBInfoHasLabel property="HOTSPOT"/>
  </OperateOnCertainResidues>
  <SelectBySASA name=core mode="sc" state="bound" probe_
  radius=2.2 core_asa=0 surface_asa=30 core=1 boundary=0 surface=0/>
  <SelectBySASA name=core_and_boundary mode="sc" state=
  "bound" probe_radius=2.2 core_asa=0 surface_asa=30
  core=1 boundary=1 surface=0/>
</TASKOPERATIONS>
```

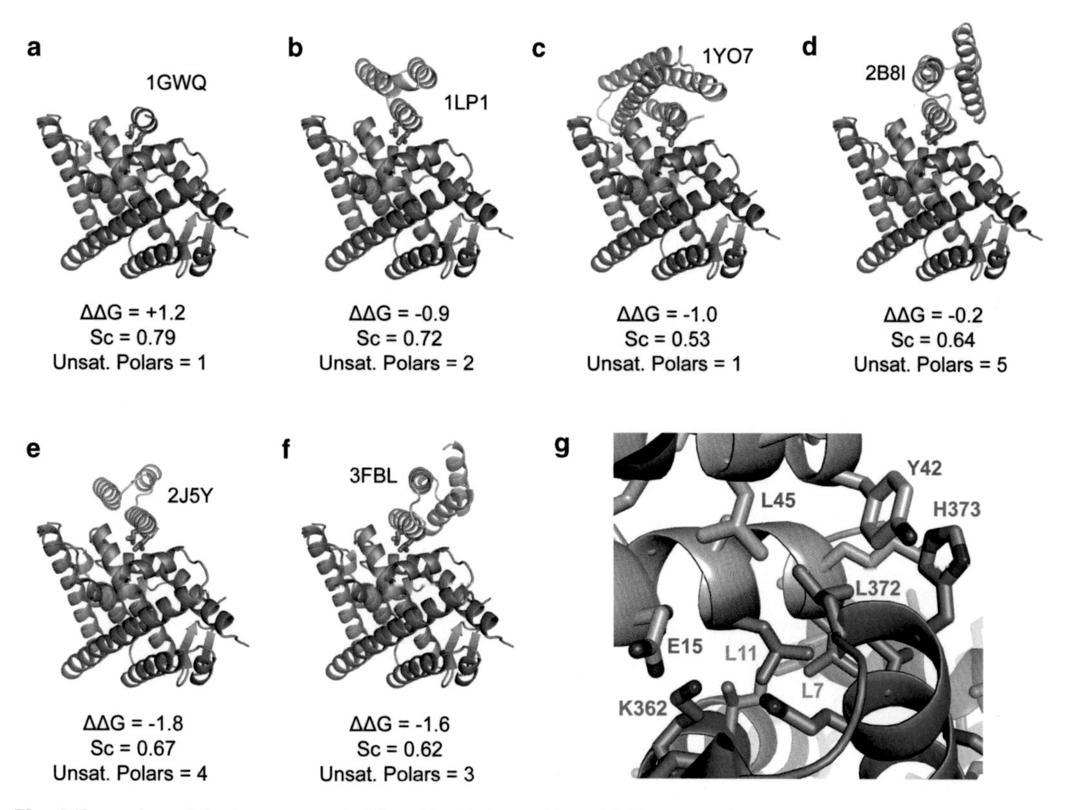

Fig. 3 Examples of designs generated by side chain grafting. (**a**) The crystal structure (PDB 1GWQ) of a LXXLL coactivator motif (*orange*) bound to ERα (*green*). Only chains A (ERα; the target) and C (LXXLL motif) are considered. The structure was energy minimized with ROSETTA and the interface was scored. (**b–f**) Five different designs generated by side chain grafting using the XML script described here. The scaffolds (*grey*; PDB codes indicated in the figure) are all helical bundle proteins. The grafted leucine hot spot residues (L690 and L694 in Fig. 2) are colored in *orange*. (**g**) The interface of the design in panel (**b**) is shown in greater detail. Designed interactions around the hot spots include hydrophobic contacts from L45, aromatic stacking between designed residue Y42 and target residue H373, and a saltbridge from E15 to K362

```
<FILTERS>
  <Ddg name=ddg confidence=0/>
  <BuriedUnsatHbonds name=unsat confidence=0/>
  <ShapeComplementarity name=Sc confidence=0/>
</FILTERS>
<MOVERS>
  <MotifGraft name="motif_grafting" context_structure=
  "context.pdb" motif_structure="motif.pdb" RMSD_toler-
  ance="1.0"   NC_points_RMSD_tolerance="1.0"   clash_
  score_cutoff="5"  clash_test_residue="GLY"  hotspots=
  "3:7"combinatory_fragment_size_delta="2:2"  max_frag-
  ment_replacement_size_delta="-8:8" full_motif_bb_align-
  ment="0" graft_only_hotspots_by_replacement="0"/>
```

```
  <build_Ala_pose name=ala_pose partner1=0 partner2=1
  interface_cutoff_distance=8.0 task_operations=hotspot_
  repack/>
  <Prepack name=ppk jump_number=0/>
  <PackRotamersMover name=design_core task_operations=
  hotspot_repack,pido_far,core/>
  <PackRotamersMover name=design_boundary task_operations=
  hotspot_repack,pido_med,core_and_boundary/>
  <PackRotamersMover name=design_interface task_operations=
  hotspot_repack,pido_near/>
  <MinMover name=sc_min bb=0 chi=1 jump=0/>
</MOVERS>
<PROTOCOLS>
  <Add mover_name=motif_grafting/>
  <Add mover_name=ala_pose/>
  <Add mover_name=ppk/>
  <Add mover_name=design_core/>
  <Add mover_name=design_boundary/>
  <Add mover_name=design_interface/>
  <Add mover_name=sc_min/>
  <Add filter_name=unsat/>
  <Add filter_name=ddg/>
  <Add filter_name=Sc/>
</PROTOCOLS>
</ROSETTASCRIPTS>
```

The first mover called in the protocols section of the XML script is MotifGraft. As with side chain grafting, options `context_structure` and `motif_structure` specify the target and motif PDB files, respectively. The `RMSD_tolerance` and `NC_points_RMSD_tolerance` are both set at 1.0 Å (the maximum recommended is 1.5 Å); during backbone grafting, these options set the maximum allowed RMSD between the motif termini and the backbone graft sites in the scaffold. A lower RMSD tolerance will enforce a better match between the motif termini and scaffold backbone, giving better results, though at the expense of more solutions. The options for `clash_test_residue`, `clash_score_cutoff`, `hotspots` and `combinatory_fragment_size_delta` are set the same as for side chain grafting. However, for backbone grafting options `full_motif_bb_alignment` and `graft_only_hotspots_by_replacement` are both turned off (i.e., set to "0"). A new option is now used; `max_fragment_replacement_size_delta="-8:8"` sets the minimum and maximum sizes of the scaffold segment that can be replaced by the motif (i.e., the resulting scaffold can vary from eight residues shorter up to eight residues longer than the original scaffold).

The protocol continues by calling a mover to mutate scaffold residues at the interface to alanine. Next, rotamers are minimized with the Prepack mover, followed by three design steps using PackRotamersMover. The first design step is restricted to scaffold residues within the hydrophobic core up to 15 Å away from the interface. Since the grafted motif is potentially very different from the scaffold segment it replaced, design of the core is necessary to stabilize the new structure. Two task operations define which residues can be designed: (1) the ProteinInterfaceDesign task operation permits design to chain 2 (the scaffold) within a distance threshold of the interface, and (2) the SelectBySASA task operation defines core, boundary, and surface residues based on solvent-accessible surface area and turns their design on or off. The second design step is restricted to 12 Å from the interface but now allows the design of core and "boundary" (i.e., partially buried) amino acids. Again, task operations define the residues for design. The third design step is now focused on optimizing all scaffold residues 8 Å from the target surface. A task operation prevents the grafted hot spot leucine residues from mutating at any stage. The final mover is a side chain minimization.

The protocol finishes with three filters to report on interface quality: the calculated binding energy, number of buried unsatisfied hydrogen-bonding atoms, and shape complementarity. Within 3 h on a laptop computer, over 200 scaffolds in the library were scanned for potential graft sites, and nearly as many designs were generated. In many of the designed proteins, helical segments of the scaffolds were swapped with the helical motif. However, in other designs, a non-helical scaffold segment was replaced; some examples are shown in Fig. 4.

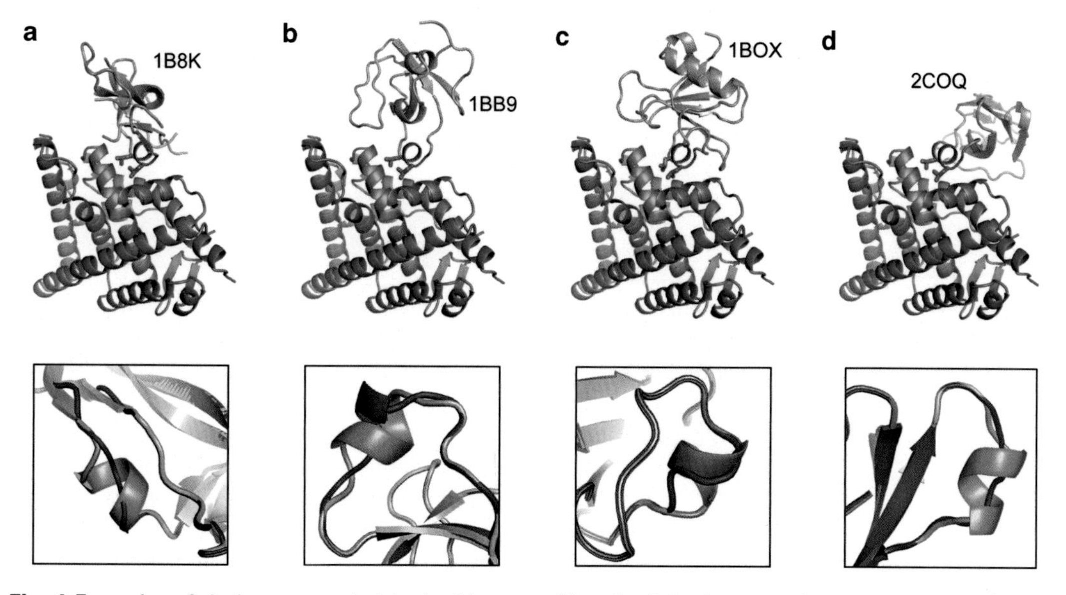

Fig. 4 Examples of designs generated by backbone grafting. (**a–d**) In the *upper* images, the target ERα is shown in *green*, the scaffold in *grey*, and the grafted motif in *orange*. The scaffold PDB is labeled. In the *lower* images, the designed proteins (scaffold and motif regions are in *grey* and *orange*, respectively) are superimposed with the original scaffold PDBs in *magenta*. Notice that scaffold loops of very different lengths and conformations were replaced with the helical motif

3.5 Selection of Designs and Optimization

To date, no computational method has been developed that can predict with perfect accuracy which designs will be functional when challenged experimentally [13]. Therefore, it is wise to proceed with designed sequences that present good metrics by multiple criteria. Designs are initially filtered based on calculated metrics for interface quality, including a favorable binding energy ($\Delta\Delta G < 0$ ROSETTA energy units, ideally the energy should be lower than the native interface from which the motif was taken), high shape complementarity ($Sc > 0.65$), and a low number of buried unsatisfied hydrogen-bonding atoms. In the XML scripts above, these filters report to a score file and will also be appended at the end of any ROSETTA output PDBs.

Once a set of designs have been selected based on the calculated metrics, it is important to perform human-guided inspection of the designed structures. There are many qualities of interfaces that are apparent to structural biologists that are not captured in standard metrics. Two common defects in ROSETTA-designed structures that are very important to avoid are buried charged residues and under-packed interfaces dominated by alanine residues (Fig. 5).

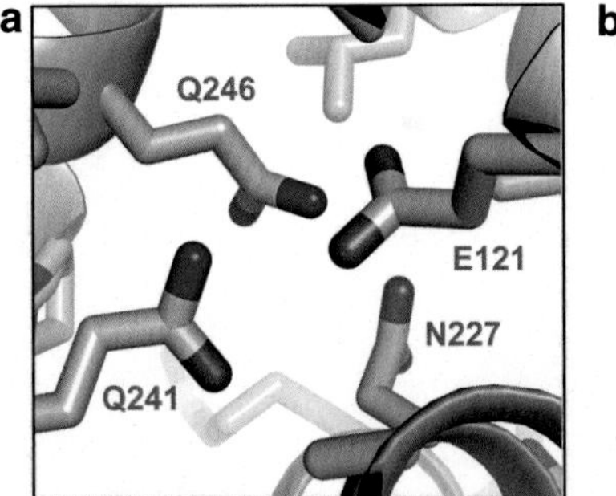

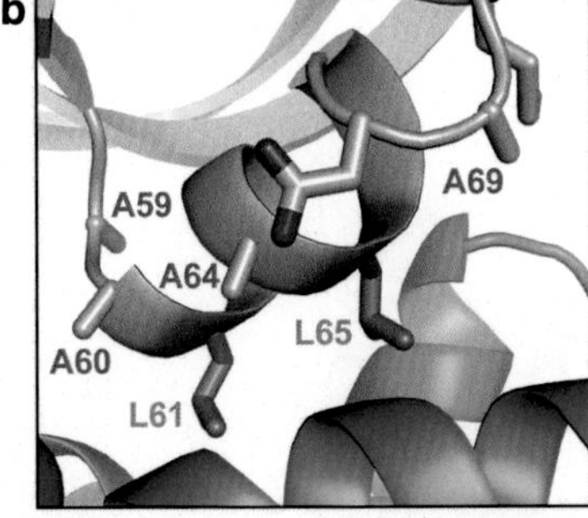

Fig. 5 Common defects in ROSETTA-designed protein binders. (**a**) After backbone grafting, the hydrophobic core of scaffold 1A0P (*grey*) was designed to support the motif. Polar and charged residues (*labeled*) were designed within the core; however, native proteins nearly always have hydrophobic cores. (**b**) Scaffold (PDB 2B29) is shown in *grey*, while the grafted leucines are in *orange* and the target ERα is *green*. The majority of designed scaffold residues at the interface (*grey sticks*) are alanines. Interfaces dominated by alanine can achieve low energies; alanine is a small hydrophobic residue that will not clash with the target surface and is therefore the "default" residue when specific interactions cannot be designed. These interfaces lack hydrogen-bonding networks and are generally under-packed

3.5.1 Reverting Designed Mutations Back to Native

It is also important to consider whether the designed scaffold will fold to its intended structure; having a spectacular interface on a computational model is irrelevant if the protein cannot fold in an experimental setting. This is particularly problematic for designed interfaces that have a large surface area dominated by hydrophobic residues. It is generally assumed that the probability of a designed sequence properly folding is inversely correlated with the number of mutations imposed on the scaffold during the design process. Therefore, it is beneficial to be conservative and make as few mutations as possible by reverting residues back to their native identities in a post-design stage. The ROSETTA application "revert_design_to_native" [2] can be used for this task; it goes through each mutated position in the scaffold, reverts to the native amino acid, and computes the change in binding energy. If the native residue scores similarly to the designed residue, then it may be safer to revert back to the native amino acid. The revert_design_to_native application requires two input PDBs: the designed PDB (containing the target (chain A) bound to the designed scaffold (chain B)) and a reference PDB that contains the target together with the native scaffold. To determine which residues have been mutated, the application sequentially compares each amino acid between the design and reference PDBs; this means the application can only be applied to designs from side chain grafting in which the two PDB files have the same number of residues. The reference PDB is easily generated by concatenating the target (context.pdb) with the scaffold PDB using the cat command:

```
#> cat context.pdb scaffold.pdb >nativecplx.pdb
```

Revert_design_to_native is run with the following command:

```
#> ${Rosetta}/main/source/bin/revert_design_to_native.
macosclangrelease    -revert_app:wt    nativecplx.pdb
-revert_app:design design.pdb -ex1 -ex2 -use_input_sc
-database ${Rosetta}/main/database/
```

3.5.2 Manually Adjusting Designs Using FoldIt

If necessary, the designed structures may be subjected to human-guided optimization. The user may wish to correct a number of frequent problematic features in ROSETTA designs, such as hydrophobic residues at the water-exposed interface edge, revert designed residues back to their native identities, mutate buried charged residues to hydrophobics, etc. There are no hard rules for manually improving designs; it is simply a matter of the designer's preference and experience. FoldIt is an excellent computational tool to perform this human-guided optimization [14]. It combines a graphic front end with molecular visualization together with many basic tools such as sequence design, rotamer repacking, and minimization (though often with creative names like "Shake" and "Wiggle"). FoldIt was developed as a protein folding and design game, bringing the advantages of crowdsourcing to solve structural biology problems [14]. The stand-alone version of FoldIt gives immediate visual and ROSETTA energy feedback, helping the user decide if any further mutations to the designed protein are warranted. The license for FoldIt Standalone is available from http://c4c.uwc4c.com/express_license_technologies/foldit, and directions will then be provided for downloading the software.

3.5.3 Filtering Designs Based on Folding Probability

Designs from backbone grafting require extra attention, as the engineering of a protein core to support the grafted motif can be challenging. Many designed sequences will not fold correctly when experimentally tested. We have found structure prediction to be a powerful filter; the designed amino acid sequences when subjected to structure prediction calculations should yield similar structures to the designed models [3]. If structure prediction returns an alternative conformation, or fails to converge on an energy minimum in a conformational landscape, then it is unlikely that the designed sequence will correctly fold. However, structure prediction is computationally expensive and not accessible on a large scale to most biochemists. Further, this evaluation method is only useful if the original scaffold sequence correctly returns the native structure; for many natural proteins, structure prediction methods are not yet able to accurately predict the known structure. Instead, designs can be relaxed with ROSETTA to determine if the designed conformation is "stable." If the designed structural model drifts, it is unlikely to occupy a low-energy conformation at the bottom of an energy funnel, and the design should either be rejected or improved using information derived from the relaxed ensemble, from which one can identify cavities and alternative conformations that should be eliminated by additional design steps. To apply this filter, first extract chain B (the designed protein) from the PDB files of the designed complexes:

```
#> for i in *.pdb; do grep " B " $i >$i.chainB; done
#> ls -1 *.chainB >monomers.list
```

Next, the designed monomers are relaxed and the RMSD to the starting structure is determined:

```
#> ${Rosetta}/main/source/bin/rosetta_scripts.macosclan-
grelease -database ${Rosetta}/main/database/ -l mono-
mers.list -use_input_sc -nstruct 1 -parser:protocol
fastrelax.xml
<ROSETTASCRIPTS>
<MOVERS>
  <FastRelax name=fstrlx repeats=4/>
</MOVERS>
<FILTERS>
  <Geometry  name=omega  omega=150  cart_bonded=100
  confidence=0/>
  <CavityVolume name=cav_vol confidence=0/>
  <Rmsd name=rmsd confidence=0 superimpose=1/>
</FILTERS>
<PROTOCOLS>
```

```
  <Add filter_name=omega/>
  <Add filter_name=cav_vol/>
  <Add mover_name=fstrlx/>
  <Add filter_name=rmsd/>
</PROTOCOLS>
</ROSETTASCRIPTS>
```

The RMSD will be low if the designed protein conformation is stable (typically ≤ 1 Å). This XML script also reports two other useful metrics prior to relaxation. The Geometry filter checks that backbone omega angles are above a defined cutoff (except for *cis*-prolines, omega angles should be close to 180°) and that Cartesian space bond angles and lengths are close to ideal (decrease the cart_bonded penalty score for a more stringent filter). The geometry at the junction points where the motif is grafted can be particularly poor, and in such cases the cart_bonded penalty score will be flagged as high and the omega angle as too low in the log report. The CavityVolume filter measures the total cavity volume in $Å^3$. This will be higher for bigger proteins and therefore should not be used as a hard filter, but any outliers with exceptionally high values likely have underpacked cores.

3.6 Experimental Validation

Despite notable advances, computational protein design has only modest success rates at the stage of experimental characterization. Hence, it is essential to have a robust and rapid experimental assay for evaluating designs. Library display methods are ideally suited to screening many designs individually or simultaneously within a mixed pool [3], and as the cost of DNA synthesis has plummeted, it is possible to screen hundreds to thousands of designs within a reasonable budget. Often initial computational designs present low affinities to the desired targets and must be optimized by targeted mutagenesis or directed evolution [1–3, 12, 15]. Experimental methods should be carefully considered before embarking on any protein design project.

3.7 Concluding Remarks

Computational design of protein–protein interactions is poised to make spectacular advancements. Fast computers, affordable DNA synthesis, and the development of tools like ROSETTA have coalesced in the past few years, such that computational design methodologies are now accessible to a wider community without requiring supercomputers or advanced programming skills. Here, we have outlined general methods for seeded interface design and encouraged readers to create new protocols tailored to their problems. Proteins made to order, once deemed science fiction, are rapidly becoming a reality.

4 Notes

1. *Formatting PDB files.* PDB files must be correctly formatted for compatibility with ROSETTA. All heteroatoms, including water molecules, should be removed. In ROSETTA "TER" statements designate different proteins in a complex, and therefore any "TER" statements within a single protein chain must be removed, such as those that are used to mark regions of missing density. While these modifications can be made in a text editor, a large number of PDB files can easily be prepared with the following UNIX command:

```
#> for i in *.pdb; do grep "ATOM " $i >$i.atoms; done
```

   This will go through all PDB files within the directory, search for all lines containing the string "ATOM", and print these lines to a new file with suffix atoms.

2. *ROSETTA energy minimization of crystallographic structures.* It may be advantageous to perform energy minimization of the structures within the ROSETTA energy function prior to matching and design. Structures from experimental data often have residues with high (i.e., energetically unfavorable) energy due to minor clashes or "imperfections," and these may be inappropriately designed by ROSETTA to alternative amino acids. This is especially problematic for backbone grafting and may lead to unnecessary sequence design of residues that should remain unchanged. Energy minimization of input PDBs generally resolves this issue. However, it is important that structures do not drift too far during the minimization protocol; after all, the original PDB files are determined from real experimental data, whereas a minimized structure will only be as real as the energy function is accurate. To perform this step, we suggest two computational protocols. First, structures can be minimized using the constrained fast relaxation protocol. To minimize a single PDB file, use option -s file.pdb in the command line. To relax all PDB files within a directory, create a list first:

```
#> ls -1 *.pdb >pdb_files.list
#> ${Rosetta}/main/source/bin/relax.macosclangrelease
-database ${Rosetta}/main/database/ -ignore_unrecog-
nized_res -relax:constrain_relax_to_start_coords -ex1
-ex2 -use_input_sc -l pdb_files.list
```

   Alternatively, structures can be minimized using RosettaScripts. A command line and example XML script are:

```
#> ${Rosetta}/main/source/bin/rosetta_scripts.maco-
sclangrelease -database ${Rosetta}/main/database/ -l
pdb_files.list -use_input_sc -ex1 -ex2 -parser:protocol
ppk_min.xml
```

Contents of ppk_min.xml:

```
<ROSETTASCRIPTS>
<FILTERS>
  <Rmsd name=rmsd threshold=1.5 superimpose=1/>
</FILTERS>
<MOVERS>
  <Prepack name=ppk jump_number=0/>
  <MinMover name=sc_bb_min bb=1 chi=1/>
</MOVERS>
<PROTOCOLS>
  <Add mover_name=ppk/>
  <Add mover_name=sc_bb_min/>
  <Add mover_name=ppk/>
  <Add mover_name=sc_bb_min/>
  <Add filter_name=rmsd/>
</PROTOCOLS>
```

In this XML script, there are two rounds of rotamer repacking and side chain/backbone minimization using the movers Prepack and MinMover. The "Rmsd" filter superimposes the minimized structure with the input PDB file; if the two differ by over 1.5 Å, then the structure is rejected and ROSETTA proceeds to the next scaffold in the list. The reasons why a structure is "unstable" during energy minimization and rejected may include inaccuracies in the ROSETTA energy function or regions of poor quality in the crystallographic models. For instance, in our initial scaffold library, we found that from 1519 protein structures, only 1419 fulfilled the filtering criteria and were included in the library to perform the modeling examples described in this manuscript.

References

1. Procko E, Hedman R, Hamilton K et al (2013) Computational design of a protein-based enzyme inhibitor. J Mol Biol 425:3563–3575. doi:10.1016/j.jmb.2013.06.035
2. Fleishman SJ, Whitehead TA, Ekiert DC et al (2011) Computational design of proteins targeting the conserved stem region of influenza hemagglutinin. Science 332:816–821. doi:10.1126/science.1202617
3. Procko E, Berguig GY, Shen BW et al (2014) A computationally designed inhibitor of an Epstein-Barr viral Bcl-2 protein induces apoptosis in infected cells. Cell 157:1644–1656. doi:10.1016/j.cell.2014.04.034
4. Leaver-Fay A, Tyka M, Lewis SM et al (2011) ROSETTA3: an object-oriented software suite for the simulation and design of macromolecules. Methods Enzymol 487:545–574. doi:10.1016/B978-0-12-381270-4.00019-6
5. Das R, Baker D (2008) Macromolecular modeling with rosetta. Annu Rev Biochem 77: 363–382. doi:10.1146/annurev.biochem.77.062906.171838
6. Fleishman SJ, Leaver-Fay A, Corn JE et al (2011) RosettaScripts: a scripting language interface to the Rosetta macromolecular modeling suite. PLoS ONE 6, e20161. doi:10.1371/journal.pone.0020161
7. Wärnmark A, Treuter E, Gustafsson J-A et al (2002) Interaction of transcriptional intermediary factor 2 nuclear receptor box peptides with the coactivator binding site of estrogen receptor alpha. J Biol Chem 277:21862–21868. doi:10.1074/jbc.M200764200
8. Savkur RS, Burris TP (2004) The coactivator LXXLL nuclear receptor recognition motif. J Pept Res 63:207–212. doi:10.1111/j.1399-3011.2004.00126.x
9. Ofek G, Guenaga FJ, Schief WR et al (2010) Elicitation of structure-specific antibodies by epitope scaffolds. Proc Natl Acad Sci U S A 107: 17880–17887. doi:10.1073/pnas.1004728107

10. Correia BE, Ban Y-EA, Holmes MA et al (2010) Computational design of epitope-scaffolds allows induction of antibodies specific for a poorly immunogenic HIV vaccine epitope. Structure 18:1116–1126. doi:10.1016/j.str.2010.06.010
11. Azoitei ML, Ban Y-EA, Julien J-P et al (2012) Computational design of high-affinity epitope scaffolds by backbone grafting of a linear epitope. J Mol Biol 415:175–192. doi:10.1016/j.jmb.2011.10.003
12. Azoitei ML, Correia BE, Ban Y-EA et al (2011) Computation-guided backbone grafting of a discontinuous motif onto a protein scaffold. Science 334:373–376. doi:10.1126/science.1209368
13. Fleishman SJ, Whitehead TA, Strauch E-M et al (2011) Community-wide assessment of protein-interface modeling suggests improvements to design methodology. J Mol Biol 414:289–302. doi:10.1016/j.jmb.2011.09.031
14. Cooper S, Khatib F, Treuille A et al (2010) Predicting protein structures with a multiplayer online game. Nature 466:756–760. doi:10.1038/nature09304
15. Whitehead TA, Chevalier A, Song Y et al (2012) Optimization of affinity, specificity and function of designed influenza inhibitors using deep sequencing. Nat Biotechnol 30:543–548. doi:10.1038/nbt.2214

Chapter 18

Computational Reprogramming of T Cell Antigen Receptor Binding Properties

Timothy P. Riley, Nishant K. Singh, Brian G. Pierce, Brian M. Baker, and Zhiping Weng

Abstract

T-cell receptor (TCR) binding to peptide/MHC is key to antigen-specific cellular immunity, and there has been considerable interest in modulating TCR affinity and specificity for the development of therapeutics and imaging reagents. While in vitro engineering efforts using molecular evolution have yielded remarkable improvements in TCR affinity, such approaches do not offer structural control and can adversely affect receptor specificity, particularly if the attraction towards the MHC is enhanced independently of the peptide. Here we describe an approach to computational design that begins with structural information and offers the potential for more controlled manipulation of binding properties. Our design process models point mutations in selected regions of the TCR and ranks the resulting change in binding energy. Consideration is given to designing optimized scoring functions tuned to particular TCR-peptide/MHC interfaces. Validation of highly ranked predictions can be used to refine the modeling methodology and scoring functions, improving the design process. Our approach results in a strong correlation between predicted and measured changes in binding energy, as well as good agreement between modeled and experimental structures.

Key words T cell receptor, Structure-guided design, Rosetta, Binding

1 Introduction

The αβ T cell receptor (TCR) is a membrane-bound heterodimer on the surface of helper or killer T cells that recognizes peptide antigens bound and displayed by major histocompatibility complex (MHC) proteins (Fig. 1). TCR recognition of peptide/MHC initiates T cell signaling and defines specificity in cellular immunity. TCR affinity for a target peptide/MHC generally correlates with in vivo potency [1, 2], which has led to the generation of many high affinity TCR variants using molecular evolution techniques such as yeast or phage display (e.g., refs. 3–6). While these methods can lead to spectacular gains in TCR affinity, there is potential to negatively impact specificity, leading to enhanced

Barry L. Stoddard (ed.), *Computational Design of Ligand Binding Proteins*, Methods in Molecular Biology, vol. 1414,
DOI 10.1007/978-1-4939-3569-7_18, © Springer Science+Business Media New York 2016

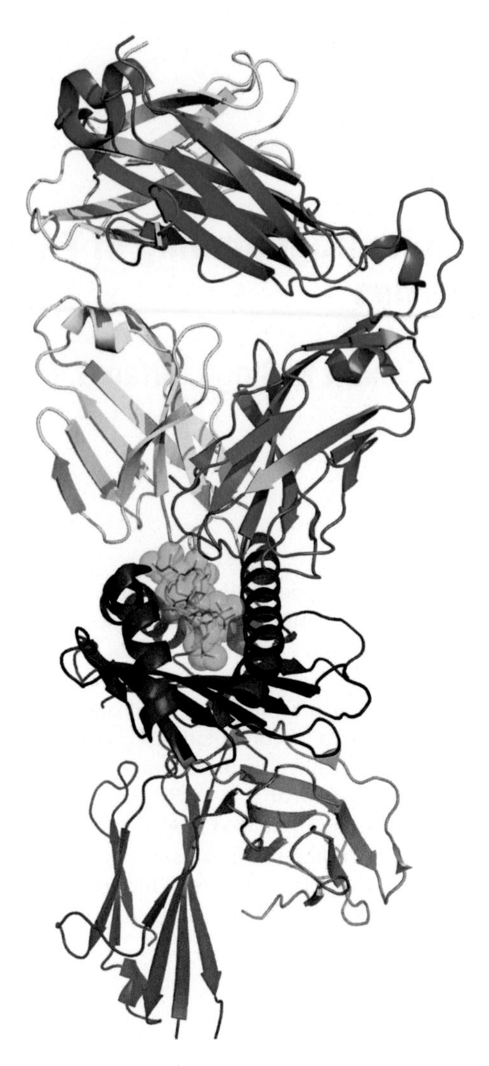

Fig. 1 Structural overview of the complex formed between a TCR (*blue*/*gold*) and peptide/MHC complex (*green*/*purple*/*orange*). The structure of the DMF5 TCR bound to the human class I MHC HLA-A2 in complex with the MART-1 ELA peptide was used for this figure [16]

cross-reactivity [7]. As potential uses for engineered TCRs include constructing genetically engineered T cells or soluble reagents to treat cancer and infectious disease [6, 8, 9], enhanced cross-reactivity could lead to dangerous autoimmunity. Further, accumulating evidence suggests that very large enhancements in affinity may lead to diminished T cell potency [2].

By incorporating structural information into the design process, computational design offers the potential to more carefully control specificity than molecular evolution. Also, computational design can permit more controlled enhancements in binding.

Computational design has been used to engineer a small number of TCRs [10–13]. While different approaches have been used, all benefit from the ability to rationalize effects on specificity and affinity through the examination of crystallographic structures and target specific regions of the interface. This latter point is crucial, as the recognition of a composite surface formed from two distinct components (the peptide and MHC protein) sets TCR recognition of pMHC apart from almost all other protein–protein interactions and requires special consideration when considering the origins of binding affinity and specificity and how they might be manipulated in productive ways [14].

Here we describe an approach to TCR computational design that recognizes the unique nature of TCR-pMHC binding and builds off our recent work with TCRs specific for viral and tumor antigens [11, 13]. Our approach uses the powerful Rosetta suite [15]. The design process models point mutations in selected regions of the TCR and ranks the resulting change in binding energy through the use of scoring functions which describe van der Waals interactions, solvation energies, hydrogen bonds, etc. As both structural modeling and energetic scoring involves trade-offs, assumptions, and known limitations, it is important to validate and if needed iteratively refine the design process with biophysical binding and structural work. For example, with the DMF5 TCR binding the MART-1 ELA and AAG peptides presented by the class I MHC protein HLA-A2 [16], we observed close agreement between predicted and measured changes in binding energy, as well as predicted and crystallographic X-ray structures, but multiple approaches for structural modeling and scoring were considered [13].

One important caveat is that in some circumstances, TCR structural properties have been shown to be surprisingly sensitive to changes in the TCR–pMHC interface (e.g., refs. 16–20). While our method attempts to accommodate some structural alterations, large conformational changes or global TCR repositioning are unlikely to be captured by the approach described here. While improvements are therefore possible, this approach can nonetheless serve as the foundation for efforts in engineering TCRs with novel binding properties.

2 Materials

1. A personal computer or high performance computing facility enabled with the latest Python and PyRosetta installations (https://www.rosettacommons.org/). The IPython command shell (http://ipython.org/) is recommended as it supports tab-completion and is useful in accessing PyRosetta functions.

2. 3 GB of available RAM and one processor core is required for each PyRosetta job. Multiple cores with accompanying RAM are required for large modeling projects (*see* **Note 1** for comments on calculation speeds).
3. Structural coordinates of the target TCR/pMHC complex; when publicly available downloadable from the Protein Data Bank (http://www.rcsb.org/pdb/home/home.do).
4. Applied mathematics software with multiple linear regression functionality, such as MATLAB (http://www.mathworks.com/products/matlab).

3 Methods

3.1 Structure-Guided Improvement of T Cell Receptor Binding

The Rosetta package includes tools for computational modeling and structure analysis and was originally designed for de novo structure prediction [15]. Rosetta is typically used to investigate research applications such as protein folding or protein design, and has been used to predict interaction energies between proteins (e.g., refs. 11, 13, 21). PyRosetta is a Python toolkit which packages the powerful Rosetta algorithms into the easily learned Python scripting language [22]. PyRosetta can be used via scripts or interactively by command line. The sections below describe a complete script which inserts, structurally adjusts, and scores point mutations at TCR–pMHC interfaces (*see* **Note 2** for a list and descriptions of variables used and **Note 3** for comments about syntax).

1. Initiate PyRosetta with the command-line flag to include additional amino acid rotamers in the design process. The additional rotamers increase sidechain sampling which may allow for lower observed energy states during design.

```
#Initialize Rosetta with additional options
from rosetta import*
init(extra_options='-extrachi_cutoff 1 -ex1 -ex2 -ex3')
```

2. Declare the score function to be used in the design process for scoring interactions. The default Talaris2013 [23] score function may be sufficient for initial design work, although other score functions such as the Rosetta 'interface' or 'ddg' functions can be examined. As highlighted below, customized score functions trained to the experimental system can lead to improved results [13].

```
#Initialize the score function
scorefxn=create_score_function('talaris2013')
```

3. Import the TCR/pMHC complex for design and store as a pose object. The structural coordinates may be stored locally or downloaded directly from the Protein Data Bank (the

example below uses the complex for the DMF5 TCR bound to the MART-1 ELA peptide presented by HLA-A2, available as the PDB entry 3QDG [16]).

```
#Download the DMF5 TCR/MHC complex from the PDB and
store as 'pose'
from toolbox import pose_from_rcsb
pose=pose_from_rcsb('3QDG')
```

4. Score the complex, then isolate and score the TCR and pMHC separately. To calculate a binding energy, subtract the TCR and pMHC scores from the complex (e.g.,: Binding $Score_{WT} = Score_{WTcomplex}\text{-}Score_{WT\text{-}TCR}\text{-}Score_{pMHC}$; *see* ref. 21) (*see* **Note 4** for comments on chain IDs and **Note 5** for comments on scoring).

```
#score the DMF5 TCR
scorefxn(pose)
import rosetta.protocols.grafting
#delete chains D and E of the complex and store
remaining coordinates as 'HLA'
HLA=Pose()
protocols.grafting.delete_region(HLA.assign(pose),
pose.pdb_info().pdb2pose('D',1), pose.total_residue())
#delete chains A, B, and C of the complex and store
the remaining coordinates as 'TCR'
TCR=Pose()
protocols.grafting.delete_region(TCR.assign(pose),
pose.pdb_info().pdb2pose('A',1),    pose.pdb_info().
pdb2pose('D',1)-1)
#calculate binding score
BindingScore=scorefxn(pose)    -    scorefxn(HLA)    -
scorefxn(TCR)
```

5. Using protein modeling software such as PyMOL (The PyMOL Molecular Graphics System, Version 1.7.4 Schrödinger, LLC.) or commands within Rosetta, scan the complex for TCR residues that are near atoms of the target peptide presented by the MHC molecule. Choosing residues close to (or contacting) the peptide is one means to help ensure peptide specificity is retained. A less restrictive approach is likely to favor improved interactions between the TCR and MHC, which could lead to undesirable enhancements in TCR cross-reactivity. When incorporated into a script, the commands below iteratively scan a TCR–pMHC interface and identify TCR residues in proximity to the peptide (*see* **Note 6** for comments on cut-off distances).

```
#measure distance between the center of mass of two
residues at positions i of the TCR and j of the
peptide.
list_of_residue_positions=[]
distance_cutoff=15
```

```
for i in range(pose.pdb_info().pdb2pose('D',1),
pose.total_residue()):
  for j in range (pose.pdb_info().pdb2pose('C', 1),
pose.pdb_info().pdb2pose('D', 1)):
distance = pose.residue(j).nbr_atom_xyz().
distance(pose.residue(i).nbr_atom_xyz())
if distance.norm<distance_cutoff:
  list_of_residue_positions.append(i)
```

6. Using the mutate_residue() command, computationally introduce the desired amino acids into each position selected in **step 5**.

```
#mutate residue i of the pose to an alanine and store
as mutant pose
from toolbox import mutate_residue
residue_list=['A','C','D','E','F','G','H','I','K',
'L','M','N','P','Q','R','S','T','V','W','Y']
for i in range(1, len(list_of_residue_positions)):
  for j in range(1, len(residue_list)):
    mutant=mutate_residue(pose, i, str(residue_
    list[j]))
  #At this point, a design could be considered com-
  plete. Either dump the pose to pdb (dump_
  pdb(mutant,'mutation_name.pdb') or continue
  repacking/refinement in subsequent steps within
  this loop.
```

7. For a simple design, the protein backbone is kept rigid and only the mutant amino acid is repacked. The results of 16 DMF5 point mutations modeled using this method are shown in Fig. 2a. While this approach is computationally inexpensive, it has potential to result in clashes and unrealistic rotamers. To

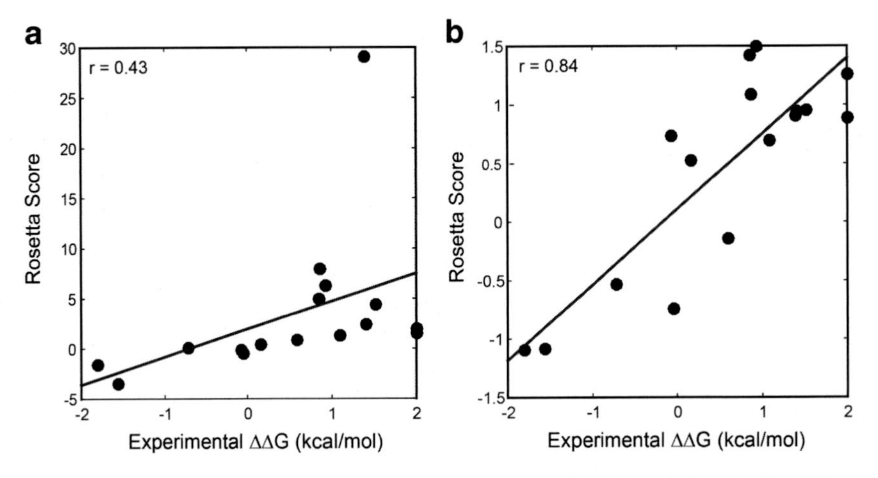

Fig. 2 Correlations between experimental values and Rosetta score of point mutations in the DMF5–ELA/HLA-A2 interface [13]. (**a**) Results when modeled with a rigid backbone and the 'interface' score function. (**b**) Results when modeled with the LoopMover_Refine () mover and the score function shown in Table 3

Table 3

New score function after stepwise multiple linear regression, removing all terms with *p* values >0.05

Term	Estimate	Standard error	*t* statistic	*p* value
fa_atr	0.33	0.07	4.52	5.83E–05
fa_elec	0.45	0.16	2.72	0.01
fa_rep	0.18	0.06	2.94	0.01
fa_sol	0.46	0.11	4.13	1.95E–4
Number of observations		42		
Root mean squared error		0.84		
R-squared		0.43		
Adjusted R-squared		0.38		

optimize the local environment around the mutated residue to minimize clashes and unfavorable interactions, residues near the mutated residue may also be repacked (*see* **Note 7**).

```
#repack the sidechain of the mutated residue to min-
imize the score from the defined score function
task=standard_packer_task(mutant)
task.or_include_current(True)
task.restrict_to_repacking()
task.temporarily_fix_everything()
task.temporarily_set_pack_residue(list_of_residue_
positions(i),True)
pack_mover=PackRotamersMover(scorefxn, task)
pack_mover.apply(mutant)
```

8. Designs can be further improved by refining the backbone of the TCR complementarity determining region (CDR) loops through a combination of cyclic coordinate descent (CCD) and Monte Carlo algorithms. Although the IMGT immunoinformatics database (www.imgt.org) [24] can be used to define TCR CDR loops, loops can also be defined by examining the structure. This may be preferable, as sequence-based definitions of loops often exclude amino acids which contact the peptide/MHC. For example, a CDR loop may be defined as occurring between residues 26 and 31 on chain 'D'.

```
#Define loop positions
start=mutant.pdb_info().pdb2pose('D',26)
cutpoint=mutant.pdb_info().pdb2pose('D',28)
end=mutant.pdb_info().pdb2pose('D',31)
```

9. A foldtree defining the flexible regions is required when manipulating the backbone. The foldtree should encompass the CDR loop and two additional residues on either side to act as "anchors."

```
#Set up a foldtree encompassing CDR1 alpha
ft=FoldTree()
ft.add_edge(1, start-2,-1)
ft.add_edge(start-2,cutpoint,-1)
ft.add_edge(start-2,end+2,1)
ft.add_edge(end+2,cutpoint+1,-1)
ft.add_edge(end+2,mutant.total_residue(),-1)
mutant.fold_tree(ft)
```

10. The LoopMover protocol in Rosetta uses a random number seed to iteratively and stochastically perturb the backbone and repack all affected sidechains. For this reason, designs vary slightly depending on the seed chosen by the LoopMover. It is suggested to perform multiple refinements and average the resulting scores to account for this variability [25]. Multiple loops may be refined at once as long as the foldtree includes the additional loops (*see* **Note 8**).

```
#Define loop and refine with cyclic coordinate descent
(CCD)
CDRloop=loop(start, end, cutpoint)
loops=Loops()
loops.add(CDRloop)
loop_refine=LoopMover_Refine_CCD(loops, scorefxn)
loop_refine.max_inner_cycles(10)
loop_refine.apply(mutant)
```

11. Refining the CDR loops multiple times increases the computational time required. The job distributor is a useful tool that can take advantage of multiple cores running the same script to generate designs/decoys in parallel. A more detailed description on the job distributor can be found in **Note 9**.

```
#create job_distributor; define number of decoys and
score function
jd=PyJobDistributor('DMF5refine'+str(list_of_resi-
due_positions[i])+'A', 3, scorefxn)
while not jd.job_complete:
  jdpose.assign(mutant)
  loop_refine.apply(jdpose)
  jd.output(jdpose)
```

12. Calculate the Binding Score for the designed complexes as described in **step 4**. Subtraction of the WT score results in a difference in energy roughly correlating to ΔΔG in kcal/mol. Negative values suggest favorable designs and possible candidates for follow-up experimental investigation.

3.2 Score Function Refinement Following Comparison With Experimental Binding Data

After many point mutations have been predicted and binding energies experimentally determined, a tailored score function may be generated to improve future predictions. An iterative approach (design, score, measure, repeat…) can optimize the "rules" for mutations in a specific interface to better predict the impacts on affinity and specificity.

1. For each mutation to be used in developing a score function, collect the values for each term available in Rosetta. This can be observed with the scorefxn.show(pose) command or within a .fasc file. The terms used in the latest release of Rosetta along with the default weights are shown in Table 1. Other terms (e.g., the Atomic Contact Energy term used in ZAFFI [13]) can be added to the score function in an attempt to improve correlation with experimental binding free energies.
2. Calculate the unweighted binding energies as described in Subheading 3.1 for all scoring terms.
3. Perform a multiple linear regression fitting all terms simultaneously to the experimental ΔΔG values. Some may be insignificant to the regression and may be removed with minimal effect. The results may be informative in understanding the biophysics

Table 1

Score function terms and weights of the Talaris2013 score function [23]

Score function term	Talaris2013 weights
fa_atr	0.8
fa_rep	0.44
fa_sol	0.75
fa_intra_rep	0.004
fa_elec	0.7
pro_close	1
hbond_sr_bb	1.17
hbond_lr_bb	1.17
hbond_bb_sc	1.17
hbond_sc	1.1
dslf_fa13	1
rama	0.2
omega	0.5
fa_dun	0.56
p_aa_pp	0.32
ref	1

Table 2

Multiple linear regression results on ΔΔG data from 42 point mutations in the A6-Tax/HLA-A2 and DMF5-AAG/HLA-A2 interface with all Rosetta terms

Term	Estimate	Standard error	*t* statistic	*p* value
dslf_fa	0	0	–	–
fa_atr	0.32	0.15	2.11	0.044
fa_dun	0.29	0.31	0.92	0.37
fa_elec	0.22	0.30	0.72	0.48
fa_intra_rep	-2.34	4.01	-0.58	0.56
fa_rep	0.03	0.13	0.26	0.80
fa_sol	0.38	0.17	2.26	0.03
hbond_bb_sc	-0.27	0.55	-0.49	0.63
hbond_lr_bb	1.38	2.06	0.67	0.51
hbond_sc	0.16	0.33	0.49	0.63
hbond_sr_bb	0.69	2.10	0.33	0.74
omega	0	0	–	–
p_aa_pp	0.27	0.61	0.44	0.66
pro_close	0.09	1.21	0.08	0.94
rama	0.69	0.89	0.78	0.44
ref	-0.02	0.37	-0.05	0.96
Number of observations		42		
Root mean squared error		0.91		
R-squared		0.50		
Adjusted R-squared		0.26		

See ref. 23 for a description of terms

within a specific TCR–pMHC interface. MATLAB's fitlm tool fits and reports weights for terms and their significance to the model (Table 2). The removeTerms tool allows a user to sequentially remove terms with a high *p* value and reweight the remaining terms to fit the model as shown in Table 3. It is important to remove terms with highest *p*-values first as some terms may become more significant as other terms are removed (e.g., fa_rep in Table 3).

```
ddg=[Experimental_ddg_values];
predictors=[array_of_predictor_values];
```

```
%Perform a stepwise multiple linear regression where
all terms in the final %output have a p value<0.05
mdl=fitlm(predictors,ddg,'Intercept' false);
```

4. To evaluate the predictive power of the regression model, consider a cross-validation approach by excluding a portion of the data (e.g., 5 %) from the training procedure. For the demonstration used here, the score function was trained to 42 data points from the A6 and DMF5 TCRs (Table 3) and 16 DMF5 data points were excluded. Once the regression model has been trained, use the resulting model to evaluate the remaining data in order to estimate the applicability of the model with future predictions. An example can be seen in Fig. 2b. The revised function eliminates a large "outlier" in the DMF5 test set that is seen with the 'interface' function, resulting in an improved fit as judged by the correlation and distribution of points around the fitted line. Without this outlier, the two functions behave similarly; however, the revised function highlights the importance of critically examining outliers and demonstrates that their exclusion may not always be appropriate.

Related to this, *N*-fold (e.g., fivefold) cross-validation is a commonly used method to assess predictive performance. To apply this method, the data are divided into *N* equally sized subsets, and for each subset a model is trained (e.g., by multilinear regression) using the points outside that set. Thus, a correlation can be produced using all points, without any training set overlapping with a test set. If the sample size is low (<100 measurements) and as much data as possible must be used in the regression model, Leave One Out Cross Validation may be used to gauge overfitting the data (see https://en.wikipedia.org/wiki/Cross-validation_(statistics)).

4 Notes

1. Rosetta performance speeds are dependent on the processing speed of the CPU core in use. Most jobs where design is limited to residue repacking can be completed in a few seconds. The LoopMovers perform complex backbone moves and calculations, and can take minutes for a single trajectory to complete.
2. Python variables in order of appearance are as follows:

```
scorefxn #holds the score function for design and
scoring
pose #holds coordinates of the full TCR/pMHC complex
HLA #holds the coordinates of the pMHC
TCR #holds the coordinates of the TCR
bindingScore #holds the binding score of the complex
list_of_residue_positions #holds a list of TCR posi-
tions as candidates for mutation
```

```
distance_cutoff #holds the specified cutoff distance
from the peptide
distance #holds distance between the center of mass
of two residues
residue_list #holds the single letter code of all 20
amino acids
mutant #holds a copy of the complex to perform
mutations.
task #holds the side chain packing settings
pack_mover #holds the mover to repack sidechains
start #holds the first position in the loop
cutpoint #holds the loop cutpoint
end #holds the end position of the loop
ft #holds the foldtree
CDRloop #holds the loop object
loops #holds all of the defined loops
loop_refine #holds the LoopMover_Refine_CCD mover
jd #holds the job distributor
jdpose #holds the pose for manipulation within the
job distributor
```

3. The commands written in Subheading 3.1 are written in Python and include the necessary variables and syntax to develop a complete script for modeling point mutations in TCRs. Commands written in Python/Pyrosetta use the # symbol to denote commented lines. Commands in Subheading 3.2 are example MATLAB commands and use the % symbol to denote commented lines.
4. The scripts assume the following chain PDB IDs: MHC heavy chain as A; β_2-microglobulin as B; peptide as C; TCR α chain as D; and TCR β chain as E.
5. The Rosetta energy unit is an arbitrary unit that loosely correlates with thermodynamic measurements. Because of this, experimental measurements may correlate best with the suggested method for calculating binding energy. Alternative approaches include scoring the entire complex.
6. A judicious cut-off distance between TCR-peptide may be useful here. We most commonly use 15 Å, although structural details and concerns about specificity and cross-reactivity may dictate smaller values.
7. Similar to the cut-off distance between TCR and peptide, we commonly repack residues within a sphere of 8 Å around the mutation. The size of the sphere may be dictated by design needs and is not necessary when using a LoopMover to refine the backbone.
8. The syntax for a foldtree encompassing multiple loops is shown below. The value of −1 indicates edges. Positive integers describe jumps where backbone regions may be manipulated without propagating throughout the rest of the structure.

```
ft=FoldTree()
ft.add_edge(1, 376-2,-1)
ft.add_edge(376-2,380,-1)
ft.add_edge(376-2,384+2,1)
ft.add_edge(384+2,380+1,-1)
ft.add_edge(384+2,408-2,-1)
ft.add_edge(408-2,411,-1)
ft.add_edge(408-2,414+2,2) #increment the positive
integer by one for each jump
ft.add_edge(414+2,411+1,-1)
ft.add_edge(414+2,mutant.total_residue(),-1)
```

9. Each core running a job distributor script and calculating a decoy will output a numeric .in_progress file. When the decoy finishes, a numbered .pdb file is created and score function information added to the .fasc file. The .in_progress file is then deleted, and the core moves on to the next trajectory that does not have a .pdb or .in_progress file in the directory.

Acknowledgements

Computational structural immunology in the authors' laboratories is supported by NIH grants R01GM103773 and R01GM067079 and an award from the Carole and Ray Neag Comprehensive Cancer Center at the University of Connecticut. TPR is supported by a fellowship from the Indiana CTSI, funded in part by NIH grant UL1TR001108.

References

1. Aleksic M, Dushek O, Zhang H et al (2010) Dependence of T cell antigen recognition on T cell receptor-peptide MHC confinement time. Immunity 32:163–174
2. Stone JD, Kranz DM (2013) Role of T cell receptor affinity in the efficacy and specificity of adoptive T cell therapies. Front Immunol 4:244
3. Holler PD, Holman PO, Shusta EV et al (2000) In vitro evolution of a T cell receptor with high affinity for peptide/MHC. Proc Natl Acad Sci U S A 97:5387–5392
4. Chlewicki LK, Holler PD, Monti BC et al (2005) High-affinity, peptide-specific T cell receptors can be generated by mutations in CDR1, CDR2 or CDR3. J Mol Biol 346:223–239
5. Li Y, Moysey R, Molloy PE et al (2005) Directed evolution of human T-cell receptors with picomolar affinities by phage display. Nat Biotechnol 23:349–354
6. Varela-Rohena A, Molloy PE, Dunn SM et al (2008) Control of HIV-1 immune escape by CD8 T cells expressing enhanced T-cell receptor. Nat Med 14:1390–1395
7. Zhao Y, Bennett AD, Zheng Z et al (2007) High-affinity TCRs generated by phage display provide CD4+ T cells with the ability to recognize and kill tumor cell lines. J Immunol 179:5845–5854
8. Morgan RA, Dudley ME, Wunderlich JR et al (2006) Cancer regression in patients after transfer of genetically engineered lymphocytes. Science 314:126–129
9. Liddy N, Bossi G, Adams KJ et al (2012) Monoclonal TCR-redirected tumor cell killing. Nat Med 18:980–987
10. Michielin O (2007) Application of molecular modeling to new therapeutic cancer approaches. Bull Cancer 94:763–768

11. Haidar JN, Pierce B, Yu Y et al (2009) Structure-based design of a T-cell receptor leads to nearly 100-fold improvement in binding affinity for pepMHC. Proteins 74:948–960
12. Malecek K, Grigoryan A, Zhong S et al (2014) Specific increase in potency via structure-based design of a TCR. J Immunol 193:2587–2599
13. Pierce BG, Hellman LM, Hossain M et al (2014) Computational design of the affinity and specificity of a therapeutic T cell receptor. PLoS Comput Biol 10:e1003478
14. Piepenbrink KH, Blevins SJ, Scott DR et al (2013) The basis for limited specificity and MHC restriction in a T cell receptor interface. Nat Commun 4:1948
15. Kaufmann KW, Lemmon GH, Deluca SL et al (2010) Practically useful: what the Rosetta protein modeling suite can do for you. Biochemistry 49:2987–2998
16. Borbulevych OY, Santhanagopolan SM, Hossain M et al (2011) TCRs used in cancer gene therapy cross-react with MART-1/ Melan-A tumor antigens via distinct mechanisms. J Immunol 187:2453–2463
17. Scott DR, Borbulevych OY, Piepenbrink KH et al (2011) Disparate degrees of hypervariable loop flexibility control T-cell receptor cross-reactivity, specificity, and binding mechanism. J Mol Biol 414:385–400
18. Adams J, Narayanan S, Liu B et al (2011) T cell receptor signaling is limited by docking geometry to peptide-major histocompatibility complex. Immunity 35:681–693
19. Burrows SR, Chen Z, Archbold JK et al (2010) Hard wiring of T cell receptor specificity for the major histocompatibility complex is underpinned by TCR adaptability. Proc Natl Acad Sci 107:10608–10613
20. Borbulevych OY, Piepenbrink KH, Gloor BE et al (2009) T cell receptor cross-reactivity directed by antigen-dependent tuning of peptide-MHC molecular flexibility. Immunity 31:885–896
21. Sammond DW, Eletr ZM, Purbeck C et al (2007) Structure-based protocol for identifying mutations that enhance protein–protein binding affinities. J Mol Biol 371:1392–1404
22. Chaudhury S, Lyskov S, Gray JJ (2010) PyRosetta: a script-based interface for implementing molecular modeling algorithms using Rosetta. Bioinformatics 26:689–691
23. Leaver-Fay A, O'meara MJ, Tyka M et al (2013) Scientific benchmarks for guiding macromolecular energy function improvement. Methods Enzymol 523:109
24. Giudicelli V, Duroux P, Ginestoux C et al (2006) IMGT/LIGM-DB, the IMGT® comprehensive database of immunoglobulin and T cell receptor nucleotide sequences. Nucleic Acids Res 34:D781–D784
25. Kellogg EH, Leaver-Fay A, Baker D (2011) Role of conformational sampling in computing mutation-induced changes in protein structure and stability. Proteins 79:830–838

Chapter 19

Computational Modeling of T Cell Receptor Complexes

Timothy P. Riley, Nishant K. Singh, Brian G. Pierce, Zhiping Weng, and Brian M. Baker

Abstract

T-cell receptor (TCR) binding to peptide/MHC determines specificity and initiates signaling in antigen-specific cellular immune responses. Structures of TCR–pMHC complexes have provided enormous insight to cellular immune functions, permitted a rational understanding of processes such as pathogen escape, and led to the development of novel approaches for the design of vaccines and other therapeutics. As production, crystallization, and structure determination of TCR–pMHC complexes can be challenging, there is considerable interest in modeling new complexes. Here we describe a rapid approach to TCR–pMHC modeling that takes advantage of structural features conserved in known complexes, such as the restricted TCR binding site and the generally conserved diagonal docking mode. The approach relies on the powerful Rosetta suite and is implemented using the PyRosetta scripting environment. We show how the approach can recapitulate changes in TCR binding angles and other structural details, and highlight areas where careful evaluation of parameters is needed and alternative choices might be made. As TCRs are highly sensitive to subtle structural perturbations, there is room for improvement. Our method nonetheless generates high-quality models that can be foundational for structure-based hypotheses regarding TCR recognition.

Key words T cell receptor, Peptide/MHC, Structure, Rosetta, Loop modeling, Docking

1 Introduction

Clonally distributed αβ T cell receptors (TCRs) recognize antigenic peptides bound and "presented" by class I or class II major histocompatibility complex proteins (pMHC; Fig. 1). TCR recognition of a pMHC initiates T cell signaling and is responsible for the specificity of T cell mediated immunity. Since initial crystallographic work in the 1990s [1, 2], dozens of new and variant TCR–pMHC structures have been reported. This structural work has significantly enriched our understanding of the principles of T cell recognition, lending insight into T cell immunobiology as well as the biophysics that underlie the remarkable binding properties of TCRs. Knowledge of TCR–pMHC structures has helped us understand the TCR's unusual dichotomy of specificity and

Barry L. Stoddard (ed.), *Computational Design of Ligand Binding Proteins*, Methods in Molecular Biology, vol. 1414, DOI 10.1007/978-1-4939-3569-7_19, © Springer Science+Business Media New York 2016

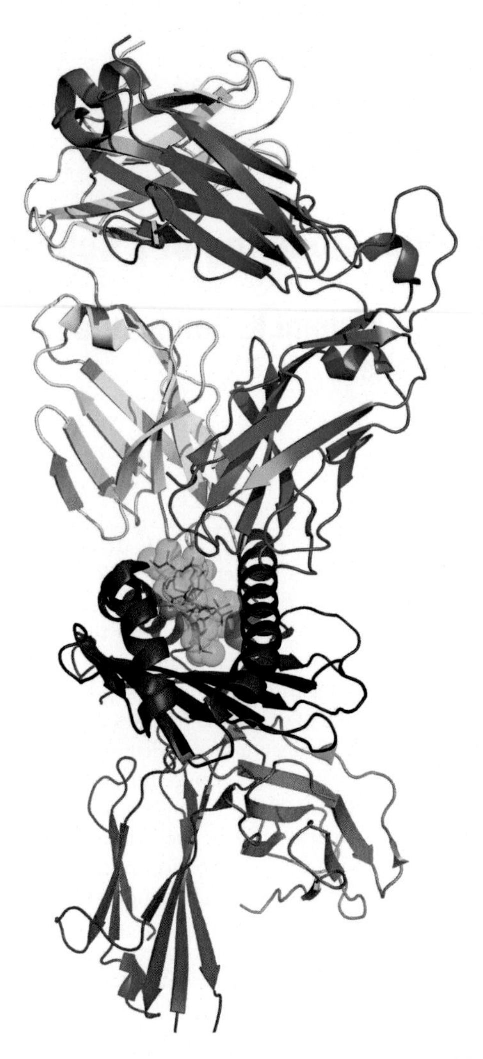

Fig. 1 Structural overview of the complex formed between a TCR (*blue*/*gold*) and peptide/MHC complex (*green*/*purple*/*orange*). The structure of the DMF5 TCR bound to the human class I MHC HLA-A2 in complex with the MART-1 ELA peptide was used for this figure

cross-reactivity; the structural basis for molecular mimicry and its potential role in autoimmunity; the principles underlying T cell recognition of tumor antigens; the basis for immune responses to pathogens and the function of "escape" mutations; vaccine design; and the mechanisms of T cell signaling (see, for example, refs. 3–8).

In addition to fundamental biophysical and immunological insights, knowledge of TCR–pMHC structures has spurred developments of novel immunologically based therapeutics. This includes T cells genetically engineered to express unique TCRs as well as soluble TCR–CD3 antibody fusions [9, 10]. These approaches redirect T cells to targets of specific interest, such as tumor or virally infected cells, and for both there is increasing interest in engineering TCR variants with enhanced recognition properties. Structural knowledge can be of obvious benefit here, permitting structure-guided computational design or helping to pinpoint which regions and amino acids should be subject to molecular evolution.

While structural information for TCR–pMHC complexes can therefore be of clear benefit for basic and applied immunology, recombinant TCRs are difficult to generate and can be challenging to crystallize (although helpful descriptions of successful, systematic approaches are available [11, 12]). While these challenges can often be surmounted, many TCR–pMHC complexes form with very weak affinities in solution [13, 14], which can further hinder crystallization. For this reason, there has been growing interest in developing procedures for modeling TCR–pMHC complexes. We recently described TCRFlexDock, a template-independent procedure which led to near-native predictions for multiple TCR pMHC complexes, in addition to TCR recognition of CD1–lipid and MR1–metabolite complexes [15, 16]. In addition to these studies, TCR–pMHC structural models have been generated by other groups to investigate specific hypotheses or facilitate structural surveys (e.g., refs. 17–21). Klausen and colleagues recently described a publicly available web server that models TCRs [22], yet this does not build a TCR–pMHC complex.

Here we outline a strategy for modeling TCR–pMHC complexes which builds on previous efforts. The procedure is designed for rapid and easy implementation and is readily extensible. Unlike TCRFlexDock, which uses unbound TCR and pMHC structures set to an average docking orientation for docking input [15, 16], this method uses a known TCR–pMHC complex as a template, utilizing the restricted pMHC binding site and the generally conserved diagonal TCR docking mode to help make predictions. As with all docking and modeling procedures, there are known caveats and areas of needed improvement, particularly given the high sensitivity of TCRs for subtle structural perturbations [23]. Nonetheless, this method is rapid and can be considered a "launching point" for generating high-quality models that permit the development of testable, structure-based hypotheses for exploring TCR–pMHC complexes.

2 Materials

1. Access to a high performance computing facility enabled with the latest Python (https://www.python.org/) and PyRosetta installations (https://www.rosettacommons.org/).
2. Access to the NCBI BLAST tool (http://blast.ncbi.nlm.nih.gov/Blast.cgi).

3. 3 GB of available RAM and one processor core is required for each PyRosetta job. Multiple cores with accompanying RAM are required for large modeling projects. A minimum of 20 cores is recommended for projects requiring thousands of decoys.
4. Sequence information of the target TCR–pMHC complex to be modeled.
5. Structural coordinates of one or more template TCR–pMHC complexes, when publicly available downloadable from the Protein Data Bank (http://www.rcsb.org/pdb/home/home.do).

3 Methods

TCR–pMHC complexes share a high level of structural homology which can be taken advantage of when modeling complexes for which there is no structural information. The procedure below assumes a "new" TCR recognizing a "new" peptide presented by the same class I MHC protein (*see* **Note 1** for comments on modeling new peptides). For demonstration, we describe modeling the complex between the antiviral TCR A6 and the Tax peptide presented by HLA-A2, using the structure of the complex between the DMF5 TCR and the melanoma-associated MART-1 AAG peptide presented by HLA-A2 [24]. Although considerable structural information is available for the A6–Tax/HLA-A2 complex, including structures for the free TCR and free pMHC [1, 3, 23, 25–29], this structural information is not utilized in the modeling procedure. Modeling the A6–Tax/HLA-A2 complex using the DMF5–AAG/HLA-A2 complex as a template is a reasonably challenging modeling task, as the two TCRs bind pMHC with different incident angles (Fig. 2). Results are also shown for modeling the DMF5–ELA/HLA-A2 and DMF5–AAG/HLA-A2 complexes using the complexes with the unrelated DMF4 TCR as templates. The latter two are also challenging modeling tasks, as the DMF4 TCR binds AAG/HLA-A2 and ELA/HLA-A2 with different docking angles [24].

3.1 Selection of Template and TCR Sequence Alignment

To begin, we first describe mapping the sequence of a target TCR onto the coordinates of the TCR in a TCR–pMHC template structure through the use of Basic Local Alignment Search Technique (BLAST). Special consideration should be given to sequence similarity and loop length when selecting the template. The most preferred template shares at least one TCR chain with the target. However, if loop sizes vary considerably (>4 residues) between target and template TCR, an alternative template might be considered. Templates with properties such as unusual docking or

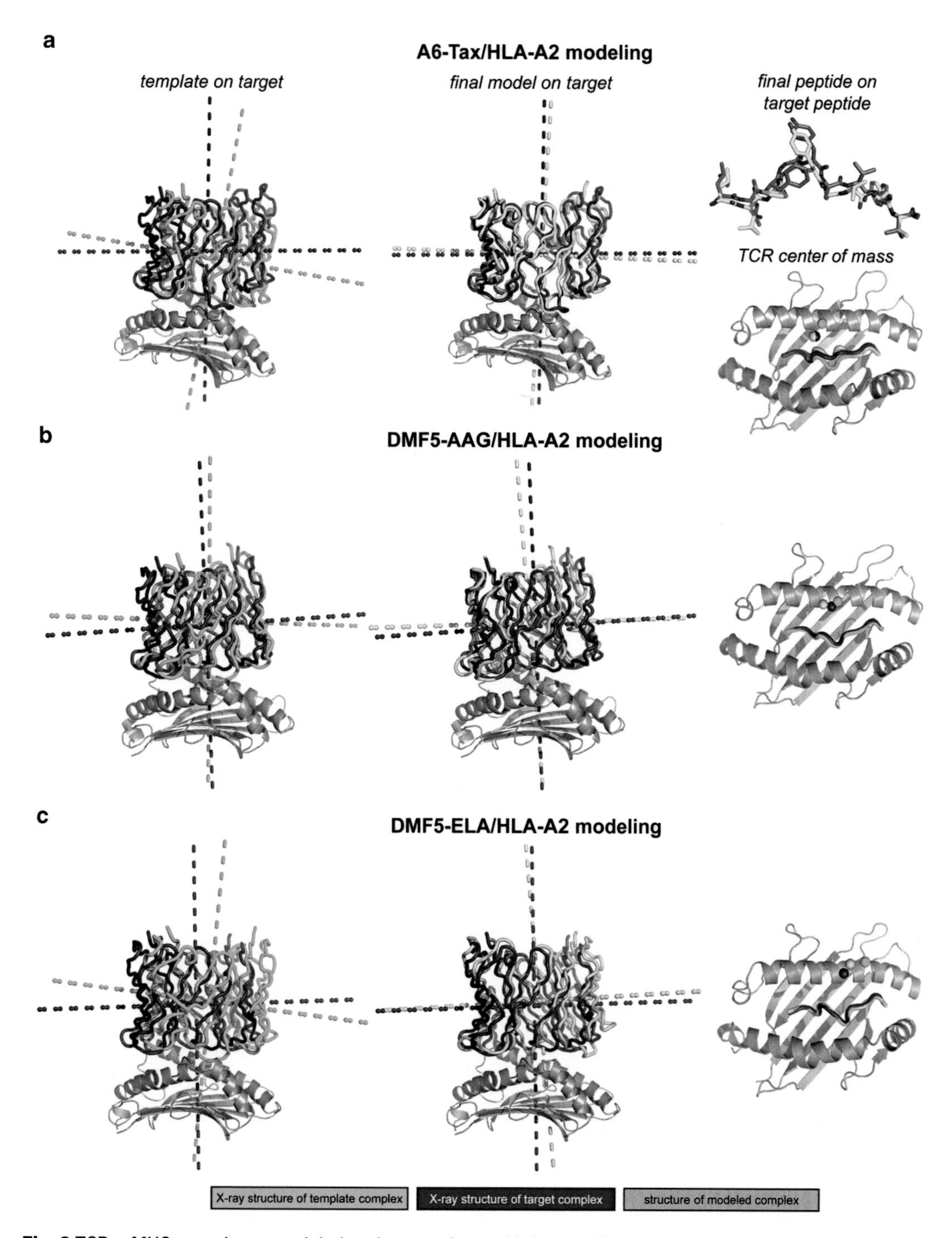

Fig. 2 TCR–pMHC complexes modeled and comparison with known X-ray structures. (**a**) Modeling of the A6–Tax/HLA-A2 complex using the DMF5–ELA/HLA-A2 complex as a template. Using the known X-ray structures, the *left panel* shows the position of the template DMF5 TCR (*green*) relative to the target A6 TCR (*blue*) when bound to pMHC, generated by superimposing the HLA-A2 heavy chain. Only the TCR variable domains and the peptide binding domain are shown. *Horizontal dashed lines* represent the TCR docking angle relative to pMHC and *vertical dashed lines* the incident angle. The *middle panel* compares the final model of the A6 complex (*yellow*) to the known structure of the target complex. The *right panel* compares the peptides from the modeled and known complexes, as well as the centers of mass of the TCRs over the MHC. (**b**) Modeling of the DMF5–AAG/HLA-A2 complex using the DMF4–AAG/HLA-A2 complex as a template. Panels and colors are the same as in (**a**). (**c**) Modeling of the DMF5–ELA/HLA-A2 complex using the DMF4-ELA/HLA-A2 complex as a template. Panels and colors are the same as in (**a**)

incident angles might also be avoided. Depending on circumstances, multiple templates might be used and the results examined for convergence on similar structural properties.

1. Align the protein sequences of the target and template TCR chains with the protein BLAST tool from NCBI, using default options. The example alignment in Fig. 3a is between the target TCR A6 α chain (Query) and template DMF5 α chain (Sbjct) sequences [30, 31].
2. The alignments in Fig. 3b demonstrates for the α chain that most of the sequence variability is at the hypervariable CDR3α loop, which is one residue longer in the A6 structure. The β chain sequences are more different, but the alignment is able to determine where the insertions should be made.

3.2 Using Rosetta to Map the Target TCR Sequence onto the Structural Template and Modify the Peptide

The Rosetta package includes tools for computational modeling and structural analysis and was originally designed for de novo structure prediction [32]. Rosetta is typically used in applications such as protein folding or protein design, and is an ideal tool here. PyRosetta is a Python toolkit which packages the powerful Rosetta algorithms into the easily learned Python scripting language [33]. PyRosetta can be used via scripts or interactively by command line. The commands used below demonstrate how PyRosetta can be used to generate an initial model of the new TCR–pMHC complex, whereby the target, "new" TCR is mapped onto the template and the peptide is altered (*see* **Note 2** for initiating Rosetta and **Note 3** for a list of variables used in the code below).

1. Load the template PDB structure (3QDJ) into PyRosetta and adjust the loops according to alignment in the previous steps. Adjusting the loop sizes can be performed manually in the pdb or using the grafting tool within Rosetta as shown below (*see* **Note 4** for comments on insertions). For amino acid insertions, glycines are used initially and backbone coordinates copied from the preceding amino acid. Insertions that are not glycine will be mutated to the appropriate amino acid in the next step. After the loops have been adjusted, backbone breaks need to be closed. This can be accomplished with the CcdLoopClosureMover, which solves the chain break through cyclic coordinate descent (*see* **Note 5** for loop definitions for closing breaks).

```
#Download the template from the PDB database and
define the last insertion site.
template_pose=pose_from_rcsb('3QDJ')
beta_insertion_site = pose.pdb_info().
pdb2pose('E',101))
#create a Glycine to insert into loop with backbone
coordinates
identical to residue 101 of chain 'E'. Insert the
Glycine(s) and close the resulting chain breaks.
```

a

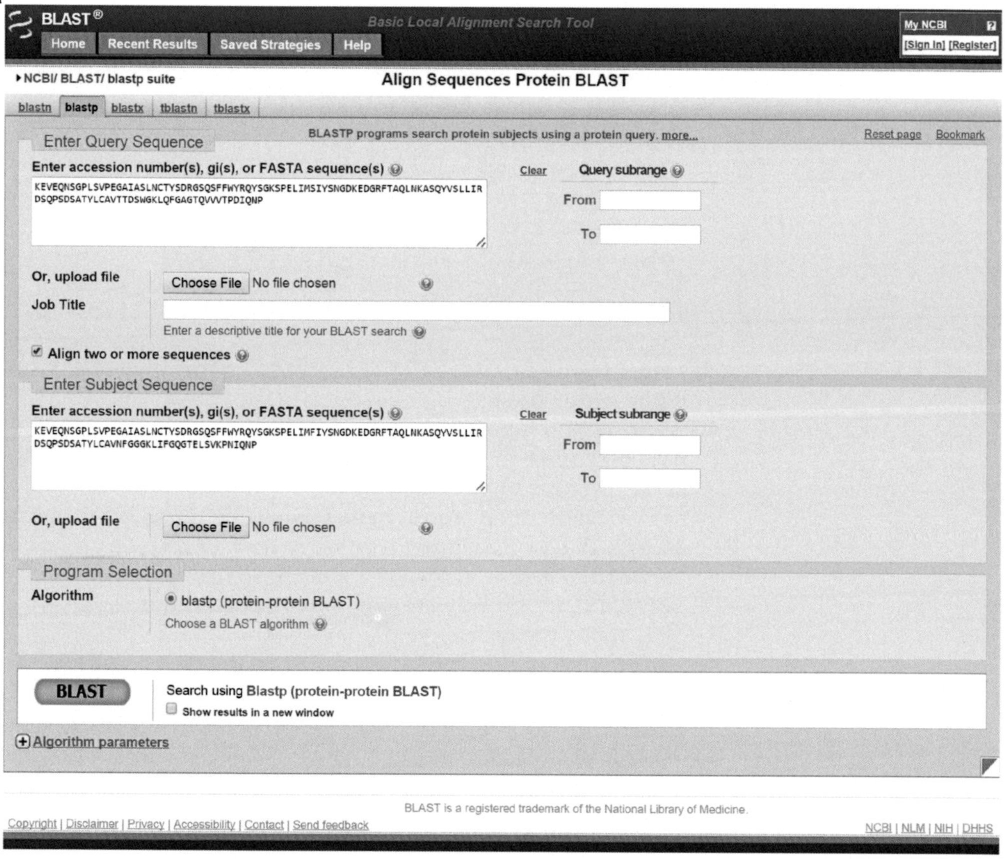

BLAST® Basic Local Alignment Search Tool

Home | Recent Results | Saved Strategies | Help

My NCBI [Sign In] [Register]

▸NCBI/ BLAST/ blastp suite

Align Sequences Protein BLAST

blastn | blastp | blastx | tblastn | tblastx

BLASTP programs search protein subjects using a protein query. more... Reset page Bookmark

Enter Query Sequence

Enter accession number(s), gi(s), or FASTA sequence(s) Clear **Query subrange**

KEVEQNSGPLSVPEGAIASLNCTYSDRGSQSFFWYRQYSGKSPELIMSIYSNGDKEDGRFTAQLNKASQYVSLLIR
DSQPSDSATYLCAVTTDSWGKLQFGAGTQVVVTPDIQNP

From

To

Or, upload file Choose File No file chosen

Job Title

Enter a descriptive title for your BLAST search

Align two or more sequences

Enter Subject Sequence

Enter accession number(s), gi(s), or FASTA sequence(s) Clear **Subject subrange**

KEVEQNSGPLSVPEGAIASLNCTYSDRGSQSFFWYRQYSGKSPELIMFIYSNGDKEDGRFTAQLNKASQYVSLLIR
DSQPSDSATYLCAVNFGGGKLIFGQGTELSVKPNIQNP

From

To

Or, upload file Choose File No file chosen

Program Selection

Algorithm blastp (protein-protein BLAST)

Choose a BLAST algorithm

BLAST — Search using Blastp (protein-protein BLAST)

Show results in a new window

Algorithm parameters

BLAST is a registered trademark of the National Library of Medicine.

Copyright | Disclaimer | Privacy | Accessibility | Contact | Send feedback NCBI | NLM | NIH | DHHS

b

```
alpha:
Query  1    KEVEQNSGPLSVPEGAIASLNCTYSDRGSQSFFWYRQYSGKSPELIMSIYSNGDKEDGRF  60
            KEVEQNSGPLSVPEGAIASLNCTYSDRGSQSFFWYRQYSGKSPELIM IYSNGDKEDGRF
Sbjct  1    KEVEQNSGPLSVPEGAIASLNCTYSDRGSQSFFWYRQYSGKSPELIMFIYSNGDKEDGRF  60

Query  61   TAQLNKASQYVSLLIRDSQPSDSATYLCAVTTDSWGKLQFGAGTQVVVTPDIQNP  115
            TAQLNKASQYVSLLIRDSQPSDSATYLCAV     GKL FG GT++ V P+IQNP
Sbjct  61   TAQLNKASQYVSLLIRDSQPSDSATYLCAVNF-GGGKLIFGQGTELSVKPNIQNP  114

beta:
Query  1    GVTQTPKFQVLKTGQSMTLQCAQDMNHEYMSWYRQDPGMGLRLIHYSVGAGITDQGEVPN  60
            G+TQ P  Q+L  G+ MTL+C QDM H  M WYRQD G+GLRLIHYS  AG T +GEVP+
Sbjct  3    GITQAPTSQILAAGRRMTLRCTQDMRHNAMYWYRQDLGLGLRLIHYSNTAGTTGKGEVPD  62

Query  61   GYNVSRSTTEDFPLRLLSAAPSQTSVYFCASRPGLAGGRPEQYFGPGTRLTVTEDLKNVF  120
            GY+VSR+ T+DFPL L SA PSQTSVYFCAS   L+ G  E +FG GTRLTV EDL  VF
Sbjct  63   GYSVSRANTDDFPLTLASAVPSQTSVYFCAS--SLSFG-TEAFFGQGTRLTVVEDLNKVF  119
```

Fig. 3 **BLAST** alignment of TCR α and β chains. (**a**) BLAST web interface for performing a protein–protein alignment, with sequences for the A6 and DMF5 α chains entered. (**b**) BLAST alignments for the A6 and DMF5 α and β chains. Residues in bold indicate locations requiring insertions or deletions

```
Gly=pose_from_sequence('G')
Gly.residue(1).set_xyz('N',    template_pose.residue
(beta_insertion_site).xyz('N'))
Gly.residue(1).set_xyz('C', template_pose.residue(beta_
insertion_site).xyz('C'))
Gly.residue(1).set_xyz('CA',   template_pose.residue
(beta_insertion_site).xyz('CA'))
Gly.residue(1).set_xyz('O', template_pose.residue(beta_
insertion_site).xyz('O'))
template_pose=protocols.grafting.insert_pose_into_
pose(template_pose, Gly, beta_insertion_site)
#close the resulting chain breaks on either end of
the insertion
loopclose=Loop(beta_insertion_site-2,beta_inser-
tion_site+2, beta_insertion_site)
movemap=MoveMap()
movemap.set_bb_true_range(beta_insertion_site-2,
beta_insertion_site+2)
set_single_loop_fold_tree(template_pose,loopclose)
add_single_cutpoint_variant(template_pose,
loopclose)
ccd=CcdLoopClosureMover(loopclose, movemap)
ccd.apply(complex)
loopclose=Loop(beta_insertion_site-2,beta_inser-
tion_site+2, beta_insertion_site+1)
set_single_loop_fold_tree(template_pose,loopclose)
add_single_cutpoint_variant(template_pose,
loopclose)
ccd=CcdLoopClosureMover(loopclose, movemap)
ccd.apply(template_pose)
```

2. Working backwards (*see* **Note 4**), complete the necessary loop adjustments for each backbone manipulation of the template TCR. For mapping A6 onto DMF5, these adjustments occur at residue 96 of chain ‘E’ and 92 of chain ‘D.’

3. Once the template has been aligned and resized to match the target sequence, ‘for’ loops for the peptide, α chain, and β chains can be set up to mutate all template residues to the target sequence. For example, a ‘for’ loop demonstrating the conversion of the DMF5 α chain sequence to A6 α chain is shown below.

```
matched_sequence=Pose()
matched_sequence.assign(template_pose)
alpha = ('KEVEQNSGPLSVPEGAIASLNCTYSDRG
SQSFFWYRQYSGKSPELIMSIYSNGDKEDGRFTA
QLNKASQYVSLLIRDSQPSDSATYLCAVTTDSWGKLQ
FGAGTQVVVTPDIQNP')
# mutate the alpha chain
for i in range(0,len(alpha)):
  matched_sequence=mutate_residue(template_pose,
  alpha_start_site+i, alpha[i])
```

4. Repeat **step 3** to map the β chain and peptide sequences onto the matched_sequence pose.
5. Build a packer task and pack_mover to repack all residues in the target structure.

```
scorefxn_high=get_fa_scorefxn()
task = TaskFactory.create_packer_task
(matched_sequence)
task.or_include_current(True)
task.restrict_to_repacking()
pack_mover=PackRotamersMover(scorefxn_high, task)
pack_mover.apply(matched_sequence)
```

6. The job distributor is a useful tool that can take advantage of multiple cores running the same script to generate designs/decoys in parallel. Before further docking or loop modeling, set up a job distributor to repack all residues in the target structure and minimize the loop backbones (*see* **Note 6** for comments on the job distributor; **Note 7** for recommended trials; and **Note 8** for loop definitions and a full FoldTree).

```
#add all CDR loops+peptide to be refined simultane-
ously. Create a FoldTree to include all loops.
peploop=Loop(376,384,380)
acdr1=Loop(408,414,411)
acdr2=Loop(433,439,435)
acdr3=Loop(475,482,479)
bcdr1=Loop(607,612,610)
bcdr2=Loop(631, 637,635)
bcdr3=Loop(675,687,680)
ft=FoldTree()
#build a complete FoldTree here.
loops=Loops()
loops.add_loop(peploop)
loops.add_loop(acdr1)
loops.add_loop(acdr2)
loops.add_loop(acdr3)
loops.add_loop(bcdr1)
loops.add_loop(bcdr2)
loops.add_loop(bcdr3)
loop_refine = LoopMover_Refine_CCD(loops,
scorefxn_high)
jd = PyJobDistributor('ready_for_docking',100,
scorefxn_high)
while not jd.job_complete:
  p=Pose()
  p.assign(matched_sequence)
  p.fold_tree(ft)
  pack_mover.apply(p)
  loop_refine.apply(p)
  jd.output_decoy(p)
```

7. Save the lowest scoring .pdb file from the job distributor for the subsequent steps.

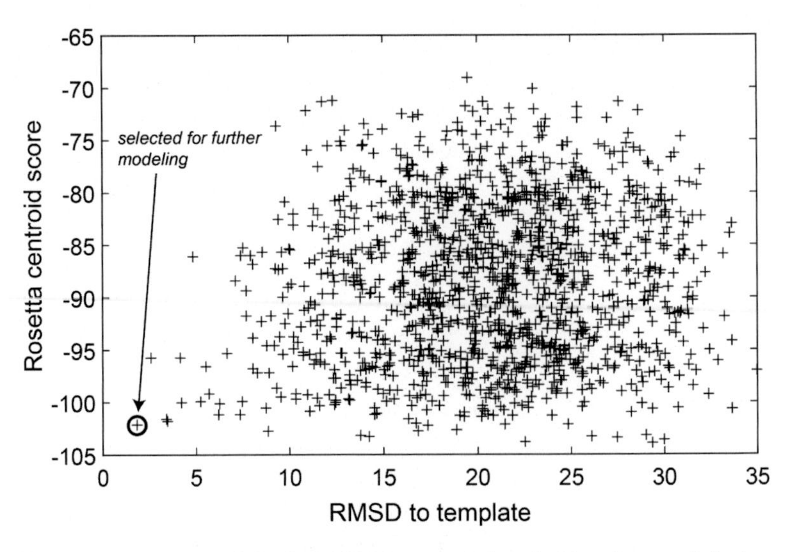

Fig. 4 Rosetta scores of the first 1000 low resolution docking decoys plotted vs. RMSD of the backbone of the decay to the template. The decoy with the lowest RMSD is selected for continued refinement, and is usually included among the lowest 20 scoring structures

3.3 Low Resolution Docking

The sections above result in an initial model of the new TCR bound to the new pMHC complex, in this case the A6 TCR bound to Tax/HLA-A2. However, although TCRs generally bind with a diagonal docking angle, there is variation in this angle, as well as the incident angle with which the TCR engages [34, 35]. Such variation is exemplified by the A6–Tax/HLA-A2 and DMF5–AAG/HLA-A2 complexes, as shown in Fig. 2.

To adjust the docking angle of the template TCR over the pMHC, the Rosetta Docking protocols are utilized [36]. The Docking movers perform rigid body translations to bring two molecules into contact and minimize scores. We use a low resolution centroid model to perform a global docking maneuver where the TCR may bind the pMHC in any orientation. Restrictions may be set in place to limit the randomization of the docking partners, but a full randomization helps remove bias and is not computationally prohibitive when low resolution centroid models are used.

The Docking protocol requires many different decoys to be generated, and many incorrect decoys may have scores similar to the true solution. To account for this, we take advantage of the diagonal docking angle found across TCR–pMHC complexes. As shown in Fig. 4, the decoy with the lowest RMSD to the template is selected for future refinement. This ensures selection of a model which retains a TCR-like binding mode, as Rosetta score alone cannot be used. Although selecting a model with the lowest RMSD to the template could potentially introduce some bias, in our experience low resolution docking is sufficiently imprecise and the variation in TCR–pMHC docking modes sufficiently small to reduce these concerns (note also that further manipulation of the docking mode is performed later by a high resolution docking maneuver). If there is still concern about template bias at this step, another TCR–pMHC complex can be used to determine RMSD values and select a model to move forward with, or multiple models generated with different templates.

1. The centroid mode within Rosetta converts amino acid sidechains into low resolution centroids. This mode is useful for sampling conformations quickly. Because the centroid is not representative of a full atom structure, there is a difference between centroid and full atom score functions, which influences how residues are repacked and manipulated. Before conversion to a centroid model, save the sidechains of the existing template to assist in repacking in subsequent steps.

```
scorefxn_low=create_score_function('interchain_cen')
recover_sidechains = protocols.simple_moves.
ReturnSidechainMover(ready_for_docking)
censwitch=SwitchResidueTypeSetMover('centroid')
centroid_complex=Pose()
centroid_complex.assign(ready_for_docking)
censwitch.apply(centroid_complex)
```

2. The docking mover requires a FoldTree that defines the TCR and pMHC as separate chains to translate independently.

```
#The jump defines where the movers are allowed to
separate the pose. "ABC_DE" is synonymous to the
complex chains (see Note 9 for information on chain
identifiers).
jump=Vector([1])
setup_foldtree(centroid_complex, "ABC_DE",jump)
```

3. The following steps involve randomizing the orientation of the TCR relative to the pMHC to escape local score minima. This is followed by sliding the two proteins into contact and a rigid body score minimization. Because of the randomization, we suggest performing the following steps within the job distributor for at least 10,000 trials.

```
jd = PyJobDistributor('docking_lowres',10000,
scorefxn_high)
jd.native_pose=full_atom_pose
while not jd.job_complete:
  p=Pose()
  p.assign(centroid_complex)
  #randomize the two partners before docking
```

```
randomize1=rigid_moves.RigidBodyRandomizeMover
(p, 1, rigid_moves.partner_upstream)
randomize2=rigid_moves.RigidBodyRandomizeMover(p,
1, rigid_moves.partner_downstream)
randomize1.apply(p)
randomize2.apply(p)
#translate the two molecules towards each other
until they come into contact
slide=DockingSlideIntoContact(jump)
slide.apply(p)
#perform rigid body score minimization
dock_lowres=DockingLowRes(scorefxn_low, 1)
dock_lowres.apply(p)
#return to full atom mode
fa_switch.apply(p)
recover_sidechains.apply(p)
pack_mover.apply(p)
jd.output_decoy(p)
```

4. As mentioned earlier, the decoy with the lowest backbone RMSD to the original template is chosen for further refinement.

3.4 Loop Modeling

We have modified the general docking protocol described by Gray et al. to include a loop modeling stage between the low and high resolution docking stages [37]. Modeling the TCR loops may be accomplished through Kinematic Loop Closure (KIC) or Cyclic Coordinate Descent (CCD) methodologies. CCD solves a chain break by finding the shortest solution to bring two termini together while KIC uses an analytical calculation to minimize the score between three pivot residues. While it is possible to model all loops simultaneously, the large number of possible loop conformations may result in discarding favorable loop conformations paired with unfavorable overall models. We prefer to model each loop consecutively to avoid this situation (*see* **Note 10** for comments on de novo loop modeling). Loops are modeled in low resolution through a Monte Carlo algorithm with the Metropolis criterion [38], and decoys are chosen based off of a steepest descent design.

1. The method outlined in Mandell et al. [38] is a well-characterized method for building loops de novo with KIC and is conveniently implemented into Rosetta as two movers for the low and high resolution stages.

```
#use the low resolution mover for one loop, and the
high resolution refinement protocol on all loops.
model_loop=Loops()
model_loop.add(aCDR1)
kic_perturb=rosetta.protocols.loops.loop_mover.
LoopMover_Perturb_KIC(model_loop)
```

```
kic_refine = rosetta.protocols.loops.loop_mover.
refine.LoopMover_Refine_KIC(loops)
```

2. Set up a job distributor to model at least 1000 decoys for the current loop.

```
jd = PyJobDistributor('loopmodel',1000,
scorefxn_high)
while not jd.job_complete:
  p=Pose()
  p.assign(current_model)
  p.fold_tree(ft)
  censwitch.apply(p)
  perturb_KIC(p)
  faswitch.apply(p)
  recover_sidechains.apply(p)
  #since backbone moved, repack sidechains
  pack.apply(p)
  kic_refine.apply(p)
  jd.output_decoy(p)
```

3. After the all of the decoys have completed, choose the lowest scoring structure and repeat the loop modeling procedure for the peptide (if modeled) and each CDR loop. Since loop conformations may be dependent on each other, continue modeling each loop until lower scores are no longer achieved (*see* **Note 10**).

3.5 High Resolution Docking

The high resolution docking procedure is similar to the low resolution procedure, but adds in full atom side chain repacking in addition to the rigid body minimization performed in Subheading 3.3. Sidechain packing allows for higher resolution discrimination of repulsive forces and charged interactions. In addition, a small perturbation is utilized to further refine the docking mode [37]. With TCRFlexDock, we used a 3 Å, 8° perturbation [16]. Since the method here incorporates a full randomization of the docking partners in low resolution docking (Subheading 3.3), a smaller perturbation could be used in the high resolution phase, although alternate values can be used to explore convergence.

1. Set up the job distributor similar to Subheading 3.2, but use the full atom movers.

```
jd = PyJobDistributor('highresdock',10000,
scorefxn_high)
jd.native_pose = loopmodel
while not jd.job_complete:
  p=Pose()
  p.assign(loopmodel)
  setup_foldtree(p, "ABC_DE",jump)
  #define the degree of perturbation between the TCR
  and pMHC (8 degree, 3 Angstrom).
  pert_mover = rigid_moves.RigidBodyPerturbMover(1,
  8, 3)
```

```
pert_mover.apply(p)
pert_mover.apply(p)
fa_slide=FaDockingSlideIntoContact(1)
fa_slide.apply(p)
dock_highres=DockMCMProtocol()
dock_highres.set_partners("ABC_DE")
dock_highres.set_scorefxn(scorefxn_high)
dock_highres.apply(p)
jd.output_decoy(p)
```

2. Similar to the low resolution docking stage, the structure with the lowest RMSD to the structure used in the docking protocol (Subheading 3.4, **step 4**) is our criteria for selection. We have found the default scores do not correlate with accurate docking (Fig. 4). Therefore, we use the template as a guide (as all TCRs bind with similar orientations) and depend on the Rosetta docking movers to modify the angle to resolve clashes.
3. If the docking angle or score changed significantly (RMS > 0.2 or abs(score) >10), additional rounds of loop modeling and docking as described in Subheadings 3.3 and 3.4 may be necessary to further refine the structure.

3.6 Analysis of Example Projects

The example above shows the modeling of the A6–Tax/HLA-A2 complex using the DMF5–AAG/HLA-A2 complex as a template. As noted earlier, this modeling task was chosen as A6 and DMF5 bind pMHC with different incident angles (Fig. 2a). We also used the above procedure to model the DMF5–ELA/HLA-A2 and DMF5–AAG/HLA-A2 complexes using the unrelated DMF4–ELA/HLA-A2 and DMF4–AAG/HLA-A2 complexes as templates. These latter two are also challenging tasks as the DMF4 TCR binds AAG/HLA-A2 and ELA/HLA-A2 with different docking angles (Fig. 2b, c) [24]. The scores and RMSDs to the target complex for each example through the course of the modeling procedure are shown in Fig. 5. This analysis demonstrates the overall level of performance (note that the RMSD to target information was not used during any stages in the modeling procedures). It also highlights a complexity observed with the DMF5–ELA/HLA-A2 complex: unlike the other complexes, the Rosetta score increased after high resolution docking, prompting additional rounds of loop refinement. Comparison with the RMSD data showed only small downward movement throughout the process, suggesting that this model may have become locked into a local energy well early in the process, possibly in the loop modeling stage. Nonetheless, the final full atom RMSD to the target complex was close at 2.4 Å (1.6 Å for only the α carbons of the TCR).

The performance for all three modeling examples are shown visually in Fig. 2. In each case, the docking/incident angles are shifted towards the correct model, yielding a good alignment as

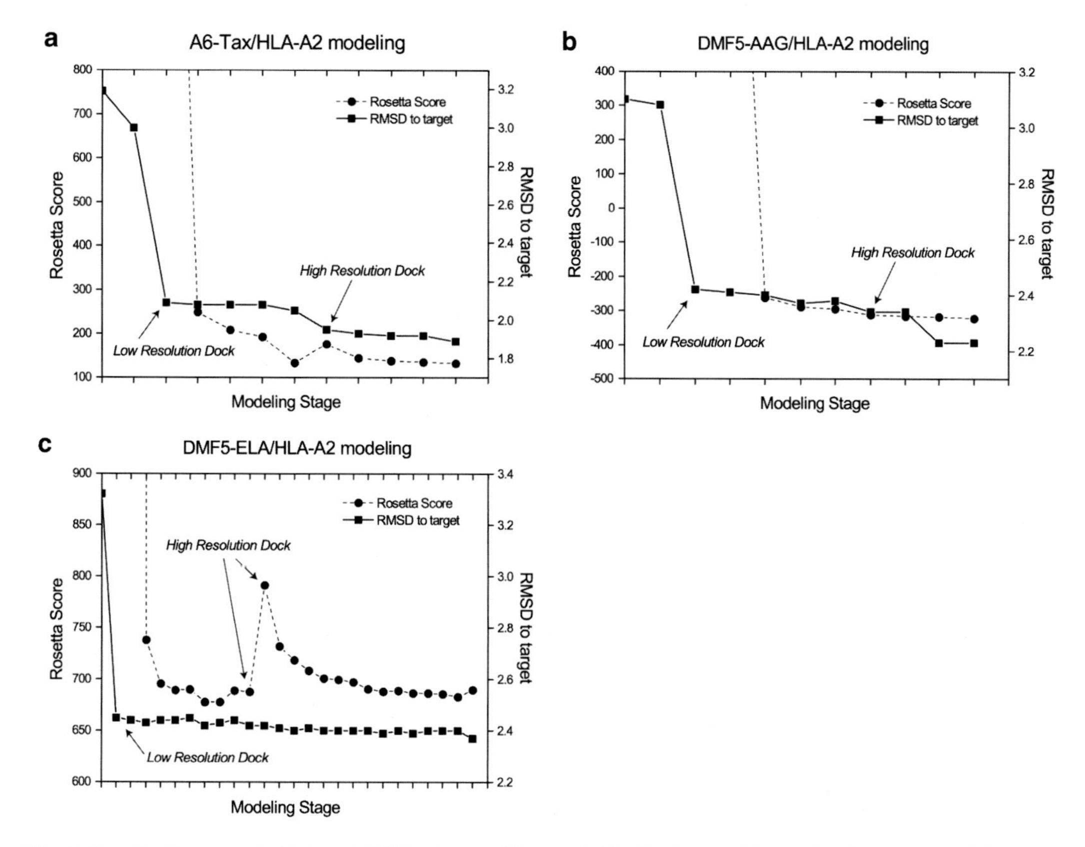

Fig. 5 Rosetta Score and all atom RMSD values of the model to the known X-ray structures as modeling progresses for (**a**) the A6–Tax/HLA-A2 complex, (**b**) the DMF5–AAG/HLA-A2 complex, and (**c**) the DMF5–ELA/HLA-A2 complex. For the DMF5–ELA/HLA-A2 complex, a second high resolution docking stage followed by further loop modeling was performed when the first stage of loop modeling failed to reduce the Rosetta score

discernible from the ribbon diagrams. The shift in docking modes is also apparent by examining the centers of mass of the template/model/known TCRs over the pMHC. In the case of the A6–Tax/HLA-A2 complex, which also involved modeling the Tax peptide from the MART-1 AAG peptide (sequences LLFGYPVYV and AAGIILTV), the peptide backbone is captured, as are the positions of key peptide side chains.

Modeling performance is quantified in Table 1. In each case, the RMSD for the backbone of the modeled TCR relative to that of the known structure is less than 2 Å, and reduced from that obtained by comparing the template to the target structures. The quality of the modeled DMF5–AAG/HLA-A2 structure seems particularly good, given that the starting model differed from the known structure by a TCR Cα RMSD of >3 Å. Details within the three TCR–pMHC interfaces are shown in Fig. 6, comparing the

Table 1

Quantitative comparison of template, X-ray structure, and modeled complexes

Template	Target (modeled complex)	Starting model to structure Cα RMSD (full complex) [Å]	Final model to structure Cα RMSD (full complex) [Å]	Final model to structure all atom RMSD (full complex) [Å]	Starting model to structure Cα RMSD (TCR only) [Å]	Final model to structure Cα RMSD (TCR only) [Å]	Final model to structure all atom RMSD (peptide only) [Å]
DMF5–AAG/HLA-A2 (3QDJ)	A6–Tax/HLA-A2 (1QRN)	2.85	1.42	1.89	1.73	1.64	1.67
DMF4–ELA/HLA-A2 (3QDM)	DMF5–ELA/HLA-A2 (3QDG)	3.16	1.96	2.37	2.02	1.64	N/A
DMF4–AAG/HLA-A2 (3QEQ)	DMF5–AAG/HLA-A2 (3QDJ)	3.38	1.76	2.23	3.09	1.49	N/A

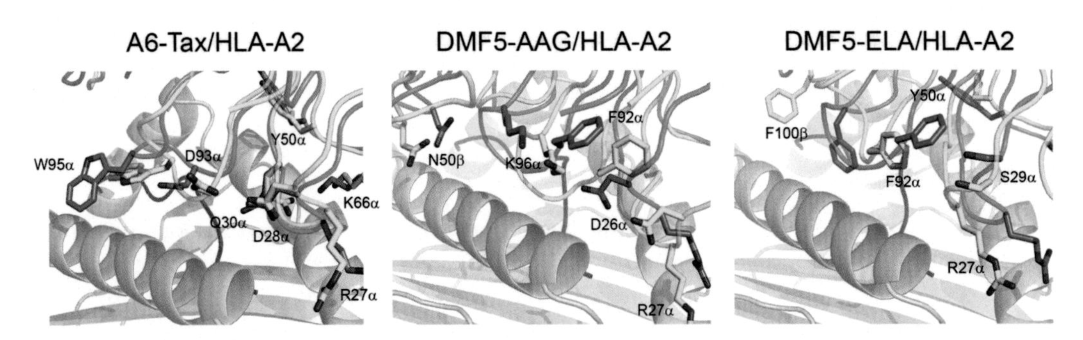

Fig. 6 Comparison of the positioning of select side chains in the known (*blue*) and modeled (*yellow*) TCR–pMHC interfaces

positions of key side chains in the models and X-ray structures. Although there is clear room for improvement in the positioning of some of the side chains, the three models demonstrate the capacity to capture interfacial features as well as general docking modes.

4 Notes

1. Modeling projects which also involve a new peptide necessitate modeling the peptide in the MHC protein, a challenging problem in itself which continues to receive considerable attention (e.g., refs. 39–42). These or other approaches could be used alongside, or integrated into, the procedures described here.
2. This project was developed with the use of multiple scripts, each with a new instance of PyRosetta and new variable declarations which may not be constant depending on the actions taken to manipulate the template (e.g., insertions change the template pdb numbering and require all new variables to be named). The following should be the header of each PyRosetta script to import all tools used in this chapter:

```
from rosetta import*
init(extra_options="-extrachi_cutoff 12 -ex1 -ex2 -ex3")
from toolbox import mutate_residue
import rosetta.protocols.grafting
from rosetta.core.pack.dunbrack import*
from toolbox import pose_from_rcsb
import rosetta.protocols.rigid as rigid_moves
from rosetta.protocols.loops.loop_mover.refine import *
from rosetta.protocols.loops.loop_closure.ccd import *
```

3. Variables used in order of appearance:

```
template_pose #holds the template pose for loop manipulation
beta_insertion_site #holds the location to perform an insertion
Gly #holds the pose for a single glycine for insertion
loopclose #holds the loop for closing chain breaks
movemap #defines the flexible backbone regions of the pose
ccd #holds the CCD mover for closing chain breaks
matched_sequence #holds the pose for mapping the target sequence onto template
alpha #holds a string of the alpha sequence
alpha_start_site #holds the pose numbering of the 1st residue of the alpha chain
scorefxn_high #holds the high resolution score function
task #holds the sidechain packing settings
pack_mover #holds the mover to repack the sidechains
peploop #holds the loop object containing the peptide
acdr1 #holds the loop object containing the αCDR1
acdr2 #holds the loop object containing the αCDR2
acdr3 #holds the loop object containing the αCDR3
```

```
bcdr1 #holds the loop object containing the βCDR1
bcdr2 #holds the loop object containing the βCDR2
bcdr3 #holds the loop object containing the βCDR3
ft #holds the foldtree of the pose
loops #holds all of the defined loops
loop_refine #holds the CCD loop refine mover
jd #holds the job distributor
p #holds the pose used for TCR modeling within the
job distributor
scorefxn_low #holds the low resolution score function
ready_for_docking #holds the pose with adjusted
loops and matched sequence
recover_sidechains #holds the mover to restore the
amino acid sidechains of the pose before conversion
to low resolution
censwitch #holds the mover for switching poses to
centroid mode
centroid_complex #holds the low resolution pose
jump #holds the jump number to identify where to per-
form the docking maneuvers
docking_lowres #holds the low resolution docked
structure
randomize1 #holds the mover to randomize the TCR
docking coordinates
randomize2 #holds the mover randomize the pMHC dock-
ing coordinates
pert_mover #holds the mover performing slight
perturbations
slide #holds the mover to bring two objects into
contact
dock_lowres #holds the mover to perform low resolu-
tion docking
fa_switch #holds the mover to convert a low resolu-
tion pose to high resolution
model_loop #holds the loop for investigating loop
conformations
kic_perturb #holds the mover to perturb a low reso-
lution loop using KIC
kic_refine #holds the mover to refine a high resolu-
tion loop using KIC
loopmodel #holds the pose used for loop modeling
fa_slide #holds the mover to slide two full atom
objects into contact
dock_highres #holds the mover to perform high reso-
lution docking
highresdock #holds the high resolution docked complex
```

4. protocols.grafting.insert_pose_into_pose inserts a pose immediately after the named residue of the template. After an insertion, residue numbering in the pdb reverts to the sequential pose numbering (e.g., residue 92 of chain D becomes residue 477). Because of this, make insertions and deletions starting

from the C termini of the beta chain and work backwards. Also, if the peptides are of different sizes, perform the same procedure on the peptide chain. However, as peptides of different lengths can take significantly different paths in the MHC binding groove [43] and peptide modeling remains challenging as noted above, consider using a template with a matched peptide length.

5. To limit the backbone perturbation of the insertion, set up the CcdLoopMover loops to include the insertion site −2 residues and the length of the insertion +1 residue.

6. Each core running a job distributor script and calculating a decoy will output a numeric .in_progress file. When the decoy finishes, a numbered .pdb file is created and score function information added to the .fasc/.sc file. The .in_progress file is then deleted, and the core moves on to the next trajectory that does not have a .pdb or .in_progress file in the directory.

7. The number of trials needed for each job distributor depends on the moves applied and the variability introduced to the template structure. For example, a full atom backbone refinement may sample all local conformations in 100 trials, a loop modeling protocol with backbone randomization may need 1000 trials, and full docking protocols will need to generate upwards of 10,000 decoys.

8. Loops may be defined using the IMGT database (http://www.imgt.org) [44] or preferably by visually inspecting the CDR loops in the template and comparing with the sequencing alignment. A loop in Rosetta is defined by two residues on either end and a 'cutpoint' to allow flexible motion without propagating through the rest of the structure. FoldTrees define independent regions of a pose and are set up to include each loop and two residues on either side to act as anchors. The syntax for a FoldTree encompassing multiple loops is shown below. The value of −1 indicates edges. Positive integers describe jumps where backbone regions may be manipulated without propagating throughout the rest of the structure.

```
ft.add_edge(1, 376-2,-1)
ft.add_edge(376-2,380,-1)
ft.add_edge(376-2,384+2,1)
ft.add_edge(384+2,380+1,-1)
ft.add_edge(384+2,408-2,-1)
ft.add_edge(408-2,411,-1)
ft.add_edge(408-2,414+2,2) #jumps increment by +1
ft.add_edge(414+2,411+1,-1)
ft.add_edge(414+2,433-2,-1)
ft.add_edge(433-2,435,-1)
ft.add_edge(433-2,439+2,3)
```

```
ft.add_edge(439+2,435+1,-1)
ft.add_edge(439+2,475-2,-1)
ft.add_edge(475-2,479,-1)
ft.add_edge(475-2,482+2,4)
ft.add_edge(482+2,479+1,-1)
ft.add_edge(482+2,607-2,-1)
ft.add_edge(607-2,610,-1)
ft.add_edge(607-2,612+2,5)
ft.add_edge(612+2,610+1,-1)
ft.add_edge(612+2,631-2,-1)
ft.add_edge(631-2,635,-1)
ft.add_edge(631-2,637+2,6)
ft.add_edge(637+2,635+1,-1)
ft.add_edge(637+2,675-2,-1)
ft.add_edge(675-2,680,-1)
ft.add_edge(675-2,685+2,7)
ft.add_edge(687+2,680+1,-1)
ft.add_edge(687+2,matched_sequence.
total_residue(),-1)
```

9. The scripts assume the following chain PDB IDs: For a class I MHC protein, heavy chain as A; β_2-microglobulin as B; peptide as C (for a class II MHC protein, the α chain would be A and the β chain would be B); TCR α chain as D; and TCR β chain as E.
10. For example, 1000 decoys of the αCDR1 loop may be remodeled. The lowest scoring decoy will be used to model 1000 decoys of αCDR2, etc. After all loops have been remodeled, it may be necessary to repeat the cycle to account for loop dependent effects. Generally, we have found three cycles (18 loop remodels) to be sufficient. We have found the peptide is often within 2 Å of the template peptide (all atom RMSD after superimposing target and template peptides) after refinement, and in our experience may not need to be randomized to identify a close-to-native conformation.

Acknowledgements

Computational structural immunology in the authors' laboratories is supported by NIH grants R01GM103773 and R01GM067079 and an award from the Carole and Ray Neag Comprehensive Cancer Center at the University of Connecticut. TPR is supported by a fellowship from the Indiana CTSI, funded in part by NIH grant UL1TR001108.

References

1. Garboczi DN, Ghosh P, Utz U et al (1996) Structure of the complex between human T-cell receptor, viral peptide and HLA-A2. Nature 384:134–141
2. Garcia KC, Degano M, Stanfield RL et al (1996) An alphabeta T cell receptor structure at 2.5 A and its orientation in the TCR-MHC complex [see comments]. Science 274:209–219
3. Borbulevych OY, Piepenbrink KH, Baker BM (2011) Conformational melding permits a conserved binding geometry in TCR recognition of foreign and self molecular mimics. J Immunol 186:2950–2958
4. Cole DK, Yuan F, Rizkallah PJ et al (2009) Germline-governed recognition of a cancer epitope by an immunodominant human T-cell receptor. J Biol Chem 284:27281–27289
5. Macdonald WA, Chen Z, Gras S et al (2009) T cell allorecognition via molecular mimicry. Immunity 31:897–908
6. Adams JJ, Narayanan S, Liu B et al (2011) T cell receptor signaling is limited by docking geometry to peptide-major histocompatibility complex. Immunity 35:681–693
7. Bulek AM, Cole DK, Skowera A et al (2012) Structural basis for the killing of human beta cells by CD8+ T cells in type 1 diabetes. Nat Immunol 13:283–289
8. Chen J-L, Stewart-Jones G, Bossi G et al (2005) Structural and kinetic basis for heightened immunogenicity of T cell vaccines. J Exp Med 201:1243–1255
9. Restifo NP, Dudley ME, Rosenberg SA (2012) Adoptive immunotherapy for cancer: harnessing the T cell response. Nat Rev Immunol 12:269–281
10. Oates J, Jakobsen BK (2013) ImmTACs: novel bi-specific agents for targeted cancer therapy. Oncoimmunology 2:e22891
11. Van Boxel GI, Stewart-Jones G, Holmes S et al (2009) Some lessons from the systematic production and structural analysis of soluble αβ T-cell receptors. J Immunol Methods 350:14–21
12. Bulek AM, Madura F, Fuller A et al (2012) TCR/pMHC optimized protein crystallization screen. J Immunol Methods 382:203–210
13. Cole DK, Pumphrey NJ, Boulter JM et al (2007) Human TCR-binding affinity is governed by MHC class restriction. J Immunol 178:5727–5734
14. Davis MM, Boniface JJ, Reich Z et al (1998) Ligand recognition by alpha beta T cell receptors. Annu Rev Immunol 16:523–544
15. Pierce BG, Weng Z (2013) A flexible docking approach for prediction of T cell receptor–peptide–MHC complexes. Protein Sci 22:35–46
16. Pierce BG, Vreven T, Weng Z (2014) Modeling T cell receptor recognition of CD1-lipid and MR1-metabolite complexes. BMC Bioinformatics 15:319
17. Xia Z, Chen H, Kang S-G et al (2014) The complex and specific pMHC interactions with diverse HIV-1 TCR clonotypes reveal a structural basis for alterations in CTL function. Sci Rep 4:4087
18. Michielin O, Luescher I, Karplus M (2000) Modeling of the TCR-MHC-peptide complex1. J Mol Biol 300:1205–1235
19. De Rosa MC, Giardina B, Bianchi C et al (2010) Modeling the ternary complex TCR-Vβ/collagenII(261–273)/HLA-DR4 associated with rheumatoid arthritis. PLoS One 5:e11550
20. Liu IH, Lo YS, Yang JM (2013) Genome-wide structural modelling of TCR-pMHC interactions. BMC Genomics 14(Suppl 5):S5
21. Leimgruber A, Ferber M, Irving M et al (2011) TCRep 3D: an automated *in silico* approach to study the structural properties of TCR repertoires. PLoS One 6:e26301
22. Klausen MS, Anderson MV, Jespersen MC et al (2015) LYRA, a webserver for lymphocyte receptor structural modeling. Nucleic Acids Res 43:W349
23. Ding YH, Baker BM, Garboczi DN et al (1999) Four A6-TCR/peptide/HLA-A2 structures that generate very different T cell signals are nearly identical. Immunity 11:45–56
24. Borbulevych OY, Santhanagopolan SM, Hossain M et al (2011) TCRs used in cancer gene therapy cross-react with MART-1/melan-a tumor antigens via distinct mechanisms. J Immunol 187:2453–2463
25. Gagnon SJ, Borbulevych OY, Davis-Harrison RL et al (2006) T cell receptor recognition via cooperative conformational plasticity. J Mol Biol 363:228–243
26. Borbulevych OY, Piepenbrink KH, Gloor BE et al (2009) T cell receptor cross-reactivity directed by antigen-dependent tuning of peptide-MHC molecular flexibility. Immunity 31:885–896
27. Piepenbrink KH, Borbulevych OY, Sommese RF et al (2009) Fluorine substitutions in an antigenic peptide selectively modulate T-cell receptor binding in a minimally perturbing manner. Biochem J 423:353–361

28. Scott DR, Borbulevych OY, Piepenbrink KH et al (2011) Disparate degrees of hypervariable loop flexibility control T-cell receptor cross-reactivity, specificity, and binding mechanism. J Mol Biol 414:385–400
29. Khan AR, Baker BM, Ghosh P et al (2000) The structure and stability of an HLA-A*0201/octameric tax peptide complex with an empty conserved peptide-N-terminal binding site. J Immunol 164:6398–6405
30. Utz U, Banks D, Jacobson S et al (1996) Analysis of the T-cell receptor repertoire of human T-cell leukemia virus type 1 (HTLV-1) Tax-specific CD8+ cytotoxic T lymphocytes from patients with HTLV-1-associated disease: evidence for oligoclonal expansion. J Virol 70:843–851
31. Johnson LA, Heemskerk B, Powell DJ Jr et al (2006) Gene transfer of tumor-reactive TCR confers both high avidity and tumor reactivity to nonreactive peripheral blood mononuclear cells and tumor-infiltrating lymphocytes. J Immunol 177:6548–6559
32. Kaufmann KW, Lemmon GH, Deluca SL et al (2010) Practically useful: what the Rosetta protein modeling suite can do for you. Biochemistry 49:2987–2998
33. Chaudhury S, Lyskov S, Gray JJ (2010) PyRosetta: a script-based interface for implementing molecular modeling algorithms using Rosetta. Bioinformatics 26:689–691
34. Rudolph MG, Stanfield RL, Wilson IA (2006) How TCRs bind MHCs, peptides, and coreceptors. Annu Rev Immunol 24:419–466
35. Miles JJ, Mccluskey J, Rossjohn J et al (2015) Understanding the complexity and malleability of T-cell recognition. Immunol Cell Biol 93:433–441
36. Chaudhury S, Gray JJ (2008) Conformer selection and induced fit in flexible backbone protein–protein docking using computational and NMR ensembles. J Mol Biol 381:1068–1087
37. Gray JJ, Moughon S, Wang C et al (2003) Protein–protein docking with simultaneous optimization of rigid-body displacement and side-chain conformations. J Mol Biol 331:281–299
38. Mandell DJ, Coutsias EA, Kortemme T (2009) Sub-angstrom accuracy in protein loop reconstruction by robotics-inspired conformational sampling. Nat Methods 6:551–552
39. Park M-S, Park SY, Miller KR et al (2013) Accurate structure prediction of peptide–MHC complexes for identifying highly immunogenic antigens. Mol Immunol 56:81–90
40. Schueler-Furman O, Elber R, Margalit H (1998) Knowledge-based structure prediction of MHC class I bound peptides: a study of 23 complexes. Fold Des 3:549–564
41. Fagerberg T, Cerottini J-C, Michielin O (2006) Structural prediction of peptides bound to MHC class I. J Mol Biol 356:521–546
42. Yanover C, Bradley P (2011) Large-scale characterization of peptide-MHC binding landscapes with structural simulations. Proc Natl Acad Sci 108:6981–6986
43. Borbulevych OY, Insaidoo FK, Baxter TK et al (2007) Structures of MART-1(26/27-35) peptide/HLA-A2 complexes reveal a remarkable disconnect between antigen structural homology and T cell recognition. J Mol Biol 372:1123–1136
44. Robinson J, Mistry K, Mcwilliam H et al (2011) The IMGT/HLA database. Nucleic Acids Res 39:D1171–D1176

Chapter 20

Computational Design of Protein Linkers

Brian Kuhlman, Tim Jacobs, and Tom Linskey

Abstract

Naturally occurring proteins often consist of multiple distinct domains joined by linker regions. Similarly, the ability to combine globular protein domains through engineered linkers would allow the creation of a wide variety of complex and useful multifunctional proteins. Recent advances in computational design of protein structures have enabled highly accurate design of novel protein structures. In this chapter we outline a computational protocol for the de novo design of protein linkers, and apply this protocol to the design of a helical linker between two rigid protein domains.

Key words Protein linkers, RosettaDesign, Molecular docking

1 Introduction

Naturally occurring proteins often consist of multiple distinct domains joined by linker regions, which can be structured or unstructured. The modularity of multi-domain proteins allows nature to create novel functions by mixing and matching functional components (domains) through evolutionary mechanisms such as gene duplication and homologous/nonhomologous end joining. Similarly, the design of linkers between existing functional domains allows the creation of novel, multifunctional proteins useful for numerous purposes. Therapeutic single-chain antibody fragments, which are used extensively as therapeutics, result from designing flexible linkers between protein domains [1]. Additionally, full-length proteins joined by flexible linkers, such as DNA binding domains fused to transcriptional activators, are ubiquitous in molecular research.

Structured linkers offer even greater functional versatility than unstructured linkers by controlling the specific spatial orientation between two protein domains. For instance, a photoactivatable protein domain was covalently joined to a GTPase implicated in cell motility using a designed structured linker. This rigid linker allows the photoactivatable domain to occlude GTPase activity in

Barry L. Stoddard (ed.), *Computational Design of Ligand Binding Proteins*, Methods in Molecular Biology, vol. 1414,
DOI 10.1007/978-1-4939-3569-7_20,

the absence of a light stimulus, creating a powerful optogenetic tool for the study of cell motility [2]. Structured linkers also enable the design of novel biomolecular structures. For example, naturally occurring homo-oligomeric proteins have been joined using a helical structured linker to form a protein nanocage [3].

Design of structured linkers between protein domains requires consideration of both the length of the linker and its amino acid sequence. Experimentally, exhaustive iteration of these metrics quickly becomes intractable, even for short linkers. However, recent advances in computational score functions and sampling methods allows for highly accurate design of specific protein structures, allowing researchers to design a set of feasible linkers in silico. Therefore, computational protein design can overcome the experimental limitations mentioned above by rapidly generating a small set of feasible candidate linkers. These candidates can then be evaluated experimentally to determine the optimal linker for the intended function.

In this chapter, we describe a method for designing a helical linker between two rigid protein domains. It is important to note that there are many methods for computational protein design, each with its own advantages and disadvantages, and that the below protocol represents only one such method. Specifically, this protocol uses the *RosettaScripts* framework included with the Rosetta molecular modeling suite [4]. RosettaScripts was chosen for its user-facing simplicity and its ability to adapt the same design protocol to a wide variety of linker design problems. Regardless of the linker design method you ultimately choose, the general steps and principles outlined here should be applicable.

A functional demo containing the full set of inputs and commands used is accessible as a Rosetta demo. A full list of Rosetta demos, along with instructions, is available at https://www.rosettacommons.org/demos/latest/

2 Materials

There are two inputs to RosettaScripts: the input starting structure(s) in Protein Data Bank (PDB) format, and a protocol file in XML format which specifies the design protocol. There are two outputs to RosettaScripts: a model of the designed protein and a file containing computational scores. The process for obtaining a starting structure is described in Subheading 2.1, and the process for writing the protocol file is described in Subheading 2.3.

2.1 Starting Structures

Design of a structural linker begins with a structure of the two domains in the desired relative orientation. A high-resolution X-ray crystal structure of a protein dimer that you would like to covalently connect with a linker is the best starting point. However, in many protein design cases, this is not available. If monomeric structures of the two domains are available, one can generate a dimeric structure using macromolecular docking software, the usage of which is outside the scope of this protocol (*see* **Note 1**). If structures of one or both monomers are not available, homology models can be generated. However, homology modeling is not perfect (it is akin to an educated guess), and incorrect models may have significant impact on the design of a rigid linker. Therefore, homology models should be treated with caution, and used only if a high-quality structural template can be identified. If your final protein dimer structure is not available in PDB format, it should be converted to this format for future steps (*see* **Note 2**).

2.2 Obtaining and Compiling the Rosetta Molecular Modeling Suite

Rosetta is a full-featured macromolecular modeling suite that can be used for a wide variety of protein design and structural prediction purposes. Rosetta is free for academic use, and licenses are available to purchase for commercial purposes. To obtain a license, visit https://www.rosettacommons.org/software/license-and-download. Once a license is obtained, download and compile the latest version available. Rosetta uses a continual-release model and new versions are available weekly. Detailed instructions for compilation can be found at https://www.rosettacommons.org/docs/latest/. Rosetta is a command-line utility, and thus, at minimum, a cursory familiarity of the command-line environment is recommended. An online tutorial with sufficient information can be found at http://cli.learncodethehardway.org/book/. Upon successful compilation of Rosetta, an executable file named starting with the word "rosetta_scripts" should exist in the Rosetta/main/source/bin directory. This executable is required for Subheading 3.

2.3 Creating Your RosettaScripts Protocol Input File

In addition to your starting PDB, an XML script describing the design protocol is necessary to run RosettaScripts. Although example scripts are contained here, it is recommended you read the documentation for the RosettaScripts syntax, which can be found at https://www.rosettacommons.org/docs/latest/scripting_documentation/RosettaScripts/RosettaScripts.

The script file is responsible for dictating the details of the linker design protocol, including the desired linker length and structure, as well as the optimization methods and allowed degrees of freedom. Typical linker design protocols, including the example below, split the task of linker design into two separate stages: a backbone construction stage, and a sidechain optimization stage (Fig. 1). This "divide-and-conquer" approach allows the complex task of linker design to be addressed as two simpler subtasks. The backbone creation stage optimizes only backbone phi/psi angles, and the sequence design stage optimizes linker amino acid sequence as well as placement of the linker side chains. In both stages, bond angles and lengths are fixed at their ideal geometric values, and

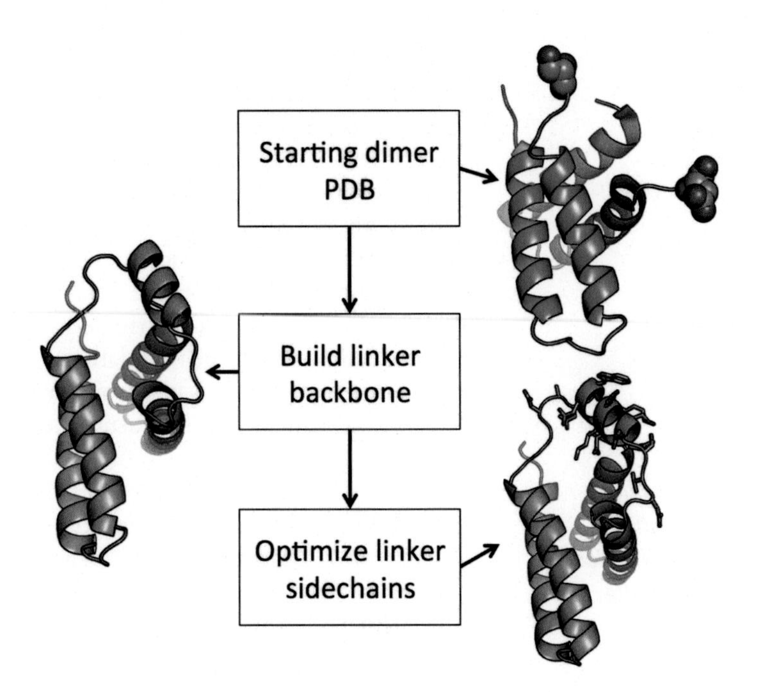

Fig. 1 Overview of the linker design protocol. In the example, a helical linker is designed between two chains of a homodimer

sidechain identities and conformation of the two domains remain constant.

RosettaScripts parses the input XML file and executes modeling operations, called "Movers", in an iterative fashion. For a complete list of available Movers, and their options, see the documentation at https://www.rosettacommons.org/docs/latest/scripting_documentation/RosettaScripts/Movers/Movers-RosettaScripts

2.3.1 Backbone Construction

The first stage of the linker design protocol is the construction of a polypeptide backbone that connects the two domains of your dimeric protein structure. In our example, we accomplish this task using the BridgeChains Mover. The BridgeChains mover constructs a polypeptide chain of the desired length and secondary structure using fragments extracted from natural protein structures [5].

In order to sample the phi/psi angles in a physically meaningful way, a placeholder "centroid" sidechain is used at every position. The placeholder in this example is a "centroid" representation of the amino acid valine. The scoring function used in this stage takes into account van der Waals attraction/repulsion (score term: vdw), the radius of gyration (score term: rg) to encourage compact structures and the favorability of the phi/psi angles (score term: rama). It does not include any terms that depend on side-chain atoms.

In addition to fragment-based backbone sampling, the BridgeChains mover is responsible for creating a closed peptide chain. The chain closure in the BridgeChains mover is accomplished using a cyclic coordinate descent algorithm [5]. The BridgeChains mover can be configured using several options, which are outlined below. Please reference the BridgeChains section of the RosettaScripts documentation for the most up-to-date list of options.

If you are building a structured linker, predominantly helical secondary structure is recommended due to stable local hydrogen bonding and reduced entropy versus loops. At the end of this stage, a newly designed linker will covalently connect your dimeric input structure if BridgeChains was successful. This linker will be composed of the specified number of residues with the input structural constraints, and will have backbone phi/psi angles based on fragments of natural proteins. However, until this point only the protein backbone has been optimized, and thus each amino acid in the linker will be the placeholder amino acid valine.

BridgeChains Options

1. 'chain1'—The chain that will be at the N-terminus in the completely linked structure. The C-terminal residue of this chain will be the beginning of your designed linker
2. 'chain2'—The chain that will be at the C-terminus in the completely linked structure. The N-terminus residue of this chain will be the end of your designed linker
3. 'motif'—The desired length and secondary structure of your completed linker. This option is specified with a string with the following format:

   <Length><SS><ABEGO>-<Length><SS><ABEGO>-...-<Length><SS><ABEGO>

   Where length is the specified number of residues; SS, or secondary structure, is specified with 'H', for helix, 'E' for strand, and 'L' for loop; and ABEGO is any of the valid ABEGO codes, which indicate allowed phi/psi angles for the residues in Ramachandran space. For more information about ABEGO codes, *see* Table 1.

Table 1
Backbone phi and psi torsion bins for ABEGO codes

Code	Min. Phi	Max. Phi	Min. Psi	Max Psi
A	-180.0	0.0	-75.0	50.0
B	-180.0	0.0	50.0	285.5
E	0.0	180.5	100.0	260.5
G	0.0	180.5	-100.0	100.0
X	-180.0	180.0	-180.0	180.0

4. 'overlap'—The number of residues to be rebuilt on each side of the new linker. Rebuilding the flanking residues of the linker will allow more flexibility in the linker positioning and will result in the creation of more linker models that are able to connect the two chains.
5. 'scorefxn'—The Rosetta score function that should be used during sampling of backbone conformations with polypeptide fragments.

2.3.2 Sidechain Optimization

The second stage of the linker design protocol is the optimization of the amino acid identities and conformations on the newly designed backbone. There are many methods for sidechain optimization available in Rosetta. Some of these protocols leave the protein backbone completely fixed and optimize only the sidechain positions, others incorporate varying amounts of backbone flexibility that allow the backbone to adjust in order to accommodate the modifications being made to the sidechain conformations. In our example, we use the PackRotamersMover, which operates on a completely fixed backbone, followed by MinMover, which performs refinement of side chain positions and backbone torsion angles.

The PackRotamersMover optimizes sidechain conformations, called rotamers, using a Monte Carlo simulated annealing algorithm [6]. The example below allows any of the 20 standard amino acids to be designed at all positions in the new linker. However, if specific amino acid identities are desired, this can be configured using a Rosetta resfile or a Rosetta Task Operation, the documentation of which can be found at https://www.rosettacommons.org/docs/latest/rosetta_basics/file_types/resfiles, and https://www.rosettacommons.org/docs/latest/scripting_documentation/RosettaScripts/TaskOperations/TaskOperations-RosettaScripts, respectively.

MinMover optimizes backbone phi/psi angles and sidechain conformations of both domains and the linker using the full Rosetta all-atom scoring function. This step results in subtle movements required to reach an energetic minimum.

Multiple cycles of design followed by minimization will typically lead to better scoring models than a single cycle alone. The ParsedProtocol and GenericMonteCarlo Movers are used to control this iteration; the full details of these Movers are available through the RosettaScripts Mover documentation.

2.3.3 Example Script

The below script will construct a linker between chain 1 and chain 2 in the input PDB. In this example, each chain is 100 amino acids. The linker will consist of 16 residues; the first three residues will be modeled with a loop secondary structure, followed by ten residues with a helical secondary structure, and then three more residues of loop. The torsion angles of the loop residues will be unrestricted, and the helix residues will be restricted to the 'A' region of

Ramachandran space ($180 \leq phi \geq 0$; $-75 \leq psi \geq 50$). Additionally, phi/psi angles of the three residues flanking the new linker will be sampled in order to allow more efficient closure of the linker (*see* **Note 3**). If a successful linker backbone is created, then the amino acid sidechains for the linker will be optimized with 20 iterations of fixed backbone design and gradient-based energy minimization.

```
<ROSETTASCRIPTS>
    <TASKOPERATIONS>
        ##############################################
        #####
        # The OperateOnResidueSubset operation, in
        conjunction
        #withthePreventRepackingResidueLevelTaskOperation
        (RLT)
        # prevents design at positions outside the new loop
        ##############################################
        #####
        <OperateOnResidueSubset name="looponly" >
            <And>
                <Index resnums="1-100" />
                <Index resnums="117-216" />
            </And>
            <PreventRepackingRLT/>
        </OperateOnResidueSubset>
    </TASKOPERATIONS>
##############################################
# The "fldsgn_cen" scorefunction is the recommended
# ScoreFunction for backbone design in which a placeholder
# amino-acid is used (in the case of this demo, that amino acid
# is valine)
##############################################
<SCOREFXNS>
            <centroid_scorefunction weights="fldsgn_
            cen" />
        </SCOREFXNS>
        <FILTERS>
        </FILTERS>
        <MOVERS>
            <BridgeChains name="connect" chain1="1"
            chain2="2" motif="3LX-10HA-3LX" over-
            lap="3" scorefxn="centroid_scorefunction" />
<PackRotamersMover name="pack" task_operations=
    "looponly" />
        <MinMover name="minimize" bb="true" chi="true" />
<ParsedProtocol name="design_and_minimize" >
```

```
            <Add mover = "pack" />
            <Add mover = "minimize" />
</ParsedProtocol>
##################################################
# Note that Talaris 2013 is the default full-atom score function in
# Rosetta, and therefore does not need to be defined in the
# SCOREFXNS section above
##################################################
<GenericMonteCarlo name = "design_mc" trials = "20" mover_
name = "design_and_minimize" scorefxn_name="talaris2013" />
        </MOVERS>
        <PROTOCOLS>
            <Add mover_name = "connect"/>
            <Add mover_name = "design_mc"/>
        </PROTOCOLS>
</ROSETTASCRIPTS>
```

3 Methods

3.1 Running RosettaScripts

To run a RosettaScript protocol, navigate to the directory containing your input PDB file and XML script in a terminal. Rosetta executables, such as RosettaScripts, can be modulated through the use of command-line flags. At a minimum, the '-s' flag is needed to specify the input PDB, and the '-parser:protocol' flag is needed to specify the XML script. Additionally, the 'nstruct' flag is used to control the number of times your complete protocol should be run. Rosetta uses random numbers during the design process, so each execution may result in a different output. Typically, it is advisable to design many candidate structures and select only the best for in vitro characterization. An example of a complete command that will run your linker design protocol 5 times is below:

```
/path/to/Rosetta/main/source/bin/rosetta_
    scripts.#distribution# \
    -s #input_PDB# \
    -parser:protocol #script.xml# \
    -nstruct 5
```

For a more exhaustive set of command-line flags used by Rosetta, visit the documentation at:

https://www.rosettacommons.org/docs/wiki/rosetta_basics/options/options-overview

3.2 Determining Appropriate Linker Length and Structure

Before running the entire linker design protocol, it is useful to run a small number of trajectories aimed at identifying an appropriate number of residues necessary to connect the two domains of your input with the desired structure. For computational efficiency, there is no need to optimize the sidechains during this step. In order to make this change, remove the "design" mover from the <PROTOCOLS> block of the example XML script from Subheading 2.3.3. Create a copy of this script for each of the linker variants you would like the test (*see* **Note 4**). For each linker variant, modify the 'motif' option for the BridgeChains Mover to represent the desired linker length and structure to be tested. Use the command from Subheading 3.1 to run each of the XML scripts. It is recommended to run at least ten trajectories (an 'nstruct' value of 10) for each linker variant.

Each trajectory can result in one of three different outcomes, each of which gives important information about the chosen linker variant. The first outcome is an error stating that the loop closure algorithm failed to close the linker. In this case, it is likely that the chosen linker is too short to bridge the gap between the two chains. In this case, additional residues may be needed. The second possible outcome is an error stating that the secondary structure and torsion angles don't match the desired specified secondary structure/ABEGO. In this case, additional residues and/or less stringent structural requirements may help. The final outcome is a successful run that produces an output PDB, which by default will be named the same as the input PDB with an additional number at the end. For example, an input PDB with the name "my_input.pdb", would result in an output named "my_input_0001.pdb" (*see* **Note 5**). A successful output indicates that a linker backbone with the specified input was created; however, it is useful to examine the output to ensure the designed linker is not unnecessarily long for the given gap.

3.3 Generation of Complete Linker Designs

Once a set of viable linker lengths and structures is determined, the full design protocol, which optimizes both the linker backbone and sidechains should be run. To generate this protocol, modify the example script from Subheading 2.3.3 to contain the desired motif you generated in Subheading 3.1. Additionally, modify the OperateOnResidueSubset TaskOperation to include only the residues outside your linker. For example, if your goal is to connect two 100-residue domains with a 10-residue linker, then your ResidueIndex selectors should include residues 1–100 and 111–210. Run the protocol using the command in Subheading 3.1. For production runs, it is recommended that at least 1000 trajectories be run for each linker variant.

3.4 Analyzing Linker Designs

The Rosetta full-atom score will be used to rank all linkers designed in Subheading 3.2. These scores are available in the Rosetta output file name 'score.sc'. This column based score file contains values for each of the score terms used by the Rosetta full-atom score function. For our purposes, only the 'total_score' and 'description' columns will be used. Sort the rows of score file by the total_score column (*see* **Note 6**). The rows with the most negative score correspond to the best output and should be selected for experimental characterization (*see* **Note 7**).

4 Notes

1. Rosetta has a built-in molecular docking protocol. For complete documentation see https://www.rosettacommons.org/docs/wiki/application_documentation/docking/docking-protocol
2. It is often useful to "clean" PDBs before using them as Rosetta input. This typically involves removal of water and other ligands that Rosetta doesn't understand by default. A python script to do this automatically can be found in Rosetta/tools/protein_tools/scripts/clean_pdb.py
3. Additional sampling restraints for the residues flanking the linker can be added using Rosetta's constraint system. For documentation see https://www.rosettacommons.org/docs/latest/rosetta_basics/file_types/constraint-file
4. As a general rule of thumb, each residue in a linker can span approximately 2 Å. So, for a 10 Å distance between the two chains you are trying to connect, a 5-residue linker is a good starting point. Helical linkers may require slightly more residues, while loop linkers may require fewer.
5. The prefix used for PDB output can be modified using the command-line flag '-out:prefix #desired_prefix#'
6. Rosetta score files can be pasted into Microsoft Excel for simplified easy sorting. Use the "text to columns" option with a space delimiter to ensure proper formatting.
7. Typical scores for well-designed linkers should typically average to less than −2.0 Rosetta Energy Units per linker residue. However, the total_score reported in the score.sc file will include the score of residues outside the linker region. The Rosetta score for each individual residue is appended to the end of each output PDB and can be used to calculate the score for only the linker residues.

References

1. Bird RE, Hardman KD, Jacobson JW, Johnson S, Kaufman BM, Lee SM, Whitlow M (1988) Single-chain antigen-binding proteins. Science 242(4877):423–426
2. Wu YI, Frey D, Lungu OI, Jaehrig A, Schlichting I, Kuhlman B, Hahn KM (2009) A genetically encoded photoactivatable Rac controls the motility of living cells. Nature 461(7260):104–108
3. Padilla JE, Colovos C, Yeates TO (2001) Nanohedra: using symmetry to design self assembling protein cages, layers, crystals, and filaments. Proc Natl Acad Sci U S A 98(5):2217–2221

4. Rohl C, Strauss C, Misura K, Baker D (2004) Protein structure prediction using Rosetta. Methods Enzymol 383(2003):66–93
5. Canutescu AA, Dunbrack R (2003) Cyclic coordinate descent: a robotics algorithm for protein loop closure. Protein Sci 12(5): 963–972
6. Kuhlman B, Baker D (2000) Native protein sequences are close to optimal for their structures. Proc Natl Acad Sci U S A 97(19):10383–10388

Chapter 21

Modeling of Protein–RNA Complex Structures Using Computational Docking Methods

Bharat Madan, Joanna M. Kasprzak, Irina Tuszynska, Marcin Magnus, Krzysztof Szczepaniak, Wayne K. Dawson, and Janusz M. Bujnicki

Abstract

A significant part of biology involves the formation of RNA–protein complexes. X-ray crystallography has added a few solved RNA–protein complexes to the repertoire; however, it remains challenging to capture these complexes and often only the unbound structures are available. This has inspired a growing interest in finding ways to predict these RNA–protein complexes. In this study, we show ways to approach this problem by computational docking methods, either with a fully automated NPDock server or with a workflow of methods for generation of many alternative structures followed by selection of the most likely solution. We show that by introducing experimental information, the structure of the bound complex is rendered far more likely to be within reach. This study is meant to help the user of docking software understand how to grapple with a typical realistic problem in RNA–protein docking, understand what to expect in the way of difficulties, and recognize the current limitations.

Key words Protein–RNA docking, NPDock, Molecular modeling, Macromolecular complexes, Structural bioinformatics, Statistical potential

1 Introduction

In almost every biological system involving protein and RNA molecules, somewhere in the process, some form of protein–RNA complex formation almost inevitably occurs. A deep grasp of the binding mechanisms that depend on both the 3D structure and interaction energies of such complexes is therefore essential to understanding the biological systems that employ them. Protein–RNA interactions have been long known to be critical in the formation of the ribosome as well as the process of protein synthesis [1, 2]. The production of small RNAs as well as the regulation of gene expression by these molecules both in prokaryotes and in eukaryotes requires numerous steps where proteins are involved [3, 4]. RNA splicing in eukaryotes is dependent on the formation

Barry L. Stoddard (ed.), *Computational Design of Ligand Binding Proteins*, Methods in Molecular Biology, vol. 1414, DOI 10.1007/978-1-4939-3569-7_21, © Springer Science+Business Media New York 2016

of protein–RNA complexes in the context of the spliceosome, a large, dynamic ribonucleoprotein machine [5].

Recently, there has been a notable growth in the number of experimentally determined structures of RNAs and protein–RNA complexes that have been solved using X-ray crystallography; however, crystallization of these macromolecules remains quite arduous and capricious [6, 7]. Hence, compared to what we would like to know, only a handful of such structures have been solved. Since it is often the case that we can only obtain the individual parts of protein–RNA complexes, where the structures of individual molecules are like the pieces of a puzzle, it is of considerable interest to develop computational tools that can predict their interactions, and assemble the puzzle [8]. Published predictions of protein–RNA complexes with the use of computational methods include examples such as ribosomes at various functional stages [9], miRNA–target–Argonaute complexes [10], and the catalytic core of the spliceosome [11].

Presently, the state of the art in our ability to model protein–RNA interactions is however very limited. The binding often involves small differences in the free energy between the complex and the separated molecules when enveloped in the surrounding solvent environment. For example, at 27 °C, a binding affinity of $k_D = 10^{-9}$ (quite strong) yields roughly −12 kcal/mol. Since the number of contacts at the RNA–protein interface may only involve a dozen residue pairs, on average, each contact may contributes less than 1 kcal/mol to the binding free energy, which is comparable to thermal energies (about 0.6 kcal/mol at 27 °C). The binding energies involve the receptor–ligand interactions themselves, the entropic effects of conformational change (particularly flexing of the RNA), entropic effects due to the formation of the complex itself, and effects related to solvent interactions [12]. In the midst of these complexities, docking programs must find some way through these uncertainties to model RNA–protein docking successfully.

In this tutorial, we perform two types of docking: bound and unbound. Bound docking involves disassembling a solved complex and attempting to put the pieces back together. Unbound docking involves starting with the independent crystallographic structures of the isolated molecules and attempting to assemble them. Clearly, the latter problem is more difficult because molecules are often somewhat plastic and tend to change shape to some extent upon binding.

For bound docking, we selected *E. coli* pseudouridine synthetase TruB bound to the T stem-loop of the RNA [13] as the target complex. For unbound docking, we attempted a rather difficult problem of identifying the binding site of yeast aspartyl tRNA-synthetase (aaRS) with aspartyl-tRNA [14]. We show how starting with a limited amount of experimental information and the individual structures, one can filter decoys (mixtures of incorrect structural poses amongst some correct poses) to obtain the approximate desired docking structure. Hence, the presentation should provide some rough idea of what one can expect of current docking methods, what sorts of strategies are required to get close (even remotely close), roughly what sorts of problems are likely to be encountered along the way and finally what directions the roads are likely to lead toward in the future.

2 Nomenclature, Materials and Software

2.1 Nomenclature

Here we introduce some docking terminology that will be used in the rest of this chapter.

- Receptor–ligand: the larger structure is typically called the receptor and the smaller one the ligand. In many cases, since RNA is quite large when compared to many proteins, the RNA is considered the receptor. Nevertheless, this is not always the case as we shall also see.
- Decoy: docking programs can generate a large number of structures, many of which are typically far from the true structure. Such structures are often called "decoys".
- Pose: a particular structural arrangement of the ligand with respect to the receptor.

2.2 RNA–Protein Structures Used in This Study

The crystal structures for these complexes were downloaded from the Protein Data Bank (PDB) at http://www.rcsb.org/ [15]. For the bound docking complex, *E. coli* pseudouridine synthetase TruB bound to the T stem-loop of RNA, we used PDB id 1K8W. For the unbound docking, the individual component of yeast aspartyl tRNA-synthetase (aaRS) and aspartyl-tRNA were used: PDB ids 1EOV [16] and 3TRA [17], respectively. In this case, the solved crystal structure of the complex is available (PDB id 1ASY) and is used as the reference to validate the docking results. Because 1ASY is a dimer complex, the selected protein and RNA chains are A and R, respectively.

2.3 ModeRNA

ModeRNA [18] is an open-source software package used for comparative modeling of RNA structures. The standalone version of ModeRNA (version 1.7.1) can be downloaded from http://www.genesilico.pl/moderna/download/ and runs on Windows (binary version) and Linux (source code). ModeRNA requires Python2.6 or higher and BioPython libraries [19]. Alternatively, the ModeRNA web server [20] can be accessed from http://iimcb.genesilico.pl/modernaserver/. Usage of the standalone version of ModeRNA is recommended for advanced users as it provides additional functionalities not available on the web server, such as multi-template modeling and removing modified nucleotides.

2.4 GRAMM

GRAMM [21] is a docking software package mainly developed for protein–protein docking, but also can be used to carry out protein–RNA docking. GRAMM is written in Fortran and can be installed on a variety of operating system platforms. The software comes with an executable file *gramm*. The user must set the path for the environment variable *GRAMMDAT* to the directory containing the data files used by GRAMM for docking. The standalone version is available for download after filling out a short registration form at http://vakser.compbio.ku.edu/main/resources_gramm1.03.download.php. GRAMM can also be accessed via a web server at http://vakser.compbio.ku.edu/resources/gramm/grammx/; however, the server currently is only set up to handle protein–protein docking.

2.5 Filtrest3D

Filtrest3D [22] is a freely available tool written in Python that helps score and/or rank 3D structures generated from a variety of other computational methods based on user-defined restraints obtained either from experimental data or computational predictions. By employing additional experimental information to help filter out some of the decoys (false positives), Filtrest3D can aid in tertiary structure prediction, macromolecular docking, etc. The filters can be weighted according to user's needs, where the default weight for any type of filter is 1.0. The user is advised to refer the online manual (http://filtrest3d.genesilico.pl/readme.html) for details. A web server version of Filterest3D can also be accessed at http://filtrest3d.genesilico.pl/filtrest3d/index.html; however, it is not able to handle a large number of structures (maximum file size < 100 Mb, roughly 1000 files). When filtering a realistic set of decoys, the standalone version of Filtrest3D is recommended, which can be downloaded at http://genesilico.pl/software/stand-alone/filtrest3d/. It requires Python 2.3 or higher and BioPython libraries ≥ 1.41. The usage of filtering restraints such as secondary structure and solvent accessibility requires the installation of external third-party software such as STRIDE [23].

2.6 DARS-RNP

DARS-RNP [24] is a coarse-grained knowledge-based potential for scoring of protein–RNA complexes. The potential can be obtained from http://genesilico.pl/software/stand-alone/statistical-potentials/. It is a standalone program that requires Python 2.6 or higher, the BioPython library version ≥ 1.45 and Numpy.

2.7 NPDock Web Server

NPDock [25] is a web server for predicting complexes of protein–nucleic acid structures. It implements a computational workflow that includes rigid body docking (with GRAMM), scoring of poses (with DARS-RNP), clustering of the best-scored models, and refinement of the most promising solutions. NPDock is available at http://genesilico.pl/NPDock/ and provides a user-friendly interface and 3D visualization of the results, without the difficulties of extensive manual processing of the results. The smallest set of input data consists of a protein structure and an RNA structure in PDB format. Advanced options are available to control specific details of the docking process and obtain intermediate results. The user only needs to prepare the protein and RNA structure of interest, submit it to the server, and, if desired, adjust some input parameters such as the number of decoys to generate.

2.8 Files

All files used in the tutorial can be downloaded from ftp://genesilico.pl/iamb/tutorial/ (file Protein-RNA docking tutorial.zip). *See* **Note 1** for the versions of software used in this study.

3 Methods

In this tutorial, we first describe a bound docking approach using the NPDock web server (which is a relatively user friendly interface) and second an unbound docking approach using manual methods (which grants the user more control over the parameters). The manual procedures are far more intricate and elaborate.

3.1 Preparation of Molecules to Be Docked

The downloaded PDB file often contains a variety of additional items in the crystal structure: for example, small molecules like water, sulfates, and nitrates that are used in the crystallization process or precipitate with the molecule, ions like Na^+, K^+, or Cl^- that bind to the molecule in the crystal structure, and additional heteroatoms which are often unique to the particular process of crystallization and are not typically required for docking. For protein–RNA docking, we are currently largely forced to focus on standard residues for the protein and RNA components, because there are few standardized conventions for naming the heteroatoms and there are a plethora of them, for which we have only a limited ability to model a few of them. Moreover, molecules like water (particularly in the unbound structures) can interfere with the docking process. Hence, all these additional items should be removed from the PDB files before using them for docking.

NPDock and GRAMM have been programmed to recognize only the PDB entries starting with the keyword "ATOM"; therefore, the user can submit the downloaded PDB structure as such. However, it is generally a good practice to prepare the separate protein and RNA files manually to be certain of what structure is being used. Therefore, we describe how to do this.

To prepare the protein structure file, this can be done using a structure visualization tool like PyMoL [26] or Chimera [27]. For this case study, open the protein file (1K8W.pdb) using PyMoL.

The file contains both water and sulfate molecules that can be removed by executing the following command in the PyMoL command line:

```
PyMOL>remove resn SO4+HOH
 Remove: eliminated 309 atoms in model "1k8w".
PyMOL>save 1K8W.pdb
 Save: wrote "1K8W.pdb".
```

As in the case of proteins, PDB files of RNA also contain many water molecules, ions, small molecules and modified nucleotides. Although NPDock and GRAMM have some ability to recognize modified bases, there are some docking programs that do not. It is very important to verify whether a modified base plays a critical role in docking. In this case, fortunately, the matter turns out to be not so serious. Therefore, for the purpose of this illustration, we assume that these modified nucleotides can be converted to their standard bases, without loss of information. Our lab has developed a tool called ModeRNA for comparative modeling of RNA structures, which also has a function for removing modifications from RNA structures. Using Python, execute the following commands to remove all heteroatoms and modifications from the RNA structure:

```
$python
>>>from moderna import * # requires moderna in the PYTHONPATH
>>>m=load_model('1K8W.pdb','B') # load chain B of 1K8W
>>>clean_structure(m) # remove all heteroatoms
Chain OK
>>>remove_all_modifications(m) # reformat to standard bases
>>>write_model(m,'na.pdb') # save the file
```

The execution of these commands results in a PDB file na.pdb containing the structural coordinates of RNA with the modified nucleotides reverted back to standard bases.

In the case of automated docking using the NPDock web server, the modified bases can be included for the docking calculation by renaming the HETATM columns in PDB file pertaining to the modified residues to ATOM.

3.2 Bound Docking

The bound docking is performed to reconstruct the geometry of a given protein–nucleic acid complex structure, where the starting point is the structure from a co-crystal; here a crystallographic structure containing both the RNA and the protein together. Bound docking is often performed to validate the accuracy of docking potentials in identifying native-like conformations, without taking into consideration any conformational changes that occur in the two macromolecules in the unbound state. In this case study, we perform bound docking on *E. coli* pseudouridine synthase

TruB bound to the T stem-loop of the RNA (PDB id 1K8W) using the NPDock web server. The steps involved in docking and identifying native-like conformations are described below.

3.2.1 Automated Protein–RNA Docking Using NPDock Web Server

After executing the procedures outlined in Subheading 3.1, navigate to the server website http://genesilico.pl/NPDock/ and click the *Submit your job* button and follow the steps described below for performing protein–RNA docking.

1. Near the top left-hand corner of the web page, enter the required job title (e.g., a name like '1K8W_test') and optionally an e-mail address for receiving a link to access the results when the job is finished. Select the *RNA-protein* option under the docking tab to specify protein–RNA docking.
2. Using the *Select file* button, upload the formatted PDB file 1K8W.pdb for the protein and enter the chain ID as 'A' under the *Select Chains* option.
3. Repeat **step 2** for the RNA and upload the formatted na.pdb file (*see* Subheading 3.1) and enter the chain ID as 'B' for the RNA chain under the *Select Chains* option.
4. To sample 50,000 conformations, change the default value from 20,000 to 50,000.
5. Leave the clustering and refinement parameters for this case of docking as default values and click the *Submit* button to start the job.

The amount of time taken to finish a particular job depends on the size of the macromolecules, the number of decoys to be sampled and other settings such as filtering criteria, clustering and refinement parameters. Once the job is completed, a web page is displayed with the IDs of the top three refined models in PDB format, which are downloadable. The web page also provides a JSmol 3D visualization tool showing the best model and a steps-vs-energy graph showing the Monte Carlo refinement of the best model. Links to the downloadable structures considered in the clustering, the results of the clustering and the raw output files from the NPDock docking pipeline are also provided for the user to perform a more rigorous analysis, if desired.

To identify native-like structures from the sampled conformations, NPDock implements a clustering algorithm proposed by Baker and coworkers [28], used successfully in protein structure prediction. The decoys are clustered based on geometrical similarity. For a given set of docking decoys, the first step is to create an all-against-all root-mean-square deviation (RMSD) matrix by calculating the RMSD for all pairs of structures. Then, the row which has the highest number of RMSD values below a given threshold (default, 5 Å cutoff) is considered to be the first cluster and removed from the matrix. The process is repeated until the number of decoys in one cluster is less than five. The three largest clusters are then considered to be the candidates that contain native-like structures and often the lowest scoring decoys from the representative clusters are identified as native-like. NPDock selects the lowest scoring models from the three largest clusters as the best models.

3.2.2 Comparison of the Docked Model to the Experimentally Observed Complex

Structural superposition is one of the most commonly used methods for assessing the quality of the docked models. If the structure of the native complex is available, then the docked model can be superimposed on the native complex. To evaluate the global similarity, the RMSD provides a criterion for judging the accuracy of the fit between the native and docked complex. One method of calculating the RMSD between the docked model and the experimentally measured structure is to perform an optimal superposition of the receptors of the two structures using a macromolecular viewer such as PyMoL. Then one calculates the RMSD of selected atoms types in the ligand; often the heavy atoms (because the ligand is the RNA molecule, this would be the phosphorus atoms, for example).

To calculate the RMSD between the best model and the reference complex, download the first model by clicking on its ID and superimpose it on the reference complex (1K8W), Fig. 1. Now, calculate the RMSD of superposition for the ligand (or RNA in this case) using the method described previously. A sample script (run_pymol.py) for calculating RMSD is provided with this tutorial. The value of superposition is found to be around 1.6 Å, implying that predicted conformation is very similar to the experimentally observed structure (the reference complex).

To identify the position of the first three clusters among the sampled conformations (as shown in Fig. 2), one must calculate the RMSD for all the decoys and plot it against the DARS scores in dars_out.txt file provided by the server in the list of raw output files.

3.3 Unbound Docking

The unbound docking involves docking of independently solved structures of a protein and RNA to identify their correct mode(s) of binding in a given protein–RNA complex. The unbound docking is significantly more difficult, because the starting point is structures in their unbound conformations and there are no reliable methods to predict conformational changes that happen during the complex formation. For this case study of unbound docking, we perform docking on yeast aspartyl tRNA-synthetase (aaRS) with aspartyl-tRNA as the target. The independently solved crystal structures of both the protein and RNA can be downloaded from PDB with ids 1EOV and 3TRA, respectively. To validate the

Fig. 1 Structural superposition of the best docking decoy on the reference complex (PDB id 1K8W). The RNA of the reference complex and the best decoy is shown in *cyan* and *magenta*, respectively

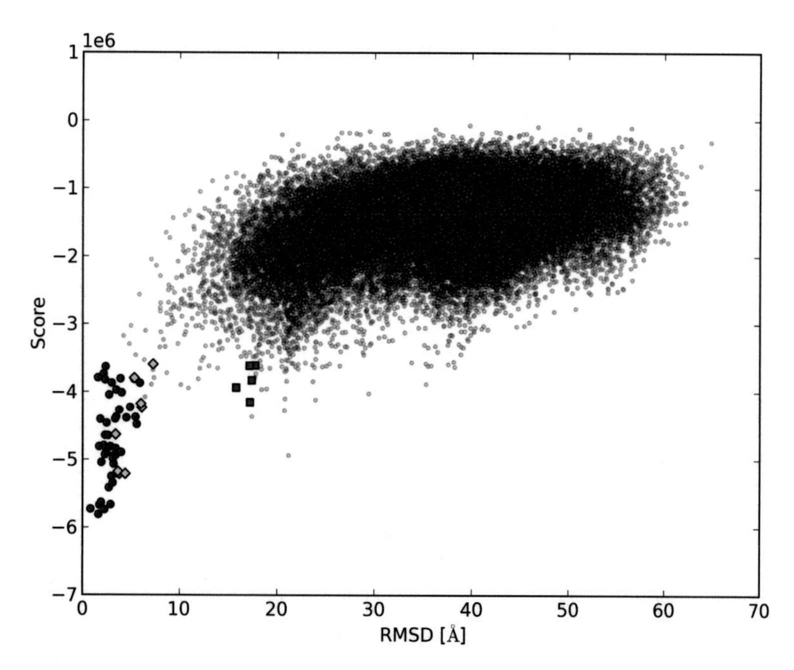

Fig. 2 Plot of the score vs RMSD for the bound docking decoys. The three largest clusters are shown in three different colors and symbols, with the first, second, and third clusters in *red circles*, *green diamonds*, and *blue squares*, respectively

docking results, we use the solved structure for this complex (PDB id 1ASY; chains A and R for the protein and RNA, respectively), Fig. 3. The steps involved in performing unbound docking and identification of the native-like conformations are described in the following subsections.

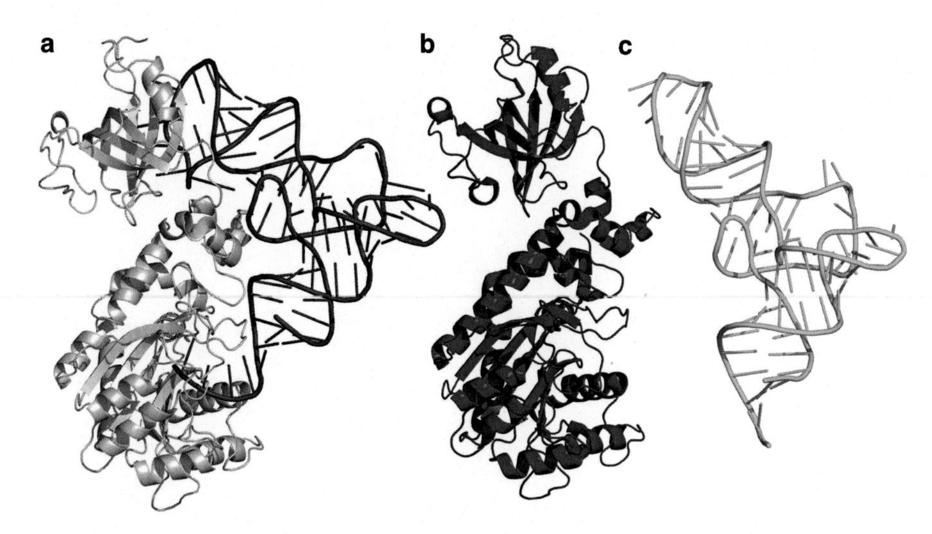

Fig. 3 Schematic representation in ribbon form for the crystal structures of the reference complex used in unbound docking and its unbound protein and RNA components. (**a**) Reference complex (pdb id 1ASY; chains A and R for protein and RNA, respectively); (**b**) unbound protein (pdb id 1EOV); (**c**) unbound RNA (pdb id 3TRA)

3.3.1 Preparation of Molecules to Be Docked

The protein structure file (1EOV.pdb) should be prepared in the same way as discussed in Subheading 3.1. However, for the RNA structure file (3TRA.pdb), the reader should consider that the structure in 3TRA.pdb lacks the CCA tail at the 3′ end of the tRNA sequence. This is a biologically relevant part of tRNA and, as it turns out, this tail is also present in the reference structure (PDB id 1ASY). Since the protein–RNA interaction depends on this tail, it should be present and one should surmise that the addition of the CCA tail is necessary to correctly guide the docking of the RNA into the CCA binding domain of the protein. Hence, further processing of the RNA structure is required.

Appending the CCA tail can be done using the ModeRNA program. To add a short tail to the PDB structure, ModeRNA requires the 3D structure coordinates (from 3TRA.pdb) as a template, and a user-defined sequence alignment between the target and the template. The alignment is constructed using a FASTA formatted file. For the first sequence in the file, add the sequence "CCA" at the end of the original RNA sequence and label it as "Target". Then, for the second sequence, align this new RNA sequence to the original sequence as shown below and save the alignment file in FASTA format as 'alignment.fasta':

```
>Target
UCCGUGAUAGUUUAAUGGUCAGAAUGGGCGCUUGUCGCGUGCCAGAUCG
GGGUUCAAUUCCCCGUCGCGGAGCCA
>3TRA template
UCCGUGAUAGUUUAAUGGUCAGAAUGGGCGCUUGUCGCGUGCCAGAUCGGGG
UUCAAUUCCCCGUCGCGGAG---
```

Since the target and the template contain exactly the same sequence (up to the CCA tail), ModeRNA copies all the atoms from the template to the target model and appends the short terminal fragment to the RNA chain. ModeRNA is also able to perform more complex modeling such as modeling of insertions and deletions for which the user is advised to refer the manual.

To clean up the RNA file 3TRA and append the CCA tail, execute the following series of commands using Python:

```
$python
>>>from moderna import *
>>>t=load_template ('3TRA.pdb','A')
>>>clean_structure(t)# resets bases, removes HOH and Mg
Chain OK
>>>a=load_alignment('alignment.fasta')
>>>m=create_model(t,a) # add the CCA tail
>>>write_model(m,"complete_RNA.pdb")
```

This generates a PDB file complete_RNA.pdb containing the modeled structure of RNA with the CCA tail included (Fig. 4).

Alternatively, one can carry out all the steps as in Subheading 3.1 (for the RNA file) by replacing 1K8W.pdb with 3TRA.pdb and na.pdb with 3TRA_clean.pdb. Then, to append the CCA tail, the following command line statement can be used:

```
$python moderna.py -t 3TRA_clean.pdb -c A -a alignment.
fasta -o complete_RNA.pdb
```

where, "-t" specifies the template structure, "-c" indicates the chain to be considered for modeling, "-a" identifies the alignment file, and "-o" assigns the name of the output file. Finally, there is yet another alternative: using the ModeRNA server, submit (separately)

Fig. 4 Schematic representation of the modeled RNA with a CCA tail. The modeled CCA tail is highlighted in *blue*

the template and alignment files in the online submission form (http://iimcb.genesilico.pl/modernaserver/submit/model/), and click the "Build Model" option to start processing the job and wait till the results are obtained.

3.3.2 Manual Protein–RNA Docking

For manual protein–RNA docking, we selected the GRAMM program to generate decoy conformations. GRAMM treats the receptor and ligand (here a protein and a RNA molecule, respectively) as rigid bodies and finds a geometric match between the two molecules by projecting the atoms on a 3D grid. The algorithm allows for softening the van der Waals interactions and permitting some degree of steric clashes that are expected to be alleviated by local conformational changes.

Protein–RNA Decoy Generation Using GRAMM

In the description of docking components, typically the larger molecule is called the "receptor" and the small one the "ligand." In this example, the size of the RNA molecule is smaller compared to the protein; therefore, the RNA will be called the ligand and the protein the receptor. GRAMM requires the following files to perform docking:

- rpar.gr—describes the docking parameters
- rmol.gr—provides the description of molecules to be docked
- wlist.gr—defines the IDs of the decoys to be extracted in PDB format.

For this case study, we perform docking in low resolution mode, as high resolution docking results in a large number of steric clashes between the protein and RNA. For this reason, select the potential range type as "grid_step" to implement low resolution docking and set the grid step radius to 3.1 Å, which is the lowest value allowed by the program for "low resolution". The repulsion parameter is set to 10 Å and the attraction double range is set to 0. The ligand is allowed to rotate at an angle of 10° and a total of 10,000 conformations are sampled. The following settings should be entered into the docking parameter file (rpar.gr):

```
Matching mode (generic/helix) ...................... mmode=generic
Grid step ........................................... eta=3.1
Repulsion (attraction is always -1) ................. ro=10.
Attraction double range (fraction of single range) ..... fr=0.
Potential range type (atom_radius, grid_step) ....... crang=grid_step
Projection (blackwhite, gray) ....................... ccti=gray
Representation (all, hydrophobic) ................... crep=all
Number of matches to output ......................... maxm=10000
Angle for rotations, deg (10,12,15,18,20,30, 0-no rot.)  ai=10
```

The boldface indicates the most important parameters to adjust for particular problems. The configuration list is defined as follows:

- mmode—defines the docking mode (generic or helix). The generic mode involves sampling for all the ligand's positions and orientations. In the helix mode, GRAMM automatically discards poses with large displacements along the helix axes and angles larger than indicated in the rmol.gr file.
- eta—step of the grid
- ro—repulsion parameter
- fr—attraction double range, used in high resolution docking.
- crang—"atom_radius" implies high resolution docking (projection of a sphere with van der Waal radius) and "grid_step" implies docking under lower resolution
- ccti—cumulative projection ("gray") is generally used in low resoluation docking, yes-no ("blackwhite") is often used in high resolution docking
- crep—switch to hydrophobic docking: in this case, it should always be "all"
- maxm—number of output structures
- ai—angle of rotation for search through rotational coordinates

These settings may vary for docking of different protein–RNA complexes. Therefore, it is advised to experiment with different combinations of parameters to obtain the best settings for docking of different complex structures.

The parameters of the rmol.gr file, which provides the information of molecules to be docked, are shown below.

```
# Filename Fragment ID Filename Fragment ID (paral/anti max.ang)
#----------------------------------------------------------------
1EOV.pdb * prot complete_RNA.pdb * na
```

Here, the usage of the asterisk (*) under the Fragment heading indicates that the entire molecule is used for docking. The user should remember that only lines with the first word "ATOM" are taken into consideration by GRAMM. GRAMM also accepts a specified fragment or region of a molecule for docking. Additionally, the chain ID (in capitals) can also be specified if docking has to be performed for a particular chain. GRAMM considers the first molecule as the receptor and the second molecule as the ligand. The ID of each fragment is used by GRAMM to name the output file (with extension *.res). For example, in this case the name of output file generated by the docking simulation will be prot-na.res.

In order to start the conformational search with GRAMM, use the command:

```
> gramm scan
```

This may take quite some time depending on the computational power of the system used for docking. It took 36 min on a machine with 24 processors (clock speed 2.8 GHz) and 24 GB RAM and running Ubuntu 14.04.1. The completion of the conformational search generates the prot-na.res file, which contains information about the 10,000 sampled conformations.

To extract the coordinates of each decoy in PDB format, GRAMM requires the wlist.gr file, which provides information about the poses to be extracted from the output file. The parameters of the wlist.gr file are shown below:

```
#File_of_predictions    First_match    Last_match    separate/
   joint+init_lig
#-----------------------------------------------------------------
prot-na.res 1 1000 separ no
```

where prot-na.res is the output file containing the information about the docked complexes. The headings "First_match" and "Last_match" denote the id of the decoys to be generated and "separate/joint" indicates whether the files should be extracted separately for each decoy or combined in a single file. The last column heading "+init_lig" specifies whether the initial conformation of the ligand should also be extracted together with each of the sampled conformations or not.

To extract the coordinates, execute the following command:

```
>gramm coord
```

See **Note 2** regarding extracting structures using this command.

Decoy Filtering According to Restraints

In a docking simulation, it is very important to obtain a representative distribution of the actual conformations, particularly those that lie in the region of the docking site. This will allow the clustering procedure to select the configuration that is also representative of the distribution that is actually found. However, it is often the case that the target interface is not easily recognized and the results of the docking simulation needs some further processing or help.

GRAMM generates a large number of conformations of the protein and RNA components where some of the poses of the RNA–protein complex may involve binding interactions that are very far away from the correct docking site. In order to obtain decoys with a reasonable native-complex-like geometry, it is advised to remove the obviously nonnative or unreasonable structures using some filtering criteria. The filtering criteria can be any information which can be obtained either from experimental data or from computational predictions. For instance, in this example, we know the CCA binding region and anticodon binding residues from the solved crystal structure of the complex. We use these restraints to filter native-like decoys from all docked conformations.

For filtering, we use Filtrest3D. The general syntax for defining any type of restraint is:

```
Restraint_type_name (
Restraint_declaration
.....
)
```

For the case of protein–RNA docking presented herein, we select the distance restraints for ranking of the decoys. We retain decoys in which the aaRS binds the anticodon loop with the anticodon binding domain, and the CCA tail with the catalytic domain, defined by two protein–RNA distance restraints: Gln138-U35 and Glu478-A75, respectively. For both these pairs we define the required distance as less than or equal to 16 Å. The decoys for which the sum of squares of deviations from a 16 Å cutoff is less than 80 are retained for further analysis. The parameters used in the restraint file (filter.filtrest) for filtering are as follows:

```
dist (
E478_near_A75: (E478) "A"-(A75) "B" (<=16)
Q138_near_U35: (Q138) "A"-(U35) "B" (<=16)
)
```

where,

- dist indicates a distance based type of restraint
- E478_near_A75 is the name of the restraint for the amino acid-nucleic acid pair E478-A75 in chains A and B, respectively.
- (<=16) specifies that the distance between any of the closest atoms between residues E784 and A75 should be less than or equal to 16 Å.

To filter the decoys using the above defined restraints, execute the following command.

```
$python filtrest3d.py --restraints filter.filtrest \
--dirfile ./structures/str_list.txt>filtrest_result.out
```

where "--restraints" specifies the restraint file, "--dirfile" indicates the path to file with list of structures to be filtered. The output is written to a file "filtrest_result.out". *See* **Note 3** for Filterest3D, if any errors are encountered.

Out of 10,000 decoys, 19 decoys are found to fulfill the filtering restraints.

Scoring of Decoys Using the DARS-RNP Potential

The next step of the docking procedure after sampling the conformations is to discriminate near-native structures. An ideal docking method should combine both sampling and scoring of decoys to identity near-native structures of protein–RNA complexes.

Unfortunately, GRAMM does not have a scoring function for protein–RNA complexes; therefore, we must use an external scoring function to identify near-native structures from the decoys generated using GRAMM.

Our team has developed a scoring potential for protein–RNA complexes called DARS-RNP [24] that has performed well in identifying near native decoys compared to the other potentials available for scoring protein–RNA decoys. To score the filtered decoys using the DARS-RNP potentials refer to the next Section.

Selection of the Most Promising Complex Model

If the number of "promising" structures (such as those selected by filtering) is large, say more than 50, it is generally advised to perform clustering to identify the largest set of similar conformations that approximate the most likely solution of the docking problem. For clustering, we use a 10 Å RMSD cutoff, as outlined at the end of Subheading 3.2.1, and we use the `DARS_potential_v3.py` script for scoring the decoys, which also clusters the best scored decoys (that in this case are the filtered structures). To perform scoring and clustering using the `DARS_potential_v3.py` script, execute the following command.

```
$python DARS_potential_v3.py -f list.txt -m 19 -c 10>DARS.out
```

where, "-f" specifies the file containing the list of structures to be scored, "-m" denotes the number of structures considered for clustering and "-c" indicates the RMSD cutoff for clustering (10 Å). The output is written to the file "DARS.out". *See* **Note 4** for the correct way of running DARS_potential_v3.py script, if errors are encountered.

This results in two clusters with the first (largest) cluster containing eight structures and the second with five structures. Open all 13 complexes in PyMoL to visualize the conformations of RNA of the two clusters (Fig. 5). Choose the first cluster to identify the near native-like docked conformation and select the lowest scoring decoy from this cluster as the final docked model.

Comparison of the Docked Model to the Experimentally Solved Structure

To assess the accuracy of the docked complex, superimpose the best docked model on the reference complex (Fig. 6). It can be clearly seen that the selected RNA pose, as well as all eight structures from the first cluster bind in a similar way to the RNA in the reference structure. However the CCA tail still does not manage to get close enough to dock into CCA binding domain of the protein. Now examine the position of the RNA in the structures present in the second cluster (Figs. 5 and 6). It is clearly visible that in addition to the CCA tail, which does not dock correctly in the CCA domain of protein, the anticodon binding loops of RNA in all five structures of the second cluster are positioned away from the binding site.

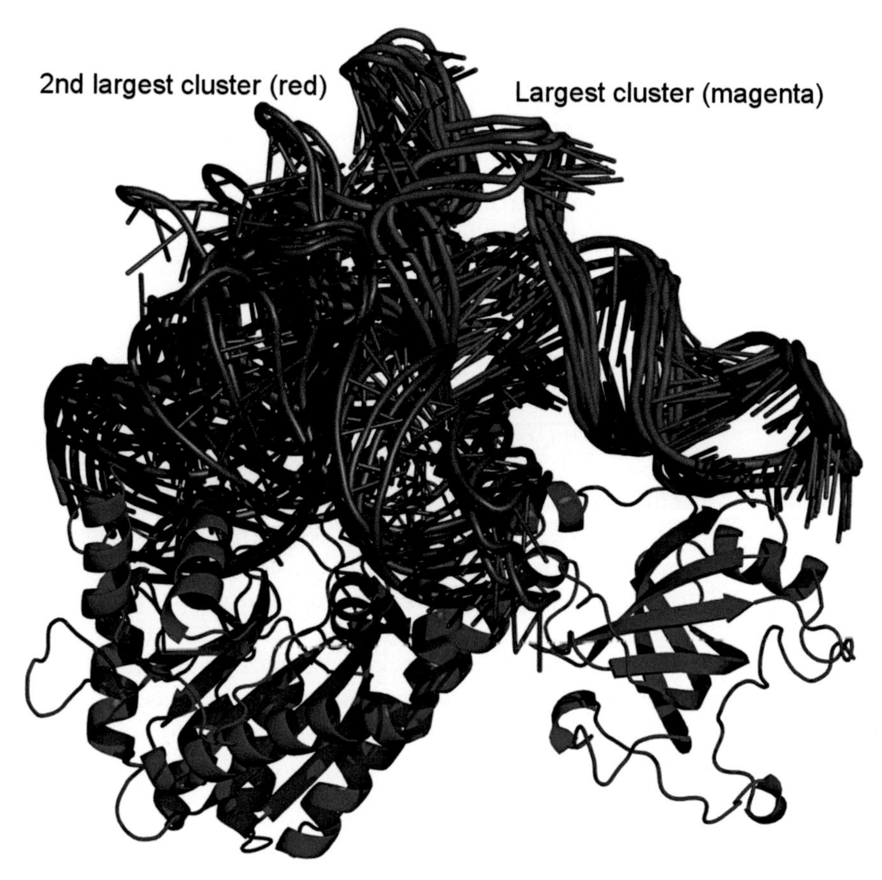

Fig. 5 Schematic representation of the conformations of RNA from the two largest clusters. The tRNAs from the first and second largest clusters are shown in *magenta* and *red* colors, respectively

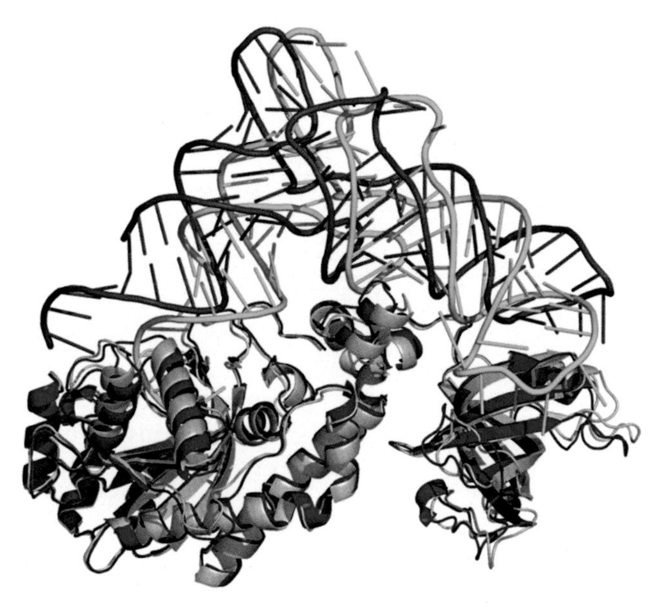

Fig. 6 Superimposed structures of the lowest scoring decoy on the reference complex. The protein and RNA of the reference complex is shown in *gray* and *cyan*, respectively. The protein and RNA of the lowest scoring decoy is shown in *blue* and *magenta*, respectively

The RMSD of the structural superposition, calculated using the method described in Subheading 3.2.2, was found to be 9.8 Å for this complex. This value indicates a much larger deviation from the reference complex than in the case of the bound docking example described earlier. One of the ways to further improve the quality of the docked model is to perform flexible optimization, for instance using Molecular Dynamics with a physics-based force field [29, 30], which may improve the quality of the model. However, such optimization is not trivial and requires extensive preparation of the system to be analyzed as well as complicated analysis of the results, which is out of the scope of this chapter.

3.4 Summary

The methods used in this study exhibit the use of computational methods to predict how protein and RNA molecules with known structures form protein–RNA complexes. We demonstrated the use of a web server NPDock for a fully automated protein–RNA complex structure modeling and the use of a workflow of various tools for a more elaborate docking, analyzing the docking results and obtaining the most promising model. Both bound and unbound docking exercises are presented for the user to understand the steps involved in performing docking for a given pair of protein and RNA components. This tutorial also explains various problems which can be encountered during the docking procedure and suggests the implementation of certain methods to overcome such problems. The analysis of the docked models in unbound docking highlights the inadequacy of the docking algorithms in sampling native-like conformations and calls for the development of better tools and algorithms for flexible macromolecular docking.

4 Notes

1. The exercises described in this tutorial were performed on a computer running Ubuntu 14.04.1 with installed Python and Biopython version 2.7.6 and 1.63, respectively. This tutorial uses the latest version of the all software mentioned in Subheading 2, available during the preparation of this manuscript.
2. The "gramm scan" command can extract only 1000 structures at a time, which means the user has to repeatedly edit the wlist.gr file to extract all 10,000 structures. To automate this process, we have provided a sample Perl script (extract_GRAMM.pl) which can be modified accordingly.
3. The user may encounter some problems while using Filtrest3D, as the program was written using old Biopython libraries. For this, we have provided the files which should be replaced, if errors are encountered.

4. The "-v" switch can be used to specify the version of Biopython used while using `DARS_potential_v3.py` script for scoring and clustering. This will save the user from running into errors, particularly in cases where this tutorial is run on a machine with installed Biopython version ≥ 1.45.

Acknowledgments

This work was supported by the European Commission (E.C. REGPOT grant FishMed, contract number 316125, to Jacek Kuźnicki in IIMCB) and by the European Research Council (ERC, StG grant RNA+P=123D grant to J.M.B). J.M.B was also supported by the "Ideas for Poland" fellowship from the Foundation for Polish Science. J.M.K., I.T., and M.M. were additionally supported by the Polish National Science Center (NCN, grants 2012/05/N/NZ2/01652 to J.M.K., 2011/03/N/NZ2/03241 to I.T., and 2014/12/T/NZ2/00501 to M.M.). The development and maintenance of computational servers was funded by the E.C. structural funds (grant POIG.02.03.00–00–003/09 to J.M.B.). Calculations were performed on a high-performance computing cluster at IIMCB, Warsaw (supported by IIMCB statutory funds). The authors are grateful to Stanisław Dunin-Horkawicz and Michał Boniecki for useful discussions and comments on the manuscript.

References

1. Moller W, Amons R, Groene JC, Garrett RA, Terhorst CP (1969) Protein-ribonucleic acid interactions in ribosomes. Biochim Biophys Acta 190(2):381–390
2. Demeshkina N, Jenner L, Yusupova G, Yusupov M (2010) Interactions of the ribosome with mRNA and tRNA. Curr Opin Struct Biol 20(3):325–332. doi:10.1016/j.sbi.2010.03.002
3. Ghildiyal M, Zamore PD (2009) Small silencing RNAs: an expanding universe. Nat Rev Genet 10(2):94–108, nrg2504 (pii)
4. Pichon C, Felden B (2007) Proteins that interact with bacterial small RNA regulators. FEMS Microbiol Rev 31(5):614–625, FMR079 (pii)
5. Hoskins AA, Moore MJ (2012) The spliceosome: a flexible, reversible macromolecular machine. Trends Biochem Sci 37(5):179–188. doi:10.1016/j.tibs.2012.02.009
6. Doudna JA (2000) Structural genomics of RNA. Nat Struct Biol 7(Suppl):954–956
7. Ke A, Doudna JA (2004) Crystallization of RNA and RNA-protein complexes. Methods 34(3):408–414
8. Tuszynska I, Matelska D, Magnus M, Chojnowski G, Kasprzak JM, Kozlowski LP, Dunin-Horkawicz S, Bujnicki JM (2014) Computational modeling of protein-RNA complex structures. Methods 65(3):310–319. doi:10.1016/j.ymeth.2013.09.014
9. Whitford PC, Ahmed A, Yu Y, Hennelly SP, Tama F, Spahn CM, Onuchic JN, Sanbonmatsu KY (2011) Excited states of ribosome translocation revealed through integrative molecular modeling. Proc Natl Acad Sci U S A 108(47):18943–18948. doi:10.1073/pnas.1108363108
10. Gan HH, Gunsalus KC (2013) Tertiary structure-based analysis of microRNA-target interactions. RNA 19:539. doi:10.1261/rna.035691.112
11. Anokhina M, Bessonov S, Miao Z, Westhof E, Hartmuth K, Luhrmann R (2013) RNA structure analysis of human spliceosomes reveals a compact 3D arrangement of snRNAs at the catalytic core. EMBO J 32(21):2804–2818. doi:10.1038/emboj.2013.198
12. Cheng LT, Wang Z, Setny P, Dzubiella J, Li B, McCammon JA (2009) Interfaces and

hydrophobic interactions in receptor-ligand systems: a level-set variational implicit solvent approach. J Chem Phys 131(14):144102. doi:10.1063/1.3242274

13. Hoang C, Ferre-D'Amare AR (2001) Cocrystal structure of a tRNA Psi55 pseudouridine synthase: nucleotide flipping by an RNA-modifying enzyme. Cell 107(7):929–939
14. Ruff M, Krishnaswamy S, Boeglin M, Poterszman A, Mitschler A, Podjarny A, Rees B, Thierry JC, Moras D (1991) Class II aminoacyl transfer RNA synthetases: crystal structure of yeast aspartyl-tRNA synthetase complexed with tRNA(Asp). Science 252(5013):1682–1689
15. Rose PW, Prlic A, Bi C, Bluhm WF, Christie CH, Dutta S, Green RK, Goodsell DS, Westbrook JD, Woo J, Young J, Zardecki C, Berman HM, Bourne PE, Burley SK (2015) The RCSB protein data bank: views of structural biology for basic and applied research and education. Nucleic Acids Res 43(Database issue):D345–D356. doi:10.1093/nar/gku1214
16. Sauter C, Lorber B, Cavarelli J, Moras D, Giege R (2000) The free yeast aspartyl-tRNA synthetase differs from the tRNA(Asp)-complexed enzyme by structural changes in the catalytic site, hinge region, and anticodon-binding domain. J Mol Biol 299(5):1313–1324
17. Westhof E, Dumas P, Moras D (1988) Restrained refinement of two crystalline forms of yeast aspartic acid and phenylalanine transfer RNA crystals. Acta Crystallogr A 44(Pt 2):112–123
18. Rother M, Rother K, Puton T, Bujnicki JM (2011) ModeRNA: a tool for comparative modeling of RNA 3D structure. Nucleic Acids Res 39(10):4007–4022. doi:10.1093/nar/gkq1320
19. Cock PJ, Antao T, Chang JT, Chapman BA, Cox CJ, Dalke A, Friedberg I, Hamelryck T, Kauff F, Wilczynski B, de Hoon MJ (2009) Biopython: freely available Python tools for computational molecular biology and bioinformatics. Bioinformatics 25(11):1422–1423
20. Rother M, Milanowska K, Puton T, Jeleniewicz J, Rother K, Bujnicki JM (2011) ModeRNA server: an online tool for modeling RNA 3D structures. Bioinformatics 27(17):2441–2442
21. Katchalski-Katzir E, Shariv I, Eisenstein M, Friesem AA, Aflalo C, Vakser IA (1992) Molecular surface recognition: determination of geometric fit between proteins and their ligands by correlation techniques. Proc Natl Acad Sci U S A 89(6):2195–2199
22. Gajda MJ, Tuszynska I, Kaczor M, Bakulina AY, Bujnicki JM (2010) FILTREST3D: discrimination of structural models using restraints from experimental data. Bioinformatics 26(23):2986–2987, btq582 (pii)
23. Frishman D, Argos P (1995) Knowledge-based protein secondary structure assignment. Proteins 23(4):566–579
24. Tuszynska I, Bujnicki JM (2011) DARS-RNP and QUASI-RNP: new statistical potentials for protein-RNA docking. BMC Bioinform 12(1):348, 1471-2105-12-348 (pii)
25. Tuszynska I, Magnus M, Jonak K, Dawson W, Bujnicki JM (2015) NPDock: a web server for protein-nucleic acid docking. Nucleic Acids Res 43:W425. doi:10.1093/nar/gkv493
26. The PyMOL Molecular Graphics System. The PyMOL Molecular Graphics System, vol Version 1.5.0.4. Schrödinger, LLC.
27. Pettersen EF, Goddard TD, Huang CC, Couch GS, Greenblatt DM, Meng EC, Ferrin TE (2004) UCSF Chimera--a visualization system for exploratory research and analysis. J Comput Chem 25(13):1605–1612
28. Shortle D, Simons KT, Baker D (1998) Clustering of low-energy conformations near the native structures of small proteins. Proc Natl Acad Sci U S A 95(19):11158–11162
29. Ditzler MA, Otyepka M, Sponer J, Walter NG (2010) Molecular dynamics and quantum mechanics of RNA: conformational and chemical change we can believe in. Acc Chem Res 43(1):40–47. doi:10.1021/ar900093g
30. Estarellas C, Otyepka M, Koča J, Banáš P, Krepl M, Šponer J (2015) Molecular dynamic simulations of protein/RNA complexes: CRISPR/Csy4 endoribonuclease. Biochim Biophys Acta 1850:1072–1090. doi:10.1016/j.bbagen.2014.10.021

Index

Barry L. Stoddard (ed.), *Computational Design of Ligand Binding Proteins*, Methods in Molecular Biology, vol. 1414,
DOI 10.1007/978-1-4939-3569-7, © Springer Science+Business Media New York 2016

G

H

I

L

M

N

P

R

S

T

U

V

X

Y

Printed in the United States
By Bookmasters